# Tourism, Diasporas and Space

Diasporas result from the scattering of populations and cultures across geographical space and time. Transnational in nature and unbounded by space, they cut across the static, territorial boundaries more usually deployed to govern tourism. *Tourism, Diasporas and Space* explores the new challenges that diasporas pose to tourism discourse.

This book introduces the main features and constructs of diasporas, and explores their implications for the consumption, production and practices of tourism. Three sets of mutually reinforcing relationships are explored: experiences of diaspora tourists, the settings and spaces of diaspora tourism, and the production of diaspora tourism. Examples are drawn from a wide spectrum of diasporic groups including the Chinese, Jewish, Southeast Asian, Croatian, Dutch and Welsh.

In a vibrant inter-disciplinary collection of essays from leading scholars in the field, *Tourism, Diasporas and Space* provides a unique navigation of the nature of the connections between tourism and diasporic populations and networks.

**Tim Coles** is Lecturer in Human Geography and University Business Research Fellow at the University of Exeter, and **Dallen J. Timothy** is Associate Professor at Arizona State University and Visiting Professor of Heritage Tourism at the University of Sunderland.

# Contemporary Geographies of Leisure, Tourism and Mobility
Series Editor: Michael Hall
*Professor at the Department of Tourism, University of Otago, New Zealand*

The aim of this series is to explore and communicate the intersections and relationships between leisure, tourism and human mobility within the social sciences.

It will incorporate both traditional and new perspectives on leisure and tourism from contemporary geography, e.g. notions of identity, representation and culture, while also providing for perspectives from cognate areas such as anthropology, cultural studies, gastronomy and food studies, marketing, policy studies and political economy, regional and urban planning, and sociology, within the development of an integrated field of leisure and tourism studies.

Also, increasingly, tourism and leisure are regarded as steps in a continuum of human mobility. Inclusion of mobility in the series offers the prospect to examine the relationship between tourism and migration, the sojourner, educational travel, and second home and retirement travel phenomena.

The series comprises two strands:

**Contemporary Geographies of Leisure, Tourism and Mobility** aims to address the needs of students and academics, and the titles will be published in hardback and paperback. Titles include:

**The Moralisation of Tourism**
Sun, sand . . . and saving the world?
*Jim Butcher*

**The Ethics of Tourism Development**
*Mick Smith and Rosaleen Duffy*

**Tourism in the Caribbean**
Trends, development, prospects
*Edited by David Timothy Duval*

**Qualitative Research in Tourism**
Ontologies, Epistemologies and Methodologies
*Edited by Jenny Phillimore and Lisa Goodson*

**Routledge Studies in Contemporary Geographies of Leisure, Tourism and Mobility** is a forum for innovative new research intended for research students and academics, and the titles will be available in hardback only. Titles include:

**1. Living with Tourism**
Negotiating identities in a Turkish village
*Hazel Tucker*

**2. Tourism, Diasporas and Space**
*Tim Coles and Dallen J. Timothy*

# Tourism, Diasporas and Space

**Edited by**
**Tim Coles and Dallen J. Timothy**

Routledge
Taylor & Francis Group

LONDON AND NEW YORK

First published 2004
by Routledge
2 Park Square, Milton Park, Abingdon, Oxon, OX14 4RN

Simultaneously published in the USA and Canada
by Routledge
270 Madison Ave, New York NY 10016

*Routledge is an imprint of the Taylor & Francis Group*

Transferred to Digital Printing 2005

© 2004 Selection and editorial matter, Tim Coles, Dallen J. Timothy;
individual chapters, the authors.

Typeset in Times by Keystroke, Jacaranda Lodge, Wolverhampton

*British Library Cataloguing in Publication Data*
A catalogue record for this book is available from the British Library

*Library of Congress Cataloguing in Publication Data*
Coles, Tim Edward, 1970–
    Tourism, diasporas and space / Tim Edward Coles.
      p. cm.—(Contemporary geographies of leisure, tourism, and mobility; 6)
    Includes bibliographical references (p.   ).
    ISBN 0–415–31124–1 (hardback: alk. paper)
    1. Tourism—Social aspects. 2. Travelers—Attitudes. 3. Ethnicity.
    4. Identity (Psychology) 5. Emigration and immigration. I. Title.
    II. Series: Routledge/contemporary geographies of leisure, tourism,
    and mobility; 6
G155.A1C5128 2004
306.4′8—dc22
    2003023380

ISBN 0–415–31124–1

Printed and bound by Antony Rowe Ltd, Eastbourne

# Contents

# Illustrations

## Plates

## Figures

**Boxes**

**Tables**

# Contributors

**Sean Carter** is Lecturer in Human Geography in the School of Geography, Archaeology and Earth Resources at the University of Exeter, UK.

**Deborah Che** is Assistant Professor at Western Michigan State University and the Chair of the Recreation, Tourism and Sport speciality group of the Association of American Geographers.

**Erik H. Cohen** is Lecturer in School of Education, Bar Ilan University, Israel, and the Scientific Director of a group of researchers and consultants at Jerusalem, Israel.

**Mara Cohen Ioannides** is a Lecturer in the English Department at Southwest Missouri State University, Springfield, Missouri, USA and sits on the Executive Board of the Midwest Jewish Studies Association.

**Tim Coles** is University Business Research Fellow and Lecturer in Human Geography in the School of Geography, Archaeology and Earth Resources at the University of Exeter, UK. He is also Honorary Secretary of the Geography of Leisure and Tourism Research Group of the Royal Geographical Society (with Institute of British Geographers).

**Noga Collins-Kreiner** is a Lecturer at the Department of Geography and Research Fellow at the Centre for Tourism, Pilgrimage and Recreation Research at the University of Haifa, Israel.

**David Timothy Duval** is a Lecturer in the Department of Tourism at the University of Otago, Dunedin, New Zealand.

**C. Michael Hall** is Professor and Head of Department, Department of Tourism, University of Otago, Dunedin, New Zealand and Honorary Professor, Department of Marketing, University of Stirling, Scotland.

**Kevin Hannam** is Principal Lecturer in Tourism in the School of Arts, Design, Media and Culture at the University of Sunderland, UK.

**Keith Hollinshead** is Professor of Public Culture in the Department of Tourism at the Luton Business School (University of Luton, UK).

**Dimitri Ioannides** is an Associate Professor of Planning in the Department of Geography, Geology and Planning at Southwest Missouri State University, Springfield, Missouri, USA and Senior Research Fellow at the Centre for Tourism and Regional Development, Bornholm, Denmark.

**Brian King** is Professor and Head of the School of Hospitality, Tourism and Marketing at Victoria University, Melbourne, Australia.

**Alan A. Lew** is Professor and Chair of the Department of Geography, Planning and Recreation at Northern Arizona University, USA.

**Joanne Maddern** is Lecturer in Human Geography at the Department of Geography at the University of Dundee, Scotland.

**Kevin Meethan** is Principal Lecturer in Sociology and Academic Director of the Social Research and Regeneration Unit at the University of Plymouth, UK.

**Nigel Morgan** is a Reader in Tourism Studies at the Welsh Centre for Tourism Research in The Welsh School of Tourism, Hospitality and Leisure Management, University of Wales Institute, Cardiff, UK.

**Thu-Huong Nguyen** is Lecturer in the School of Hospitality, Tourism and Marketing at Victoria University, Melbourne, Australia.

**Dan Olsen** is a PhD Candidate in the Department of Geography at the University of Waterloo, Canada.

**Annette Pritchard** is a Reader in Tourism Studies and Director of the Welsh Centre for Tourism Research in The Welsh School of Tourism, Hospitality and Leisure Management, University of Wales Institute, Cardiff, UK.

**Marcus L. Stephenson** is Senior Lecturer in the Sociology of Tourism and is the Director of Postgraduate Research in the Department of Business and Service Sector Management at London Metropolitan University, UK.

**Victor B. Teye** is Associate Professor in the Department of Recreation Management and Tourism at Arizona State University, Tempe, Arizona, USA.

**Dallen J. Timothy** is Associate Professor in the Department of Recreation Management and Tourism at Arizona State University, Tempe, Arizona, USA and Visiting Professor of Heritage Tourism at the University of Sunderland, England.

**Alan Wong** is a Lecturer in the School of Hotel and Tourism Management at Hong Kong Polytechnic University.

# Preface

Rather fittingly, this book is the result of a transnational enterprise over the last two years. It is a celebration of the power of e-mail and the speed of telecommunications, of transgressing time zones and transcending geopolitical boundaries, and of an intricate multi-nodal social network with hubs in south-west England and in the south-western corner of the USA. It is the endpoint of a long and often arduous journey for both of us. It started with the realisation that tourism and diaspora are two prolific subjects of contemporary inter-disciplinary academic enquiry. In no small measure, their popularity as objects of their respective academic gazes stems from their position as defining features and conditions of the fin-de-millennium condition. Tourism, leisure and culture have become increasingly implicated within, reflections of, and trans-formed by, the restructuring of contemporary society and economy. Diasporas have rightfully been described as exemplars of transnationalism and the contribution of globalisation to the conduct of diasporic communities has been duly acknowledged. Somewhat surprisingly, among the burgeoning corpuses of attendant work, the establishment of explicit conceptual and theoretical linkages between the two themes appeared elusive. Although scholars of diaspora espoused the importance of routes and roots in the mediation of diaspora and diasporic identities, paradoxically they appeared reticent to explore the fuller implications of tourism for diaspora and vice versa. Equally taciturn were those in tourism studies who, by and large, overlooked diasporas as 'travelling cultures' in every sense of the term.

Or, so it seemed at the time. Since the start of this project we have uncovered reassuringly insightful, yet relatively fledgling interest in diaspora among tourism research workers. Like the concept itself, contributions on diaspora and tourism have been widely scattered among the literature, often to be found in the most unexpected and far-flung locations, and frequently taking unexpected, hybridized forms by lending theory, concept and method from a number of sources and inspirations. Diaspora is a topic area with which tourism academics have engaged, but one which has for the most part been bypassed and sidelined in the interests of other allegedly more relevant and critical debates. In producing this collection we contend that diasporas should occupy a more privileged position in tourism discourse. Diasporas are major communities and they challenge the hegemonic position of the nation-state in global society through their cross-border relations and mobilities, articulated not least through travel and tourism. A much deeper understanding of diasporic travel and tourism is clearly key towards a fuller understanding of mobilities in contemporary global society. Diasporas are also emblematic of the need to deploy new conceptual toolkits and fluid, reflexive approaches through which deeper, more relevant readings of modern-day social motivations for travel and tourism may be constructed.

With this volume we hope to achieve two important goals: first, to raise the profile of diasporas in tourism studies, and thereby to point to their pivotal importance in establishing richer conceptual linkages between tourism and mobility; and second, to provide a platform from which to induce further critical research on tourism and diasporas. The approach we adopt is an inter-disciplinary one. Our contributors are from a diverse array of backgrounds. As befits tourism studies more widely, among the authors are those with backgrounds in history, geography, sociology and anthropology as well as tourism. The essays presented here represent a synthesis of the major developments in current research on diaspora tourism. As one of the contributors queries: is tourism studies ready to embrace the challenges of investigating diasporas? The answer may be that tourism research has already embarked on its journey of diasporic discovery and enlightenment. However, it is not a case of 'the more we know, the less we understand', rather 'the more we know, the more we appear to have to learn'.

Tim Coles, Exeter, UK
Dallen J. Timothy, Gilbert, Arizona, USA
October 2003

# Acknowledgements

Inevitably with a book of this size and scope, production has been a major task. We would like to thank Andrew Mould (Commissioning Editor at Routledge) and Michael Hall (Series Editor) for their interest and enthusiasm in this project from a very early stage. Melanie Attridge at Routledge kept us on an even keel during the editorial process. We should like to thank the support staff in the Department of Geography at the University of Exeter for their excellent contribution to the manuscript. Cathy Aggett heroically fought her way through the text and diligently worked to make sense of our annotations to produce a crisp manuscript. Helen Jones turned our doodles into the excellent maps, diagrams and figures in this volume. Andrew Teed produced the photographic plates to his usual immaculate standards.

Several contributors to this volume desire to acknowledge people who have influenced their work and lives. Keith Hollinshead acknowledges the diligent word processing and punctilious proofing services of Janet H. (and the budding Ian David H.) in the compilation of his chapter. 'Thank you. The Desborough Two, indeed'. David Duval wishes to thank Paul Wilkinson, Bill Found, Bonnie Kettel, Ray Rogers, David Trotman, David Telfer, Nick Kontogeorgopoulos and Victor Teye for comments on the arguments presented in his chapter. Thanks are also due to Mike Hall and Loretta Baldassar for useful discussions on migration and tourism and to Dave Palmer for comments on an earlier version of his chapter. For their joint chapter, Mike Hall and David Duval wish to recognize Colin Smithies, Department of Economics, University of Otago, for his insight into the economic history of New Zealand, particularly as it affected international migrants from the Pacific Islands and Mel Elliot of the Department of Tourism for her assistance with drafting some of the tables in their chapter. Dallen Timothy and Victor Teye would like to express their gratitude to the College of Public Programs, Arizona State University, for partial funding for the research upon which their chapter is based and to the people of the Cape Coast and Elmina Castle who made their work possible. Erik H. Cohen sends his thanks and best regards to Feuershtein Hadas, Educational Director of the Exodus Project; Danny Levine, Senior Educational Consultant of the Exodus Project; Doubi Szwarc, Director of the UK groups; Dany More, Director of the Israel Experience programmes; Jean-Charles Zerbib (FSJU) and Reuven Naamat (Youth and Hechalutz Department), Co-directors of the France groups; and Eva Labi (Regional Director, FSJU), who agreed to be interviewed at length. Thanks also go to Allison Ofanansky, who helped Erik Cohen organize and edit his chapter. Alan Lew and Alan Wong wish to recognize funding for their study by the School of Hotel and Tourism Management of the Hong

Kong Polytechnic University and Northern Arizona University. A previous version of Tim Coles' chapter was presented as a paper at the Annual Conference of the Association of American Geographers in Los Angeles (March 2002) and at the Tourism Society's *Family History Tourism* meeting in Dorchester, UK (November 2002). Participants at both venues provided helpful and constructive feedback, as did David Duval, Gareth Shaw and his co-editor, Dallen Timothy. Tim recognizes Brian Hay of *VisitScotland* for very kindly making available strategy and policy documentation on genealogical tourism. Finally, Deborah Che wishes to thank Tim Coles and Dallen Timothy for their constructive comments on an earlier draft of her chapter; Margaret Pearce for her cartographic assistance in drafting the early versions of the figures; and the Joint Archives of Holland, Michigan, for their kind assistance with her research program. Thanks to all.

# 1  'My field is the world'

## Conceptualizing diasporas, travel and tourism

*Tim Coles and Dallen J. Timothy*

### Tourism, migration and mobility: a missing piece of the jigsaw?

The quotation in the title is inspired by a photograph in Alan Kraut's (1982: 112) monograph *The Huddled Masses: The Immigrant in American Society, 1880–1921.* Dated circa 1900, the photograph depicts a scene in the departure hall of a German steamship company. Written in German and painted on the wall in bold Latin typeface for all to see, emigrants were offered this thought to reflect upon as they queued to secure passage on a Hamburg-Amerika steamer. In many respects, it encapsulates the themes and issues addressed by this book as well as the situation confronting the future *émigré* just prior to departure: the world of opportunities for travel and migration; the widespread reach and development of communications systems; spatially-extended communities linked by complex social networks articulated through major global nodes; for better or worse, new migrants' experiences along the way and wherever they may finally settle; the possibility of return; and, finally, the unfolding impact of each of these (and other) aspects on the migrants' identities as their journeys are recalled, appraised and acted upon.

Over a century has elapsed since the photograph was taken. Time and space have compressed; communications have become more straightforward, rapid and efficient not least through the Internet; and more extensive, intricate transnational social networks have emerged. As Urry (2000: 154) observes, 'most societies are not nations, let alone nation-states'. Instead, the world is now characterized by the proliferation of 'nation peoples'. These groups are defined by varying kinds and degrees of displacement and ambiguous location and, according to Urry, many may be regarded as diasporic. According to Mitchell (1997a: 534), 'diaspora' has been used by most scholars in a working sense to describe 'the situation of a people living outside of their traditional homeland'. Barber's (2001: 178) equally brief definition views diasporas as 'communities that define themselves by reference to a distant homeland from which they once originated'.

The aim of this book is to explore the contemporary connections and relationships between diasporas and tourism. It focuses on diaspora tourism, or tourism primarily produced, consumed and experienced by diasporic communities. Here, the intention is to concentrate on the relationship between the diasporic condition and the production and consumption of tourism for diasporas themselves rather than diasporas as exotic Others to be gazed upon (Urry 1990). In particular, we focus on three sets of themes that are beginning to emerge in tourism studies of diaspora: namely, diaspora experiences of tourism; the spaces occupied by diaspora tourists; and the production of tourism for and by diasporic communities. The book attempts to bridge the disciplinary divide between diaspora and tourism. According to Edward Bruner (1996: 290), 'the literature on diaspora and hybridity has on the whole neglected tourism, perhaps because tourist visits are thought to be temporary and superficial'. In his view this is a regrettable position because,

> travellers such as migrants, refugees, exiles, expatriates, émigrés, explorers, traders, missionaries and even ethnographers may also travel for limited periods of time. To develop travelling theory, we need to know more about all patterns of travel (Clifford 1989), including tourism (Bruner and Kirshenblatt-Gimblett 1994).
>
> (Bruner 1996: 290)

Irrespective of whether metaphors such as 'travel' and 'journeys' (cf. Clifford 1997) are deployed to unravel diasporic identities, diaspora studies has by and large bypassed tourism as a consideration in the mediation and sustainability of diasporic communities. While Bruner's criticism places much of the blame for the estrangement squarely at the door of diaspora studies, tourism studies is equally culpable of having overlooked diasporas. To date, there has been just tacit recognition of the relevance of diasporic communities. This is notwithstanding their relevance as a key type of community and hence a basic constituency to be acknowledged in contemporary tourism management (Richards and Hall 2000: 2–3). Equally axiomatic has been the treatment of the distinctive cultural capital diasporas offer for commodification in place imagery and destination marketing (van Hear 1998; Richards and Hall 2000; Klemm 2002) and the role played by some diasporic migrants in servicing the tourism and hospitality sectors in cities as reserves of relatively low cost, non-militant, often unskilled labour (Eade 2000; Williams and Hall 2000a,b; Church and Frost 2004).

This schism is reflective of a similar separation between tourism and migration. Although both talk to the same basic theme of mobility, as Williams and Hall (2000a,b) contend, tourism and migration as subject areas have been uneasy companions until recently. Put more emphatically, they argue that,

> the largely discrete literatures on tourism and migration have, at best, served to mark out the core areas of their research concerns. The failure to conceptualize adequately and define their fields of enquiry has . . . [led to] very few attempts to disentangle the changing relationships between tourism and migration . . . [which represent] an increasingly important component of the new forms of mobility.
>
> (Williams and Hall 2000b: 7)

According to Feng and Page (2000: 247), one of the reasons for the ring-fencing was that population geography and migration research were not valorized as key issues within the dominant tourism research agenda (cf. Hall and Page 1999; Shaw and Williams 2002, 2003). While more mundane, functional definitions conceptualized tourism as temporary or short-term migration away from home (Cooper *et al.* 1998: Hall and Page 1999; Shaw and Williams, 2002), paradoxically there was an unwillingness to engage in a more sustained, theoretical debate to explore the increasingly mutually implicated natures of tourism and migration in the late twentieth century. In this context, scattered populations of migrants were relegated primarily as the subjects for ethnic tourism and as travellers likely to undertake religious and secular pilgrimages practically as socio-cultural rites of passage (Shair and Karan 1979; Hudman and Jackson 1992; Park 1994; Vukonić 1996; Hall 2002; Jutla 2002; Olsen and Timothy 2002). Tourism was primarily portrayed as a lens through which visitors could gaze on exotic Other ethnic communities and indigenous groups (Urry 1990; MacCannell 1992; King 1994). Critical debates attended such issues as the authenticity and alleged perversion of local cultures in the face of pressure from tourists (Adams 1997; Wood 1998) and the ethnic politics of tourism development (Pitchford 1995; van der Berghe 1995; Callahan 1998; Jamison 1999; Wall 1999).

Détente has characterized the more recent relationship between tourism and migration. Of late, two collections in particular, have explored the interactions between tourism and migration at the turn of the twenty-first century (Williams and Hall 2000a; Hall and Williams 2002). One of their central messages is that globalization has stimulated new forms of travel, tourism and migration whose production and consumption are intricately bound together (Williams and Hall 2000b; Oigenblick and Kirschenbaum 2002). In one sense, they concur with Franklin and Crang's (2001: 11) clarion call that '. . . tourism should search for links with other mobilities such as commuting, mobile labour markets, migration and Diasporas [sic]'. Notwithstanding, we would contend that, among these groups, diasporas have not been afforded the consideration that their status in contemporary transnational, global society would merit. Rather, they have been marginalized in recent discourses on tourism and mobility in favour of such themes as second-home ownership and retirement migrations (Williams *et al*. 1997; Tomljenovic and Faulkner 2000). The contributions presented in this book attempt to energize greater discussion about, and debate over, the connectivities between diasporas and tourism. Diasporas are complex entities. Almost inevitably, it is impossible here to review in any great detail the full intricacies and nuances of the discourses on diaspora. In what remains, we attempt to contextualize the subsequent chapters by offering an introduction to diasporas and their linkages with tourism consumption and production.

## Towards conceptualization of diaspora

Diaspora is a word with long and rich historical lineage. For Helmreich (1992: 245), the etymology of word 'diaspora' may be traced back to the Greek word for 'dispersion' from the words for 'through' and 'sow or scatter' and originates in the Greek translation of the 'Book of Deuteronomy' in the Bible. Braziel and Mannur (2003) note that through its religious significance, the term was pervasive in medieval rabbinical writings about the Jewish diaspora and the predicament of Jews living outside Palestine.

Definitions and conceptualizations of diaspora are fluid and contested and have been the focus of considerable debate. Diasporas are groups of people scattered across the world but drawn together as a community by their actual (and in some cases perceived or imagined) common bonds of ethnicity, culture, religion, national identity and, sometimes, race. 'Diaspora suggests a dislocation from the nation-state or geographical location origin and a relocation in one or more nation-states, territories, or countries' (Braziel and Mannur 2003: 1). Several writers note the importance of the original point of dispersal, the 'homeland', as occupying a focal point in the mediation of diasporic identity (Safran 1991). For instance, Sheffer (1986: 3) regards modern diasporas as 'ethnic minority groups of migrant origins residing and acting in host countries but maintaining strong sentimental and material links with their countries of origin – their homelands'. Although diasporic communities vary greatly, Cohen (1997: ix) contends that, irrespective of their historical trajectories and experiences, all 'acknowledge that the "old country" – a notion buried deep in language, religion, custom or folklore – always has some claim on their loyalty and emotions'. Moreover, 'a member's adherence to a diasporic community is demonstrated by an acceptance of an inescapable link with their past migration history and a sense of co-ethnicity with others of a similar background'. James Clifford (1994, 1997), in contrast, warns of the problems of over-emphasizing origin and return. He draws attention to the extent of scattering; the lateral reach and complexity of intra-diasporic networks; and the geopolitical juxtapositions of diasporas. For him,

> diasporas usually presuppose longer distances and a separation more like exile: a constitutive taboo on return, or its postponement to a remote future. Diasporas also connect multiple communities of a dispersed population. Systematic border crossings may be a part of this interconnection, but multilocale diaspora cultures are not necessarily defined by a specific geopolitical boundary.
>
> (Clifford 1997: 246)

Diasporic processes and communities are not always the outcomes of voluntary actions. Robin Cohen (1997: ix) acknowledges that 'when applied to humans, the ancient Greeks thought of diaspora as migration and colonization' but for several groups – Jews, Palestinians and Armenians notable among them – diaspora has had much more sinister historical connotations, signifying as it does a sense of group identity resulting from collective trauma, banishment and exile. Paul Gilroy (1993) underscores the horror and cruelty of slavery in mediating the black Atlantic diaspora (see also Bruner 1996). Cohen (1997: 27) points to the origins of the Armenian diaspora in trade and commerce, only for brutal treatment at the hands of the Turks to lead to their forced displacement from 1915 to 1916. A similar scenario, he contends, was played out by the Irish at the hands of the British as migration followed the famine of 1845 to 1852.

Based on comparative readings of diaspora histories, several authors have attempted to define diaspora not by any single meta-statement, but rather based on a series of common characteristics and principal components (Safran 1991; Cohen 1997; Shuval 2000). Safran (1991: 83–4) postulated six attributes that captured the essence of diasporic communities. Not entirely satisfied with what he terms 'Safran's desiderata', Cohen (1997: 23) argued that there was too great an emphasis on the relationship between the diaspora and its homeland. Instead, he reworked the schematic principally to orientate it more towards the condition of the diaspora beyond the homeland; that is, in terms of scattering for aggressive or voluntarist reasons, the positive virtues of retaining a diasporic identity while abroad and the power of collective identity expressed not just with the homeland but also in the place of settlement and with

---

***Box 1.1* Robin Cohen's nine common features of a diaspora**

1 Dispersal from an original homeland, often traumatically, to two or more foreign regions
2 Alternatively, the expansion from a homeland in search of work, pursuit of trade or to further colonial ambitions
3 A collective memory and myth about the homeland, including its location, history and achievements
4 An idealization of the putative ancestral home and a collective commitment to its maintenance, restoration, safety and prosperity, even to its creation
5 The development of a return movement that gains collective approbation
6 A strong ethnic group consciousness sustained over a long time and based on a sense of distinctiveness, a common history and belief in a common fate
7 A troubled relationship with host societies, suggesting a lack of acceptance at the least or the possibility that another calamity might befall the group
8 A sense of empathy and solidarity with co-ethnic members in other countries of settlement
9 The possibility of a distinctive creative, enriching life in host countries with a tolerance for pluralism.

Source: adapted from Cohen (1997: 26).

---

co-ethnic members in other countries. The result was a definitional scheme for diaspora based on nine common characteristics (Box 1.1).

This diagnostic is an idealized one and one which has been compiled by reference to several diasporas. As the listing is a composite, Cohen recognizes that not all the characteristics have to be evident in every contemporary diasporic grouping. Similarly, the exact assemblages and strengths of the characteristics will vary among different sub-groups and intra-diasporic constituencies. A similar logic is asserted by Judith Shuval (2000) with her definitional schematic (Table 1.1). Responding to a concern that diaspora 'encompasses a motley array of groups such as political refugees, alien residents, guest workers, immigrants expellees, ethnic and racial minorities and overseas communities' (Shuval 2000: 41), she proposes a general framework, the attributes of which are intended to allow robust and structured comparison between different types of diasporas. For her, diasporas may also be defined more clearly by reference to the characteristics of, and within, host society and its disposition(s) towards diaspora groups.

Beyond checklist approaches, Cohen (1997) proposes a five-fold typology of diasporas based on commonalities of experiences and the structural processes mediating diaspora. Victim Diasporas such as the Jews, the African diaspora, the Armenians and the Palestinians are typified by their forced and traumatic displacement from a territory, not least resulting from (nation-)state formation or denial. In contrast, Cohen (1997: 57) argues that Labour Diasporas, as exemplified by the Indians under British Rule, arise from scattering in pursuit of work. He notes, however, that not all groups who migrate internationally need necessarily be described as diasporas. The British are regarded as the quintessence of Imperial Diasporas. Like the Spanish, Portuguese, Belgians, French and Germans, the British scattered to further their colonial

*Table 1.1*  Judith Shuval's theoretical paradigm of diasporas

| I. Characteristics of the diaspora group | II. Characteristics of homeland | III. Characteristics of host |
|---|---|---|
| a. Chronology of group<br>b. Causes of dispersion toward ethnic groups<br>c. Differentiation (to sub-groups?)<br>d. Retention of ethnic culture<br>e. Location, links and relations among members<br>f. Quality of relations among members<br>g. Attitudes and feelings to homeland | a. Level of reality<br>b. Legitimacy<br>c. Attitude of residents and government in homeland to diaspora and returnees<br>d. Behaviour to returnees<br>e. Behaviour of returnees | a. Structural features<br>b. Cultural-ideological stance<br>c. Behaviour of government and sub-groups toward ethnic groups<br>d. Relevance of homeland to host government and sub-groups in host society |

Source: abridged from Shuval (2000: 50).

ambitions. Similarly, Trade Diasporas refer to extended networks of merchants, traders and entrepreneurs who carry out their business by buying, selling, trading and marketing their goods and services over long distances. These are exemplified by the reach of Chinese traders in southern and east Asia and Lebanese merchants in West Africa and the Americas. Finally, Cohen (1997: 127) proposes Cultural Diasporas as an attempt to address the postmodernists' fascination with the 'collective identity of homeland and nation [which] is a vibrant and constantly changing set of cultural interactions that fundamentally question the very ideas of "home" and "host"'. For Hague (2001: 145), a cultural diaspora exists where connections between people are not so much based on shared historical experiences or movement to return home, but rather they are grounded in the belief of common ethnic and cultural origins. Although Urry (2000: 155) asserts that all diasporas are by definition in part inherently cultural, in a strong echo of Gilroy's (1993) ideas and Bhabha's (1994) postulates (see also Ch. 2), Cohen (1997) explores the way in which Caribbean peoples are cemented as much by literature, political ideas, religious convictions and life-styles as permanent migration under conditions of postcolonialism. While it may be tempting to pigeon-hole each diaspora into one of these groupings, it is clear that the boundaries between the individual groupings are somewhat blurred. Equally, it is possible for an individual diaspora to have dual or multiple presence in more than one of the groupings. For instance, indentured Indian labour, which may reasonably be typical of a Labour Diaspora may in fact in certain instances also be categorized under the heading Victim Diaspora.

## The diasporic condition and the 'hype of hybridity'

According to Shuval (2000: 43), in essence, the critical components of such definitions are a history of dispersal, myths and memories of the homeland, alienation in the host country, desire for eventual return, ongoing support of the homeland and a collective identity defined by these relationships.

For many commentators, however, such apparently rigid approaches towards definition are unacceptable, perhaps even quasi-imperialist taxonomical exercises, reducing as they do a concept of inherent complexity and fluidity to a series of distinct, discrete and stiff criteria (cf. Bhahba 1994; Hall 1990, 1996; Hollinshead 1998; Braziel and Mannur 2003). For Mitchell (1997a,b), one of the alluring reasons for the study of diaspora is that it challenges prior orthodox narratives of fixity and mobility. The propensity in earlier discourse was to reduce the world to a series of banal binary oppositions (Soja 1996). Diasporas, as metaphors for social and cultural analysis at large and as entities in themselves, suggest that instead of strict, sclerotic, bi-partite divisions, more effective modes of explanation are plurality, compromise and nego-tiation (cf. Anthias 2001). In short, as Lisa Lowe (1991) emphasizes, diasporic communities are notable for their hybridity, heterogeneity and multiplicity and, lending from Stuart Hall, she asserts that diasporic identity is a matter of 'becoming' and 'being'. Simply put, diaspora identity is creolized or hybridized (Featherstone 1995; Friedmann 1999; Nurse 1999); it is shaped by a melange of influences and constraints – cultural, social, political, economic – mediated through articulated through such themes as ancestral inheritance, the process of migration, the experience in the host space and further subsequent influences from the homeland to the remote diasporic communities (Mitchell 1997; Urry 2000).

Thus, through their roots and their routes, diaspora identities are multi-faceted and composed of complexly inter-woven strands of ethnicity, religion and ancestry. Diasporic communities have specific geographies and histories, they have multiple loyalties, they move between regions, do not occupy a single cultural space and, perhaps most importantly, operate exterior to state boundaries and their cultural effects (Mitchell 1997a). Interstitial positions are occupied by diasporas for whom there is a growing sense of their location in-between different cultures (Mirzoeff 2000: 2) which, in the case of these 'halfway populations' (Hollinshead 1998; see Ch. 2), may often be expressed by feelings of unease, ambiguity and ambivalence. More rigid forms of definition simply favour particular diasporic groups which in turn become the preferred objects of the academic gaze (Cohen 1997; Braziel and Mannur 2003). Readings grounded in fixed notions of home, identity and exile are also accompanied by the propensity to,

> privilege the geographical, political, cultural and subjective spaces of home-nation as an authentic space of belonging and civic participation, while devaluing and bastardizing the states of displacement and dislocation, rendering them inauthentic places of residence.
>
> (Braziel and Mannur 2003: 6)

Earlier interpretations are further compromised insofar as some groups such as Haitian, Cuban, Vietnamese and Khmer refugees may wish to bury deeply in their sub-conscious their troubled memories and recollections of the complex reasons and turbulent times that precipitated their departure from the home country. In this instance, looking back nostalgically may not be a primary action or defining feature of diaspora. Other complex histories may similarly frustrate definitional approaches based on the dualism of host country and homeland mediated by flow. For instance, Falzon (2003: 662) documents how Hindu-Sindhis left Sind in newly created Pakistan in 1947 and settled in Bombay in India. Today, Bombay, not Sind, functions as, what he terms, the 'cultural heart' of the diaspora, 'the node that connects and organizes translocality'. Thus, as

Braziel and Mannur (2003: 19) suggest, many members of diasporic communities 'may not know where *home is in order to stay there*' (emphasis in original). They draw on the work of Caren Kaplan (1996: 7) who concluded that,

> For many of us there is no possibility of staying at home in the conventional sense – that is, the world has changed to the point that those domestic, national or marked spaces no longer exist.
>
> (Quoted in Braziel and Mannur 2003: 19)

Instead, alternative conceptualizations have been proposed which attempt to embrace the complexity and plurality of diaspora. Braziel and Mannur (2003: 4) argue that 'once conceptualized as an exilic or nostalgic dislocation, diaspora has attained new epistemological, political and identitarian resonances as its points of reference proliferate'. For instance, Brah (1996: 180) argues that diasporas should be understood as 'historically contingent genealogies in the Foucauldian sense'. By exploring the historical trajectories of diaspora, she provides a critique of the fixed origins thesis such that a 'homing desire' may be identified, but this is entirely different to a desire for a 'homeland' (see also Falzon 2003). The distinction is crucial because it alerts us to the fact that not all diasporas are motivated to return. In this respect, the metaphor of the rhizome may be usefully deployed in so far as diasporas may be perceived as rootless (sometimes even schizophrenic). Pnina Werbner (2002: 119) describes diasporas as '*chaorders*, chaotic orders, which are inscribed both materially and imaginatively in space, time and objectifying practices'. From dislocated positions in their multi-nodal networks, although organizationally chaotic, diasporas are notable for their shared sense of co-responsibility, in particular as articulated in material gestures across space and in the struggle for enhanced citizenship rights for themselves and fellow diaspora members elsewhere.

## Diasporas, citizenship and transnationalism

The term 'hyphenated community' as an alternative to diasporic community has resulted from the semantic coupling of the homeland and the host state. For example, people originally of Irish origin who have settled and lived in the USA are referred to as 'Irish-Americans', persons of Asian descent in the UK are often described as 'British-Asians' and Russians with German ancestry from Volgaland who subsequently migrated to the USA are German-Russian-Americans.

Hyphenation in this manner presents commentators with a series of analytical opportunities as well as potential pitfalls. As Soja's work (1996) intimates, it is effectively the hyphen in 'hyphenated community' that is a crucial first step towards understanding diasporic identity. This is because it implies the resolution of the contemporary act of 'being' with the historical process of 'becoming'. Effectively, the hyphen demarcates the diasporic community as a distinct social group in the host state while simultaneously distinguishing it from other similar groups scattered in the diaspora but originating from the same homeland. Thus, although professing a common bond to the homeland as well as accepting some common historical antecedents, Scottish-Americans, Scottish-Canadians, Scottish-New Zealanders and Scottish-South Africans will, for this reason, inevitably have similar, yet contrasting identities, shaped, as they have been and will continue to be, by the alternative narratives and stimuli in the receiving countries.

The hyphen emphasizes that diaspora is a byword for compromise, negotiation and differentiation, even instability and metamorphosis. Unfortunately, for some critics, 'diaspora' and associated hyphenations have been indiscriminately used in under-theorized and even untheorized ways (Cohen 1997; Braziel and Mannur 2003). Here, the hyphenated designation may obscure plurality rather than expose it fully. For instance, labels such as 'African-American', 'Asian-American' and 'British-Asian', or descriptors such as 'the African diaspora' or the 'Black Atlantic diaspora' may be used (too) casually, almost for convenience's sake, to distinguish particular groups without full forethought of the implications. 'Catch-all' terms of this manner mask important differences within wider diasporic communities as well as obscuring the particularities and complexities of trajectories and episodes of past identity formation. Lowe (1991) and Radhakrishnan (2003) depict important internal fissures inside groups described broadly as 'Asian-Americans' and 'Indian-Americans', while Paul Gilroy (1993) warns against essentializing narratives of the 'African diaspora' by arguing that important cleavages exist within this 'group' in terms of social, cultural, economic lineaments. As such, Braziel and Mannur (2003: 3) warn that 'theorization of diaspora should not be divorced from historical and cultural specificity'. Stuart Hall's (1990) reading of cultural identity and diaspora extends this logic. He reads the Caribbean as triply traversed by a *Présence Africaine*, *Présence Européenne* and *Présence Américaine*, (as well as several other cultural presences such as the Indian, Chinese and Lebanese among others) that over time both mediate and position, as well as re-negotiate and relocate, Caribbean identities. In this respect, it is useful to reflect that diasporas do not exist in 'splendid isolation', practically hermetically sealed away from other diasporic communities and groups in host society. Abstractions that deal with dual host and homeland may overlook that multiple diasporic landscapes may be superimposed on one another in space. Instead of singularity and exclusivity of spatial occupation, diasporic populations exist side-by-side in many countries, cities and neighbourhoods.

In its favour, Hague (2001: 145) observes that the hyphenation highlights the crucial duality of ethnicity and citizenship which is imbued in each diasporic community. Hague argues that it is significant that ethnic identity (usually) precedes citizenship in the hyphenated construction. Ethnicity is especially important in fashioning self-identity, but the coupling of ethnic self-identification with a citizenship affiliation mediates a much stronger identity. However, we would contend that, although not without merit, such a view downplays the significance of citizenship in diasporic identity forming. Citizenship may precipitate further troubling dilemmas that add to the feelings of destabilization, uncertainty and ambiguity that so characterize the diasporic condition. By provoking the issue of affiliation (to a state or states), diasporas are forced to confront their roots and routes and how these mediate sense(s) of belonging. The most obvious and immediate dilemma is, as Scheffer (1995: 13) recognizes, where to take citizenship. Scheffer's view is that ideally homeland governments would prefer migrants to retain their original citizenship, with only temporary status when away and regular contacts with home. Should they decide to settle away permanently, home governments would prefer the migrants to remain as 'incipient diasporas' because, as an interim stage, this does not preclude the possibility of return; it presupposes reasonably strong contact with social, cultural and political institutions at home; political control over the diasporic communities is made much easier; and diasporic organizations are less likely to reflect the host country's interests.

As Clarence (1999: 202) reminds us, citizenship refers to more than membership of a particular state. Rather, 'citizenship is a status bestowed on those who are full members of a community. All who possess the status are equal with respect to the rights and duties with which this status is bestowed' (Marshall 1992: 18, cited in Marshall and Bottomore 1992). Thus, citizenship incorporates issues of participation and access as well as rights and obligations of the citizens themselves. As a basis for discussion, Clarence invokes Marshall's (1992) triadic conceptualization of civil (rights to secure individual freedom and justice), political (rights to participate in elections to institutions that exercise power) and social citizenship (rights to economic welfare, security, social heritage and socially acceptable way of life). Delanty (2000: 14) notes that Marshall's ideas, originally published in 1950, marked the shift from a previous market-based model of civil society to a state-based model, thereby reflecting a gradual confluence of liberalism with social democracy.

Criticized now as dated, Anglo-centric, lacking in universality, failing to address gendered and ethnic inequalities and underestimating the power of the state (Clarence 1999; Delanty 2000; Urry 2000; Pearson 2002; Murphy and Harty 2003), Marshall's ideas serve two purposes here: they allow us to confront the prior orthodoxy of 'entitlement' and its relationship to diaspora; and they introduce more recent, radical alternative conceptualizations of citizenship resulting from transnationalism. Although transnationalism is a highly contested concept (Hannerz 1996; Portes *et al.* 1999; Vertovec 1999; Delanty 2000; Faist 2000; Papastergiadis 2000; Kivisto 2001), the working definition adopted here is Braziel and Mannur's (2003: 8). For them, trans-nationalism is 'the flow of people, ideas, goods and capital across national territories in a way that undermines nationality and nationalism as discrete categories of identification, economic organization and political constitution'. They differentiate 'diaspora from transnationalism . . . in that diaspora refers specifically to the move-ment – forced or voluntary – of people from one or more nation-states to another. Transnationalism speaks to larger, more impersonal forces – specifically those of globalizations and global capitalism'. Faist (2000: 197) adds the caveat that 'diasporas tend to constitute a specific type of transnational community' and in his view they 'can only be called transnational communities, if the members also develop some significant social and symbolic ties to the receiving country', although these ties need not necessarily be concrete.

Thus, in a world characterized by dynamism, flows across borders and enhanced mobilities of goods, services, knowledges, risks, cultures and travellers, older constructs of citizenship are challenged by the 'exemplary communities of the transnational moment', diasporas (Tölölyan 1991: 4–5). As Cohen (1997: ix) observes, the old dogma that 'immigrants would identify with their adopted country in terms of political loyalty, culture and language can no longer be taken for granted'. In other words, a former, very static view, whereby to qualify for citizenship diasporic members as immigrants had to assimilate or integrate over a long period, has been largely superseded by alternatives such as ethnic pluralism and the border crossings of social spaces. This is notwithstanding the concession that assimilation may be a more powerful force for the second and subsequent generations (Portes 1999, cited in Kivisto

2001: 563). Instead, new forms of citizenship have emerged which reflect the erosion of the power of the state, the increasing importance of sub-state groups, such as diasporic communities, and their claims for the same political and democratic rights as majority (national) groups (Tambini 2001; Hindess 2002; Murphy and Harty 2003). Globalization and transnationalism have mediated a situation whereby states have been compelled to move 'into a realm of global citizenship where rights and duties and forms of participation and identity, operate in a "post" or "de" nationalized and border-less world of labour, capital and knowledge movements' (Pearson 2002: 991–2).

The implications for the relationship between diasporas and tourism are profoundly important. Levitt and de la Dehesa (2003) identify a more erudite approach on the part of homeland states to their relationships with their diasporas. Instead of a more suspicious, ambiguous and cynical relationship of the type articulated by Sheffer (1995), heightened globalization may forge stronger ties between migrants and their home states. Increasingly, states are willing to de-couple residence and citizenship. By effectively extending the state boundaries to incorporate those living overseas, states are prepared to allow migrants to participate in the national development process. As discussed later (Chs 12–16), tourism is a vital, but critically disregarded framework through which overseas citizens can exercise their rights to participate and by which they may be encouraged to do so by institutions at home. Thus, tourism represents a vital medium by which post-national and post-sovereign social relations may be resolved because it acts practically as a strong socio-cultural glue which bonds the home state with 'its' migrants. Moreover, as David Duval (2003; Ch. 3) argues, tourism is one major mechanism by which the de-territorialization of culture functions. Increasingly, as Papastergiadis (2000: 115) puts it, 'people now feel they belong to various communities despite the fact that they do not share a common territory with all other members'. In these 'pluri-local' or 'hetero-local' (Zelinsky 2001) social networks, people feel connected with one another across geopolitical boundaries and sometimes vast distances by imagined and/or tangible common bonds. Through the return visit, tourism becomes an embodiment of and facilitator for, these widespread social practices. Faist's (2000) thesis provides further support for such a valorization of tourism. He argues the terms 'transnational social spaces' and 'transnational communities' are often used practically synonymously. For him, a more nuanced view of international migration, in fact, reveals that there are three types of transnational social spaces: transnational kinship groups, transnational circuits and transnational communities. These are the outcomes of three primary mechanisms of integration that operate in transnationalism: reciprocity in small groups, exchange in circuits and solidarity in communities. Transnational social spaces, which may be occupied by diasporas, operate on different scales from families and kinship groups to circuits and networks of interest (perhaps in trade) and to collectives and communities (such as the diaspora *per se*). In each of these cases, tourism is a crucial structural framework through which the agencies of these three types of transnational social space function and are articulated. It provides a means of connecting people as the basis for reciprocity in kinship visits; it facilitates the performance of exchange in the development of trade circuits and networks; and it provides a platform for the mobilization of the collective where solidarity is the objective.

## Ethnicity, diaspora and tourism

Without wishing to become embroiled in the intricacies of definition (Banton 2001), an operating definition of the concept of ethnicity may be the 'process by which individuals allude to a sense of belonging to groups with similar socio-cultural traits and normative behaviour' (Drury 1994, cited in Stephenson 2002: 379). From this perspective, Stephenson (2002) notes that ethnicity has become a frequently discussed component in tourist motivation. In an early paper, King (1994) identified two forms of 'ethnic tourism' in which ethnicity is a primary determinant. The first and perhaps more predictable form is evident in Smith's (1978) and Graburn's (1978) early work among others; namely, ethnic tourism is manufactured from a desire to seek out the cultural exoticism of other ethnic groups and societies (McIntosh and Goeldner 1990: 139–40). Exotic 'others' become the primary focus of the tourist gaze (Urry 1990; MacCannell 1992). For example, the cultures of indigenous peoples in Australia (Hollinshead 1996; Zeppel 1998; Moscardo and Pearce 1999), Canada (Li 2000) and New Zealand (Barnett 1997; Ryan 1997) have been heavily commodified (Butler and Hinch 1996). Ethnic tourism becomes a means by which another culture may be experienced and interpreted by outsiders. According to Li (2000), it is effectively an antidote to the rationalizing discourses of western white culture identified by Dean MacCannell (1992), albeit the strength of ethnicity as a motivation varies notably among visitors to ethnic attractions (Moscardo and Pearce 1999; Ryan and Huyton 2000a,b).

King's (1994: 173–4) second and less frequent application of the term applies to travel movements whose primary motivation is ethnic reunion. He notes that,

> [t]his travel could be motivated by a desire to delve into family histories through travel to the relevant country. It might or alternatively might not involve actually staying with family . . . and this type of ethnic tourism has tended to be regarded as virtually synonymous with the visiting friends and relatives or VFR traffic.
>
> (King 1994: 174)

Here, he argues, the emphasis is not on contrast or on the exotic as in the first form. Rather, the search for similarity, belonging and group identification is a primary motivation. Esman (1984) noted that some ethnic groups use travel and tourism to the 'home country' to (re)assert, reaffirm and perform their heritage (cf. Timothy 2002a). Thanopoulos and Walle (1988) recorded that 30 per cent of Greek-Americans are potential travellers back to Greece, while in high summer 1989, 38 per cent of visitors to Poland were Polish-born (Ostrowski 1994). However, there are subtle, yet significant variations among ethnic tourists. Some may be motivated by familial piety and obligation as practically ethnic pilgrimages to ancestral homes (Cohen 1974), some may be motivated by temporary returns as expatriate migrant workers and others may even pave the way for remigration of members of the community (King and Gamage 1994; Nguyen and King 1998; Kang and Page 2000; Feng and Page 2000). Travel among and within ethnic groups is uneven depending on the structural framework of social, cultural and economic conditions in which an ethnic group is embedded (Stephenson and Hughes 1995).

Of course, this latter reading of ethnic tourism forms a starting point for much of the subsequent attention here to diasporas and tourism. Scattering of ethnic groups around the globe is an obvious precondition for this type of ethnic tourism. Thus, from Wood's (1998: 218) review essay, it is hardly a blinding revelation that the three principal conceptual strands that bind tourism with ethnicity, ethnic relations and ethnic identities are also applicable to diasporic groups: first, tourism becomes a form of ethnic relations (in this case between members of the diaspora and/or with members of other ethnic groups) (van der Berghe 1980, 1994); second, tourism plays a role in the development of touristic ethnic (i.e. diaspora) cultures, in which interaction with tourism becomes an integral part of the construction of ethnic (i.e. diaspora) identity; and third, through the de-differentiation of the tourist realm, touristic modes of visualization and experience become characteristics of the expression and consumption of ethnicity (see also Picard and Wood 1997).

Where diasporas differ from other ethnic groups and hence warrant more detailed consideration with respect to tourism, is in their distinct assemblages of characteristics and attributes, their temporal and spatial experiences, their contemporary geographical juxtapositions and their social and cultural constructs. We would contend that diasporas have been under-valorized in tourism discourse because the potency of the mutually implicated relationships between tourism and the dual conditions of 'being' and 'becoming' have yet to be fully recognized (cf. Ch. 2). As the next section identifies, on a more functional level particular patterns and processes of tourism consumption and production precipitate from the diasporic condition. However, travel and tourism have crucial roles to play reflexively in the processes of learning and self-discovery that define the fluid, constantly unfolding nature of diasporic identities (Hollinshead 1998). Tourism does not just represent a vehicle for straightforward, practically automatic voyages of self-discovery and identity affirmation. Visits to homelands or elsewhere into the diaspora may result in troubling, disconcerting and ambiguous experiences as well as new-found ambivalences (Stephenson 2002; Duval 2003). Tourism contributes to the construction of contemporary narratives of diasporic heritages which articulate to members of diasporas, as Lowe (1991) may put it, who they are and how they came to be.

## Spaces and places of diaspora travel and tourism

Given the complexities and nuances of the relationship between tourism and diaspora, it is none the less three of the central, most frequently mentioned and widely accepted characteristics of diaspora that have immediate resonances for tourism enquiry: namely, the duality of the 'home' and the 'host' country in the consciousness of diaspora members; the myths, nostalgia, imagined and actual histories of the group and the home; and perhaps, most importantly, that identities, behaviour and cultures in diasporic communities 'abroad', although similar to the 'homeland' and elsewhere in the diaspora, are inevitably distinctive and contrasting due to the infusions and conflations borne of their interstitial existence. When teased apart further, these three facets either alone or in combination suggest that there are six distinctive patterns of travel and tourism associated with the spaces and places occupied and travelled through by diasporas. Each results in quite individual encounters and visitor experiences and each has major consequences in terms of the production of tourism products and packages as well as place more widely.

First and perhaps most predictably, members of diasporic communities make trips in search of their roots and their routes with aims of reaffirming and reinforcing their identities. Most commonly, these are associated with trips back to their original homelands, but they may also include, as a second variant, trips to visit co-members of the extended community beyond the homeland. These trips, which often take the form of secular pilgrimages, are practised by diaspora members in the vain hope of discovering more about themselves, their ancestry, their heritage, their families and their extended communities. Stephenson (2002) describes how members of a UK Caribbean community travel to ancestral lands is mediated in no small measure by mothers' and grandmothers' encouragement to maintain links with their place of origin. Matriarchal as well as peer group networks contribute to the creation of particular place narratives and the generation of aspiration. Duval (2003) charts the return visits of Toronto's Eastern Caribbean communities. Their experiences revealed that visits were used as a means of retaining social histories as well as contextualizing social and cultural backgrounds after migration. Importantly, his study pointed to the ambivalences of experience encountered by some diaspora tourists which were sometimes compounded by their discomfort at their ambiguous reception in the homeland (cf. Stephenson 2002: 409). Bruner (1996) explores visits to Ghana by African-Americans and their meetings with local Akan-speaking Fanti at Elmina Castle, a major staging post in the mid-Atlantic slave trade (see also Ch. 7). Considerable differences are evident in the readings of slave castles between indigenous West Africans and African-Americans with the latter described as 'too emotional' by the former. This state of enhanced sentiment is ascribed by Bruner (1996: 293) to the 'almost mythic image of Africa as Eden. For black American men . . . a return to Africa is a return to manhood, to a land where they feel they belong, where they can protect their women and where they can reconnect with their ancestry'. Epstein and Kheimets (2001) focused on the concepts of dis- and re-connection with diasporic homelands in their study of the visits made by Russian Jews to Jerusalem in the post-Soviet era. They drew similar conclusions by identifying a 'double pilgrimage': in the first element, their trips comprise visits to King David's capital and the foundations of the original and ancient Jewish state; as part of the second they visit the roots of Christian civilization (Via Delorosa, the Garden of Gethsemane and the Holy Sepulchre). Greatest understanding of the tourists' roots was obtained from Yad VaShem, the 1953 Holocaust commemoration. The significance of the double pilgrimage is in its appeal to post-Soviet perceptions of self-identity which they read as the need to embrace the heritage of Grand Russian culture, an essentially Christian meta-narrative and the Jewish legacy, a feature which was denied by Soviet censorship.

The search for roots and routes has also manifested itself in the rise of so-called 'genealogical' (Nash 2002; Meethan 2002; Ch. 9), 'ancestral' (Fowler 2003) or 'family history' tourism. This form of travel may be both domestic and international depending on the family's routes and roots. Increasingly, visitors are travelling longer distances and over longer periods to retrace the footsteps and experiences of their ancestors. Genealogical tourism may comprise several components, some of which overlap with ethnic reunion tourism. Visiting friends and relatives in extended families and communities to reaffirm bonds of kinship may be accompanied by visits to poignant sites in the personal heritages of individuals and communities. As Fowler (2003) points out, these are increasingly being supplemented by the visitor's search for documented evidence and tangible artefacts of a forebear's existence. More structured, targeted research trips to local libraries, archives and government offices for 'official documentation' are being built into private and commercially marketed trips. Once the domain of local history societies, in recent time 'family history' has become one of the most commonly practised recreational pursuits throughout the world with tourism joining the Internet as the means for an individual to develop a richer understanding of his or her personal heritage (Timothy 1997; Fowler 2003). As subsequent chapters demonstrate, those searching for their roots and routes represent potentially fruitful market segments in an increasingly competitive global (cultural) tourism market place (Liu *et al.* 1984; Thanopoulos and Walle 1988; Morgan *et al.* 2002).

The third pattern practically represents the first in reverse. Residents of the original 'homeland' may make a trip to diaspora spaces to discover how co-members of the diaspora, perhaps even their friends and relations, have adapted to life and conditions in another place. Although many of these visits may also be routine VFR exercises (Feng and Page 2000; Kang and Page 2000), many are centred on the consumption of experiences, events, spectacles and festivals in their particular manifestation beyond 'home' in the diaspora (see Ch. 17). For instance, weekend city packages to Boston, New York and Chicago to experience the St Patrick's Day parades, pageants and events are popular short break products in the Irish market. Similarly, Scottish-Americans celebrate 6 April as Tartan Day (Hague 2001) and the 'Juneteenth Celebrations' in the USA attract many African-American visitors commemorating, as they do, General Granger's proclamation in Galvaston (Texas) on 19 June 1865 that all slaves were free (Janiskee 2002). In 2001 there were 285 Juneteenth celebrations in 46 states and most were held in Texas and California.

Spectacles like Juneteenth are not exclusive to, or possibly even dominated by, the consumption of diaspora tourists (see Zelinsky 2001). Thus, as Hoelscher's (1998) work on Swiss-Americans in New Glarus (Wisconsin) makes clear, in a variation of the above, diasporic communities also become the object of a wider tourist gaze. Diasporic destinations become notable attractions and features on 'mainstream', non-diaspora tourists' vacation itineraries; in effect, they come under a particular lens of 'ethnic tourism' to gaze on exotic Others. Local commodification of unique imagined and/or real diasporic heritage(s) may help produce local place distinctiveness in an increasingly competitive global market otherwise characterized by thematic replication and serial reproduction (Short and Kim 1999; Coles 2003). Either deliberately or unintentionally, the melange of cultural and ethnic influences in diasporic spaces produce distinctive place products and experiences which appeal to non-diasporic cultural (or ethnic) tourists. Ukrainian, Polish and Swedish neighbourhoods in Chicago warrant mention in most guidebooks (Given 2001), while the Polish Museum of America is second in the top 25 attractions in the city (Sinclair 2002). Patagonian tourism development has benefited greatly from the cultural capital imbued in the landscape by nineteenth-century Welsh migrants. In this part of southern Argentina, the peculiarities of afternoon tea, an annual Eisteddfod (festival) and the Welsh architectural style combined with more recent Argentinean cultural heritage have conspired to engender a vacationscape of great appeal to domestic visitors and the overseas Welsh (Schlüter 1999). In Neu Braunfels (Texas), German heritage is privileged in order to differentiate the the town in the visitor market place (Adams 2002; cf. Hoelscher 1998). This commodification creates ironies and tensions in two respects. First, the Hispanic population is growing rapidly and is marginalized in the tourism commodification process. Second, although not explicitly settled by Bavarians, as indeed much of Texas was not, the local community has still chosen to use the iconography and cultural references of southern Germany to fashion place identity. Such a deliberately selective approach is not untypical in the USA (Zelinksy 2001).

The themes of travelling, mobility and movement and transit spaces in the process of diasporic scattering are the basis for the fifth form. For many European-Americans, Ellis Island and the Statue of Liberty have become one of the most important attractions managed by the US Park Service. Equally, for many Asian-Americans, Ellis Island, although not directly implicated in their diasporic episodes, has come to symbolize (indirectly) their migration to and entry in to the USA (see Ch. 10; Kraut 1982; Kirshenblatt-Gimblett 1998). The European port towns of Rotterdam, Bremen, Hamburg, Liverpool, Southampton, Cork and Omagh have recently collaborated to develop a network of common heritage attractions to celebrate their roles as nodes in the mass migrations of the late-nineteenth and early-twentieth century (Richards and Bonink 1995: 177; see also Hoerder 1993). Spaces of transit do not necessarily have to include points of departure or entry, disembarkation or administrative processing such as port, quays, immigration depots and customs houses. Sites of 'dark tourism' or 'thanatourism' often recall dislodgements, dislocations and dispossessions in the collective histories of diasporas (Lennon and Foley 2000; Dann and Seaton 2001; Butler 2001; Essah 2001; Seaton 2001). Concentration camps and other sites of Nazi atrocities in the Holocaust have become regular features on Jewish travellers' itineraries to Europe (Kugelmass 1993, 1994; Ashworth 1996; Gruber 1999, 2002). In the case of Jewish-Americans' travel in the USA, they include important – in some cases former – Jewish neighbourhoods in major cities (Ioannides and Cohen Ioannides 2002; see also Ch. 6). Brooklyn and the Lower East Side of Manhattan offer subsequent generations the opportunity to walk the streets their forebears once trod and to imagine the conditions in which they lived (cf. Riis 1890). As Conforti (1996) has recently argued, ghettos have become popular tourist attractions. Urban 'ethnic tourism enclaves' (Timothy 2002b), or 'ethnic villages and showplaces' (Zelinksy 2001: 94), such as Chinatowns, Little Italies and Little Indias (Conforti 1996; Henderson 1999; Chang 2000; Eade 2000), have been heavily developed and deliberately commodified by public and private capital to attract and to cater for large volumes of visitors. The existence of these enclaves and hence their potential roles as tourist attractions is, however, uncertain as many face considerable threats from the forces of contemporary urbanization and urbanism (Buzzelli 2000; Eade 2000; Gabaccia 2000; Timothy 2002b).

Given the process of post-arrival colonization, the final form of travel flows and tourism spaces generated by diasporas is to destinations, resorts, retreats and vacation spaces which they have fashioned for themselves in the host state. For example, the Jewish community in the North East USA developed and congregated at the Catskill mountain retreat (Brown 1998; Ioannides and Cohen Ioannides 2002). Similarly, much of the capital invested in the early development of the resort of Sosua in the Dominican Republic was from exiled German Jews who arrived in 1941 (Cameron 2000).

## Structure of the book

We have placed diasporic spaces and how they are mediated for and negotiated by the diaspora tourist at the centre of the book's organization. By definition, diasporas exist scattered across space, tourism consumes space and place and the mutually reinforcing relationships between diaspora and tourism are played out in highly particularized spaces. The book is divided into three sections between which there is a degree of overlap. Briefly put, these are concerned with how diaspora tourists consume and experience space; the types of spaces and settings occupied by diaspora tourists; and the mechanics of commodifying diaspora and stimulating diaspora tourism.

### Diasporic experiences of tourism

In the first section of the book, we aim to explore diasporic experiences of tourism. The emphasis is on the interaction between tourism experience and identity; how identity helps to figure the selection and choice of tourism and travel experiences and episodes; and how the tourist experience may be reflected upon, or reflexively shape the fluid, constantly unfolding identities of diasporic groups and their individual members.

Keith Hollinshead offers an intricate reading of the connectivities between tourism and diaspora as well as a critique of current tourism engagement with diasporas. Inspired by post-colonial discourse and the work of Gilroy and Bhabha in particular, he presents a detailed exposé of two approaches to conceptualizing diaspora to augment the discussion above. His contribution articulates the inherent complexity and multiplicity of diasporic populations and questions whether it is possible to know and understand them in a full sense. His argument echoes Braziel and Mannur's (2003: 3) warning against an 'uncritical, unreflexive application of the term "diaspora"'. One of the key issues raised particularly in Chapter 2 and elsewhere in his work (Hollinshead 1996, 1998), is the ontological foundations of the subject matter and their epistemological challenges for tourism studies. Tourism clearly may impact on diasporic identity and vice versa, but this relationship resolves in highly complex and deeply subtle ways. Not surprisingly in light of the intricacy of most writings on diaspora, somewhat provocatively Hollinshead questions whether members of the tourism academy are equipped to interpret and decode relationships between tourism and diaspora and their attendant processes of mediation and negotiation. This, he contends, is not possible until tourism researchers appreciate more sympathetically the full dimensions of the fluidity, dynamism and interstitiality that define diasporic groups.

Subsequent chapters in this section take up Hollinshead's call to arms. Beyond his elaborate hypothecations, other contributors delve into specific connotations of diaspora discourse for understanding diaspora tourism. The common denominator is the mutually implicated nature of the experience of tourism and diasporic identity. David Duval adopts a transnationalist perspective to conceptualize the return visits of members of the Eastern-Caribbean diaspora living in Toronto (Ch. 3). Duval stresses the positive role of tourism as a discrete social practice in enabling transnational social networks to function. By bringing diaspora members into physical contact with one another, tourism cements the social relevance of the extended community for individual members while renewing, reiterating and reinforcing their cultural norms and values. Duval concedes that ambiguities and ambivalences may also be evident in individuals' tourism experiences. In contrast, in his account of tourism, racism and the UK Afro-Caribbean diaspora (Ch. 4), Marcus Stephenson addresses the internal and external limits placed on tourism by diasporic identity. He argues that identity may impact on tourism patterns and practices, not so much by determining the places to visit, but rather by suggesting destinations to avoid or from which diaspora members may be excluded (cf. Philipp 1994). Tourism may deliver diasporas socially alienating encounters with other social groups. In this respect, the search for cultural familiarization and the identification with others within the diaspora through tourism assumes an altogether different meaning.

Michael Hall and David Duval investigate the links between tourism and migration among Pacific Islanders in New Zealand (Ch. 5). Travel experiences of American Jews are discussed by Dimitri Ioannides and Mara Cohen Ioannides (Ch. 6). In their separate ways, these chapters make two common observations. First, both expose the influence of post-migration conditions on post-migration tourism. Conditions in the new home and how these came to be mediated strongly direct motivations to travel and patterns of tourism consumption. For instance, Hall and Duval point to the reassuring role of diasporic networks in helping migrants from the Pacific Islands adjust to New Zealand. Conversely, when economic conditions in their new home worsened in the 1980s, such networks also allowed the migrants to return to the islands temporarily. As a second observation, both warn against the problems of employing broad diaspora descriptors. Umbrella designations such as 'Jewish-Americans' and 'Pacific Islander-New Zealanders' are convenient labels; they define, by means of common attributes, diasporas and hence differentiate them against other social and ethnic groups. However, they mask important internal variations resulting from distinctive migrational trajectories, cultural and social practices and experiences in the host country. In this respect, Ioannides and Cohen Ioannides note that many of the journeys made by Jewish-Americans could be termed 'modern-day pilgrimages'. They are not necessarily driven by religious observance, but rather acts of nostalgic devotion at the sites of their ancestors. Their voyages reflect what it means to be Jewish in a more complete cultural sense. Thus, the existence of distinctive cultural ('sub-')groups among the Jewish-American population will almost inevitably precipitate distinctive travel and tourism patterns by virtue of their different roots and routes.

The search for personal and collective memory forms the subject of the final three chapters in the section. However, the focus switches to the experiences of individual diaspora tourists. In Chapter 7, Dallen Timothy and Victor Teye document the visits of African-Americans to West Africa. Visits to Ghana are read as laden with nostalgia and a means by which visitors may confront their troubled past and assert their heritage which is often denied them in the USA. Similar themes are evident in Erik Cohen's account of the Exodus Program (Ch. 8). This boat trip is designed to preface longer visits made by young diaspora Jewish travellers to Israel (Cohen 1999; Cohen *et al.* 2002). The Exodus Program is a short-term, carefully choreographed, stage-managed, quasi-simulation of a defining moment in Israel's history. Cohen argues that the Exodus Program is important as both a diaspora-building and identity-forming exercise and that it fosters enhanced appreciation of subsequent visits to Israel. Not only does it offer a chance to bond with other travellers, it provides crucial opportunities for young people to question and to reflect upon, their identity, their forebears' identities and the existence of the State of Israel. Finally, Kevin Meethan presents early findings of research on the linkages between genealogy and tourism (Ch. 9). Like Cohen, Meethan emphasizes the importance of tourism as contextualization for members of diasporic communities. Where Cohen emphasizes historical and geopolitical narratives as media for self-discovery, Meethan's respondents, who are drawn from diverse ethnic backgrounds and national affiliations, are concerned with highly localized, family-specific texts and artefacts for constructing their self-identities and asserting their personal heritages. While such 'roots tourists' may not necessarily be concerned with highly profound questions such as their position within and membership of, major global diasporas, genealogical investigations driven by tourism allow a much richer tapestry of family roots and routes to be assembled strand by strand, albeit often through mundane and banal discoveries. For roots tourists, place specificity and material traces of ancestors, however parochial, legitimize kinship. As Sean Carter reflects (Ch. 12), small differences are important in establishing the sense of us rather than them (Thrift 2000: 384).

## *Settings and spaces for diaspora tourism*

The second section of the book considers the settings and spaces in which diaspora tourism is performed. Diaspora tourism is not an activity that takes place in isolation, divorced from other structures and agencies. Diaspora tourists occupy time and space with other tourists. They routinely compete with non-diaspora tourists for access to resources, attractions, amenities and services. They contest placial representations with non-diasporic tourists as well as with tourists from other distinctive diasporic affiliations. Thus, the objective of this section is to explore the more literal as well as abstract (cultural, social, economic and political) spaces in which diaspora tourism takes place. As the contributions to this section illustrate, tourism is seen as a catalyst to wider processes of change stimulated by and for diasporas, while concurrently diaspora tourism is impacted on by over-arching imperatives and meta-narratives in the cultures, societies, economies and political systems in which diasporas are embedded.

The Ellis Island Immigration Museum in New York is the subject of Joanne Maddern's essay (Ch. 10). As the gateway to the USA for the 'moving millions' of the nineteenth- and twentieth-century migrations, Ellis Island and the nearby Statue of Liberty have become collectively symbolic of the centrality of diasporic communities in US culture and society. Maddern unpicks the multiple strands of discourse surrounding the production of the Ellis Island experience and draws wider lessons on the negotiation of visitor attractions in multi-ethnic cities (Caffyn and Lutz 1999). She notes that Ellis Island is a space that simultaneously includes and excludes diasporic groups, although like other flagship museums it aspires to present a common heritage (cf. Golden 1996). This results in 'interpretive mismatches' (Craik 1998: 115) between different diasporic groups as well as between the intended audience and the producers. Ellis Island privileges (Central and Eastern) European migrations which form the basis for mainstream narratives of immigration into the USA. At the same time, other groups such as African-Americans and Japanese-Americans – that is, major forerunners and successors of the European migrants – are less visible and are forced to use Ellis Island as a means of attacking rationalizing discourses and orthodoxies.

Maddern's concerns about the multiple ways in which tourism spaces may be read is echoed in subsequent chapters on the Vietnamese community in Australia by Nguyen and King (Ch. 11) and the global Chinese diaspora by Alan Lew and Alan Wong (Ch. 13). Obvious patterns of travel associated with family and national, religious and cultural practices in Vietnam are highlighted. Nguyen and King point, though, to the importance of positionality in tourism research on diaspora. As intimated by Hollinshead (Ch. 2), there is a propensity to view diaspora from western perspectives. The irony of an apparently insidiously imperialist orthodoxy pervading a supposedly post-imperialist enterprise is not lost in wide-ranging critiques of diaspora studies (Braziel and Mannur 2003). It should, however, be acknowledged in tourism studies of diaspora that there must be greater sensitivity on the part of the investigator to the role his or her position may have ultimately on the nature of the diasporic reading (Hall 1990; Chow 2003). Alternative readings of the role, meaning and construction of tourism experiences are required for different diasporic groups given the unique collisions of their social, cultural, political and economic settings. Constructs such as 'Orientalism' (Said 1995) and 'Balkanization' (Meštrović 1994) are usefully informed by and contribute to further, our understanding of the relationships between tourism and diaspora (cf. Ch. 12). With an altogether more modest objective, in developing their migrant (tourism) consumption model, Nguyen and King argue for a more complete understanding of the structures in both the 'home' and host countries as potential push and pull factors on diasporic travel patterns. In particular, they point to Buddhism in configuring social and cultural practices both at home and in the diaspora and hence its role in mediating tourism.

Lew and Wong deploy the concept of social capital as a means of exploring the role of diaspora tourism in the political economy of China. They target the social settings, spaces and practices of tourism and contend that tourism, in general, has the potential to develop social capital, but in many cases it fails to do so. In contrast, diaspora presents an institutional framework through which tourism facilitates the assembly of intricate social networks which satisfy wider aims and objectives. Based on a reading of eastern social and cultural codes, they argue that Guanxi and Confucianism induce distinctive travel patterns and practices and thus, in turn, tourism enables diaspora to function more smoothly and to overcome the friction of distance.

Sean Carter (Ch. 12) explores how the Croatian diaspora has been connected with home through tourism since the break up of Yugoslavia. Although Croatia had been one of the principal destinations in the former Yugoslavia (Fox and Fox 2000; Partridge 2003), new forms of tourism consumption have been mediated and actively encouraged by the state. His conclusions echo Lew and Wong's on the operation of diaspora over space, although he is mainly concerned with the mobilization of diaspora by political institutions for the purpose of building and galvanizing support for an independent Croatian state. Diasporas transgress political spaces, not least in their patterns of tourism consumption. Accordingly, he argues that there is a need to progress beyond the orthodox ethnically based readings of diasporas and tourism.

### Mobilizing diasporas for tourism

Visits made by diaspora tourists may take several forms. Diaspora tourists may visit friends and relatives or participate in structured tours and packages or both. They travel independently, often aided by information gleaned from the Internet. Some travel agents and tour operators offer dedicated products to diaspora tourists (cf. Butler *et al.*

2002; Klemm 2002; Klemm and Kelsey 2002). State tourism organizations have started to engage with 'their' diasporas overseas in order to tap into markets which they perceive to be culturally close and hence already sympathetically predisposed to a trip 'home'.

The final section of the book focuses on the particularities of the production of tourism for diasporas. The contributors set out to chart the deliberate efforts made to attract diaspora visitors through the commodification of their distinctive themes, motifs, attributes and conditions. These supply-side accounts describe the emotional devices and nostalgic triggers, policies and strategies, products and marketing campaigns intended to attract members of the diaspora to consume destinations. They address the extent to which the complex characteristics and intricately woven narratives of diaspora are acknowledged, interpreted and employed by the multifarious stake-holders in diaspora tourism production. In the process they question the role played by public and private sector actors as agents in the mediation of diasporic identities through the products, experiences and marketing themes they convey.

Three chapters explore the attempts made by state tourism organizations to mobilize diasporas for tourism. In his discussion of recent efforts to draw Jewish-Americans to Germany (Ch. 14), Tim Coles argues that producers must develop a more detailed understanding of diasporas if they are to access these markets effectively. This requires not only an understanding of diasporic conditions and of what should (and indeed may) be commodified for diaspora tourism, but also an appreciation of the different types of diaspora visitor and their varying degrees of diasporic motivation. In an echo of prior contributions to this volume, diaspora tourism marketing is often linked with wider politico-economic ambitions. Diaspora campaigns may appear as adjuncts to, and take place against the backdrop of, larger tourism marketing initiatives and place promotions. While the business case for this approach is obvious, it runs the risk that appeals to diasporic tourists will be largely unheard among the multiple, competing messages of other heritage and cultural tourism products and experiences. Moreover, producers should reflect that inevitable compromises and commercial realities deliver hybridized products and that there are more subtle outcomes generated by selective commodification.

This is a theme pursued by Nigel Morgan and Annette Pritchard in their decon-struction of the Welsh Tourist Board's *Homecoming 2000 – Hiraeth 2000* campaign (Ch. 15). Introduced in time for the millennium celebrations, the aim was to capitalize on nostalgic yearnings among diaspora members to make trips back to Wales. They painstakingly analyse the promotional devices, iconography and imagery involved. As they contend, not only have the mechanics of marketing destinations to diasporas been overlooked, there has been little interest in the nature and relevance of the iconography, narratives and straplines to diaspora markets. Welsh national identity, they note, is composed of three major strands, but paradoxically what they term 'Welsh Wales' is marginalized in the campaign. Instead, they lament that 'British-Wales', or the metropolitan view from Cardiff, which they argue is rootless, placeless and lacking the vibrancy of 'Welsh-Wales', is privileged in order to conform with the dominant messages of devolved Welsh governance.

Kevin Hannam detects a similarly contradictory approach by the Indian government to diaspora and tourism (Ch. 16). Although similar to the Chinese diaspora because social and cultural obligations shape relations between communities in India and overseas, as yet the state has struggled to mobilize effectively diasporic peoples for tourism in spite of the potential economic and social opportunities they may deliver. More structured attempts to access diaspora have been grounded in the cultural politics of the Indian state, nation-building after independence and Partition in 1947 and the manufacture of a new 'secular', collective diaspora to which many overseas Indians do not wish to subscribe and ironically to which the Indian government itself is ambivalent. With investment in marketing at a premium, he argues that a more intelligent, culturally-sensitive approach to India's many diasporas is required by the federal and state governments.

Deborah Che (Ch. 17) unpacks the history of Tulip Time in Holland (Michigan), a local celebration of diasporic identity rooted in the cultural practices of migrant-settlers from the Netherlands. She charts transformations in the festival over time and explores the drivers behind transition. Change and authenticity are the watchwords of her contribution, concentrating as she does on the relevance of the experience to local tourism stakeholders and visitors alike. To survive, like other ethnic celebrations and spectacles, Tulip Time has had to evolve to reflect conditions locally and in visitor markets. This, though, has created tensions as some sections of society have argued its authenticity has been compromised. By recognizing that ethnicity and, in particular, authenticity are negotiated terms lacking absolute conditions (Friedmann 1999), she identifies two forms of diasporic spectacle: ones that represent static snapshots of yesteryear; and more iterative, evolving events such as Tulip Time. Although the latter are culturally distant from original versions, they reflect more faithfully the constant re-negotiation of diasporic identities over time, the contemporary cultural manifestations of diaspora and the prominent role such spectacles exert reflexively in identity mediation.

Finally, Noga Collins-Kreiner and Daniel Olsen (Ch. 18) document the different types of products and experiences brought to Jewish markets by private tour operators and travel agents via the Internet. They develop a typology of nine different product types. These may lean heavily towards Jewish heritage, culture, religion and leisure. However, in keeping with Ioannides and Cohen Ioannides, culture more generally and not necessarily the exclusive concepts of Israel or religion, directs the nature of the products. The variety of products is a function of the internal heterogeneity of diaspora markets as well as tourists' individualized tastes and preferences for touristic activities and diasporic reference points. Fragmentation into multiple niches, each of which is attended by distinctly assembled products, is indicative, they argue, of the pursuit of competitive advantage among producers who realize that heritage and cultural tourism markets are increasingly congested.

## References

Adams, J. (2002) 'The "Bavarianization" of German Texas: ethnic tourism development in Neu Braunfels', paper presented at the Annual Conference of the Association of American Geographers, Los Angeles.

Adams, K.M. (1997) 'Ethnic tourism and the renegotiation of tradition in Tana Toraja (Sulawesi, Indonesia)', *Ethnology* 36(4): 309–20.

Anthias, F. (2001) 'New hybridities, old concepts: the limits of "culture"', *Ethnic and Racial Studies* 24(4): 619–41.

Ashworth, G. (1996) 'Holocaust tourism and Jewish culture: the lessons of Krakow-Kazimierz', in M. Robinson, N. Evans and P. Callaghan (eds) *Tourism and Culture: Towards the 21st Century*, London: Athenaeum Press.

Banton, M. (2001) 'Progress in ethnic and racial studies', *Ethnic and Racial Studies* 24(2): 173–94.

Barber, B.J. (2001) *Jihad vs. McWorld. Terrorism's Challenge to Democracy*, New York: Ballantyne Books.

Barnett, S.J. (1997) 'Maori tourism', *Tourism Management* 18(7): 471–74.

Bhabha, H. (1994) 'Frontlines/borderposts', in A. Bammer (ed.) *Displacements: Cultural Identities in Question*, Bloomington, IN: Indiana University Press.

Brah, A. (1996) *Cartographies of Diaspora. Contesting Identities*, London: Routledge.

Braziel, J.E. and Mannur, A. (eds) (2003) *Theorizing Diaspora. A Reader*, Malden, MA: Blackwell Publishing.

Brown, P. (1998) *Catskill Culture: A Mountain Rat's Memories of the Area*, Philadelphia, PA: Temple University Press.

Bruner, E.M. (1996) 'Tourism in Ghana: the representation of slavery and the return of the black diaspora', *American Anthropologist* 98: 290–304.

Bruner, E.M. and Kirshenblatt-Gimblett, B. (1994) 'Maasai on the lawn: tourist realism in East Africa', *Cultural Anthropology* 9: 435–70.

Butler, D.L. (2001) 'Whitewashing plantations: the commodification of a slave-free antebellum south', *International Journal of Hospitality and Tourism Administration* 2(3/4): 163–75.

Butler, D.L., Carter, P.L. and Brunn, S.D. (2002) 'African-American travel agents: travails and survival', *Annals of Tourism Research* 29(4): 1022–35.

Butler, R.W. and Hinch,T. (eds) (1996) *Tourism and Indigenous Peoples*, London: Thompson I.B.P.

Buzzelli, M. (2000) 'Toronto's postwar little Italy: landscape change and ethnic relations', *Canadian Geographer* 44(3): 298–305.

Caffyn, A. and Lutz, J. (1999) 'Developing the heritage tourism product in multi-ethnic cities', *Tourism Management*, 20(5): 213–21.

Callahan, R. (1998) 'Ethnic politics and tourism. a British case study', *Annals of Tourism Research* 25(4): 818–36.

Cameron, S. (2000) *Footprint Handbook: Dominican Republic*, 1st edn, Bath: Footprint Books.

Chang, T.C. (2000) 'Singapore's Little India: a tourist attraction as a contested landscape', *Urban Studies* 37(2): 343–66.

Chow, R. (2003) 'Against the lures of diaspora: minority discourse, Chinese women and intellectual hegemony', in J.E. Braziel and A. Mannur (eds) *Theorizing Diaspora. A Reader*, Malden, MA: Blackwell Publishing.

Church, A. and Frost, M. (2004) 'Tourism, the global city and the labour market in London', *Tourism Geographies* 6(2) (forthcoming).

Clarence, E. (1999) 'Citizenship and identity: the case of Australia', in S. Roseneil and J. Seymour (eds) *Practising Identities. Power and Resistance*, Basingstoke: Macmillan.

Clifford, J. (1989) 'Notes on travel and theory', *Inscription* 5: 177–88.

Clifford, J. (1994) 'Diasporas', *Cultural Anthropology* 9: 302–38.

Clifford, J. (1997) *Routes. Travel and Translation in the Late Twentieth Century*, Cambridge, MA: Harvard University Press.

Cohen, E. (1974) 'Who is a tourist? A conceptual clarification', *Sociological Review* 22(4): 527–55.

Cohen, E.H. (1999) 'Informal marketing of Israel Experience Educational Tours', *Journal of Travel Research* 37(3): 238–43.

Cohen, E.H., Ifergan, M. and Cohen, E. (2002) 'A new paradigm in guiding: the *madrich* as a role model', *Annals of Tourism Research* 29(4): 919–32.

Cohen, R. (1997) *Global Diasporas*, London: Routledge.

Coles, T.E. (2003) 'Urban tourism, place promotion and economic restructuring: the case of post-socialist Leipzig', *Tourism Geographies* 5(2): 190–219.

Conforti, J.M. (1996) 'Ghettos as tourist attractions', *Annals of Tourism Research* 23(4): 830–42.

Cooper, C., Fletcher, J., Gilbert, D., Wanhill, S. and Shepherd, R. (1998) *Tourism: Principles and Practice*, 2nd edn, Harlow: Pearson Education.

Craik, J. (1998) 'Interpretive mismatch in cultural tourism', *Tourism, Culture and Communication* 1(2): 115–28.

Dann, G.M.S. and Seaton, A.V. (2001) 'Slavery, contested heritage and thanatourism', *International Journal of Hospitality and Tourism Administration* 2(3/4): 1–29.

Delanty, G. (2000) *Citizenship in a Global Age. Society, Culture, Politics*, Buckingham: Open University Press.

Drury, B. (1994) 'Ethnic mobilization: some theoretical considerations', in J. Rex and B. Drury (eds) *Ethnic Mobilization in a Multi-Cultural Europe*, Aldershot: Avebury.

Duval, D. (2003) 'When hosts become guests: return visits and diasporic identities in a Commonwealth eastern Caribbean community', *Current Issues in Tourism* 6(4): 267–308.

Eade, J. (2000) *Placing London. From Imperial City to Global Capital*, New York: Berghahn Books.

Epstein, A.D. and Kheimets, N.G. (2001) 'Looking for Pontius Pilate's footprints near the Western Wall: Russian Jewish tourists in Jerusalem', *Tourism, Culture and Communication* 3(1): 37–56.

Esman, M.R. (1984) 'Tourism as ethnic preservation: the Cajuns of Louisiana', *Annals of Tourism Research* 11: 451–67.

Essah, P. (2001) 'Slavery, heritage and tourism in Ghana', *International Journal of Hospitality and Tourism Administration* 2(3/4): 31–49.

Faist, T. (2000) 'Transnationalism in international migration: implications for the study of citizenship and culture', *Ethnic and Racial Studies* 23(2): 189–222

Falzon, M.-A. (2003) '"Bombay, our cultural heart": rethinking the relationship between homeland and diaspora', *Ethnic and Racial Studies* 26(4): 662–83.

Featherstone, M. (1995) *Undoing Culture. Globalization, Postmodernism and Identity*, London: Sage.

Feng, K. and Page, S.J. (2000) 'An exploratory study of the tourism, migration–immigration nexus: travel experiences of Chinese residents in New Zealand', *Current Issues in Tourism* 3(3): 246–81.

Fowler, S. (2003) 'Ancestral tourism', *Insights* March: D31–D36.

Fox, J. and Fox, R. (2000) 'Croatia and cultural tourism: a social and economic imperative', *Revue de Tourisme – The Tourist Review* 2000(1): 14–18.

Franklin, A. and Crang, M. (2001) 'The trouble with tourism and travel theory', *Tourist Studies* 1(1): 5–22.

Friedmann, J. (1999) 'The hybridization of roots and the abhorrence of the bush', in M. Featherstone and S. Lash (eds) *Spaces of Culture. City–Nation–World*, London: Sage.

Gabaccia, D. (2000) *Italy's Many Diasporas*, London: Routledge.

Gilroy, P. (1993) *The Black Atlantic: Modernity and Double Consciousness*, London: Verso.

Given, T. (2001) *Chicago Condensed*, 3rd edn, Footscray, IL: Lonely Planet Publishing Ltd.

Golden, D. (1996) 'The museum of the Jewish diaspora tells a story', in T. Selwyn (ed.) *The Tourist Images. Myths and Myth-making in Tourism*, Chichester: Wiley.

Graburn, N. (1978) 'Tourism: the sacred journey?', in V. Smith (ed.) *Hosts and Guests. The Anthropology of Tourism*, Oxford: Blackwell.

Gruber, R.E. (1999) *Jewish Heritage Travel: A Guide to East-Central Europe*, Northvale, NJ: Jason Aronson Inc.

Gruber, R.E. (2002) *Virtually Jewish. Reinventing Jewish Culture in Europe*, Berkeley, CA: University of California Press.

Hague, E. (2001) 'The Scottish diaspora. Tartan Day and the appropriation of Scottish identities in the United States', in D.C. Harvey, R. Jones, N. McInroy and C. Milligan (eds) *Celtic Geographies. Old Culture, New Times*, London: Routledge.

Hall, C.M. (2002) 'ANZAC Day and secular pilgrimage', *Tourism Recreation Research* 27(2): 83–7.

Hall, C.M. and Page, S.J. (1999) *The Geography of Tourism and Recreation. Environment, Place and Space*, London: Routledge.

Hall, C.M. and Williams, A.M. (eds) (2002) *Tourism and Migration: New Relationships between Production and Consumption*, Dordrecht: Kluwer.

Hall, S. (1990) 'Cultural identity and diaspora', in J. Rutherford (ed.) *Identity: Community, Culture and Difference*, London: Lawrence and Wishart.

Hall, S. (1996) 'Politics of identity', in T. Ranger, Y. Samad and O. Stuart (eds) *Culture, Identity and Politics: Ethnic Minorities in Britain*, Aldershot: Avebury.

Hannerz, U. (1996) *Transnational Connections: Culture, People, Places*, London: Routledge.

Helmreich, S. (1992) 'Kinship, nation and Paul Gilroy's concept of diaspora', *Diaspora* 2(2): 243–49.

Henderson, J. (1999) 'Attracting tourists to Singapore's Chinatown: a case-study in conservation and promotion', *Tourism Management* 21: 525–34.

Hindess, B. (2002) 'Neo-Liberal Citizenship', *Citizenship Studies* 6(2): 127–43.

Hoelscher, S. (1998) 'Tourism, ethnic memory and other-directed place', *Ecumene* 5(4): 369–98.

Hoerder, D. (1993) 'Introduction. Special issue: European ports of emigration', *Journal of American Ethnic History* 13(1): 3–5.

Hollinshead, K. (1996) 'Marketing and metaphysical realism: the disidentification of Aboriginal life and traditions through tourism', in R. Butler and T. Hinch (eds) *Tourism and Indigenous Peoples*, London: Thompson I.B.P.

Hollinshead, K. (1998) 'Tourism and the restless peoples: a dialectical inspection of Bhabha's halfway populations', *Tourism, Culture and Communication* 1(1): 49–77.

Hudman, L.E. and Jackson, R.H. (1992) 'Mormon pilgrimage and tourism', *Annals of Tourism Research* 19: 107–21.

Ioannides, D. and Cohen Ioannides, M.W. (2002) 'Pilgrimages of nostalgia: patterns of Jewish travel in the United States', *Tourism Recreation Research* 27(2): 17–25.

Jamison, D. (1999) 'Tourism and ethnicity: the Brotherhood of Coconuts', *Annals of Tourism Research* 26(4): 944–67.

Janiskee, R. (2002) 'Juneteenth on the rise: the geography of America's "Second Independence Day"', paper presented at the Annual Conference of the Association of American Geographers, Los Angeles.

Jutla, R. (2002) 'Understanding Sikh Pilgrimage', *Tourism Recreation Research* 27(2): 65–72.

Kang, S. K.-M. and Page, S.J. (2000) 'Tourism, migration and emigration: travel patterns of Korean-New Zealanders in the 1990s', *Tourism Geographies* 2(1): 50–65.

Kaplan, C. (1996) *Questions of Travel: Postmodern Discourses of Displacement*, Durham, NC: Duke University Press.

King, B. (1994) 'What is ethnic tourism? An Australian perspective', *Tourism Management* 15(3): 173–6.

King, B.E.M. and Gamage, M.A. (1994) 'Measuring the value of the ethnic connection: expatriate travellers from Australia to Sri Lanka', *Journal of Travel Research* 33(2): 46–9.

Kirshenblatt-Gimblett, B. (1998) *Destination Culture: Tourism, Museums and Heritage*, Berkeley, CA: University of California Press.

Kivisto, P. (2001) 'Theorizing transnational immigration: a critical review of current efforts', *Ethnic and Racial Studies* 24(4): 549–77.

Klemm, M. (2002) 'Tourism and ethnic minorities in Bradford: the invisible segment', *Journal of Travel Research* 41: 85–91.

Klemm, M. and Kelsey, S.J. (2002) 'Catering for a minority? Ethnic groups and the British travel industry', paper presented at Tourism Research 2002 – An Interdisciplinary Conference in Wales, Cardiff.

Kraut, A. (1982) *The Huddled Masses. The Immigrant in American Society, 1880–1921*, Wheeling, IL: Harlan Davidson Inc.

Kugelmass, J. (1993) 'The rites of the tribe: the meaning of Poland for American Jewish tourists', in J. Kugelmass (ed.) *YIVO Annual*, vol. 21, *Going Home*, Evanston, IL: Northwestern University Press.

Kugelmass, J. (1994) 'Why we go to Poland: Holocaust tourism as secular ritual', in J.E. Young (ed.) *The Art of Memory: Holocaust Memorials in History*, Munich: Prestel.

Lennon, J. and Foley, M. (2000) *Dark Tourism. The Attraction of Death and Disaster*, London: Continuum.

Levitt, P. and de la Dehesa, R. (2003) 'Transnational migration and the redefinition of the state: variations and explanations', *Ethnic and Racial Studies* 26(4): 587–611.

Li, Y. (2000) 'Ethnic tourism: a Canadian experience', *Annals of Tourism Research* 27(1): 115–31.

Liu, J., Timur, A. and Var, T. (1984) 'Tourism income multipliers for Turkey', *Tourism Management* 5(4): 280–7.

Lowe, L. (1991) 'Heterogeneity, hybridity, multiplicity: marking Asian American differences', *Diaspora* 1(1): 24–44.

MacCannell, D. (1992) 'Empty meeting grounds', *The Tourist Papers*, London: Routledge.

Marshall, T.H. and Bottomore, T. (1992) *Citizenship and Social Class*, London: Pluto Press. (Updated edn. Original, T.H. Marshall, published under the same title in 1950, Cambridge: Cambridge University Press).

McIntosh, R.W. and Goeldner, C.R. (1990) *Tourism. Principles, Practices and Philosophies*, 6th edn, New York: Wiley.

Meethan, K. (2002) 'Tourism, "Roots" and the Internet', paper presented at Tourism Research 2002 – An Interdisciplinary Conference in Wales, Cardiff.

Meštrović, S.G. (1994) *The Balkanization of the West. The Confluence of Postmodernism and Postcommunism*, London: Routledge.

Mirzoeff, N. (2000) 'The multiple viewpoint', in N. Mirzoeff (ed.) *Diaspora and Visual Culture. Representing Africans and Jews*, London: Routledge.

Mitchell, K. (1997a) 'Different diasporas and the hype of hybridity', *Environment and Planning D: Society and Space* 15: 533–53.

Mitchell, K. (1997b) 'Transnational discourse: bringing geography back in', *Antipode* 29(2): 101–14.

Morgan, N., Pritchard, A. and Pride, R. (2002) 'Marketing to the Welsh diaspora: the appeal of *hiraeth* and homecoming', *Journal of Vacation Marketing* 9(1): 69–80.

Murphy, M. and Harty, S. (2003) 'Post-Sovereign Citizenship', *Citizenship Studies* 7(2): 181–97.

Moscardo, G. and Pearce, P.L. (1999) 'Understanding ethnic tourists', *Annals of Tourism Research* 26(2): 416–34.

Nash, C. (2002) 'Genealogical identities', *Environment and Planning D: Society and Space* 20(1): 27–52.

Nguyen, T.H. and King, B. (1998) 'Migrant homecomings: Viet kieu attitudes towards travelling back to Vietnam', *Pacific Tourism Review* 1: 349–61.

Nurse, K. (1999) 'Globalization and Trinidad carnival: diaspora, hybridity and identity in global culture', *Cultural Studies* 13(4): 661–90.

Oigenblick, L. and Kirschenbaum, A. (2002) 'Tourism and immigration. Comparing alternative approaches', *Annals of Tourism Research* 29(4): 1086–1100.

Olsen, D. and Timothy, D. (2002) 'Contested religious heritage: differing views of Mormon heritage', *Tourism Recreation Research* 27(2): 7–15.

Ostrowski, S. (1991) 'Ethnic tourism: focus on Poland', *Tourism Management* 12(2): 125–31.

Park, C. (1994) *Sacred Worlds. An Introduction to Geography and Religion*, London: Routledge.

Papastergiadis, N. (2000) *The Turbulence of Migration. Globalization, Deterritorialization and Hybridity*, Oxford: Polity Press.

Partridge, C. (2003) 'How to make a Split decision: the war put tourists off the delights of this coast, but now the tide is turning', *The Observer*, 20 April 2003 (Cash section): 16.

Pearson, D. (2002) 'Theorizing citizenship in British settler societies', *Ethnic and Racial Studies* 25(6): 989–1012.

Philipp, S.F. (1994) 'Race and tourism choice. A legacy of discrimination?', *Annals of Tourism Research* 21(3): 479–88.

Picard, M. and Wood, R.E. (eds) (1997) *Tourism, Ethnicity and the State in Asian and Pacific Societies*, Honolulu: University of Hawaii Press.

Pitchford, S. (1995) 'Ethnic tourism and nationalism in Wales', *Annals of Tourism Research* 22(1): 35–52.

Portes, A. (1999) 'Conclusion: toward a new world – the origins and effects of transnational activities', *Ethnic and Racial Studies* 22(2): 463–77.

Portes, A., Guarnizo, L.E. and Landholt, P. (1999) 'The study of transnationalism: pitfalls and promise of an emergent research field', *Ethnic and Racial Studies* 22(2): 217–37.

Radhakrishnan, R. (2003) 'Ethnicity in an age of diaspora', in J.E. Braziel and A. Mannur (eds) *Theorizing Diaspora. A Reader*, Malden, MA: Blackwell Publishing.

Richards, G. and Bonink, C. (1995) 'Marketing cultural tourism in Europe', *Journal of Vacation Marketing* 1(2): 173–84.

Richards, G. and Hall, D. (eds) (2000) 'Introduction', *Tourism and Sustainable Community Development*, London: Routledge.

Riis, J. (1890) *How the Other Half Lives*, New York: Charles Scribners' Sons; reprinted (1997), Harmondsworth: Penguin.

Ryan, C. and Huyton, J. (2000a) 'Aboriginal tourism – a linear structural relations analysis of domestic and international tourism demand', *International Journal of Tourism Research* 2(1): 1–15.

Ryan, C. and Huyton, J. (2000b) 'Who is interested in Aboriginal tourism in the Northern Territory, Australia? A cluster analysis', *Journal of Sustainable Tourism* 8(1): 53–88.

Ryan, C. (1997) 'Maori and tourism: a relationship of history, constitutions and rites', *Journal of Sustainable Tourism* 5(4): 257–78.

Safran, W. (1991) 'Diasporas in modern societies: myths of homeland and return', *Diaspora* 1(1): 83–99.

Said, E. (1995) *Orientalism. Western Conceptions of the Orient*, London: Penguin Books, reprinted abridged edition, originally published in 1978.

Schlüter, R. (1999) 'Sustainable tourism development in South America: the case of Patagonia, Argentina', in D.G. Pearce and R.W. Butler (eds) *Contemporary Issues in Tourism Development*, London, Routledge.

Seaton, A.V. (2001) 'Sources of slavery – destinations of slavery: the silences and disclosures of slavery heritage in the UK and US', *International Journal of Hospitality and Tourism Administration* 2(3/4): 107–29.

Shair, I.M. and Karan, P.P. (1979) 'Geography of the Islamic Pilgrimage', *Geojournal* 3(4): 599–608.

Shaw, G. and Williams, A.M. (2002) *Critical Issues in Tourism*, 2nd edn, Oxford: Blackwell.

Shaw, G. and Williams, A.M. (2004) *Tourism, Tourists and Tourism Spaces*, London: Sage (forthcoming).

Sheffer, G. (1986) 'A new field of study: modern diasporas in international politics', in F. Sheffer (ed.) *Modern Diasporas in International Politics.* London: Croom Helm.

Sheffer, G. (1995) 'The emergence of new Ethno-national diasporas', *Migration. A European Journal of International Migration and Ethnic Relations* (No. 28), 1995(2): 5–28.

Short, J.R. and Kim, Y-H. (1999) *Globalization and the City*, Harlow: Longman.

Shuval, J.T. (2000) 'Diaspora migration: definitional ambiguities and a theoretical paradigm', *International Migration* 38(5): 41–57.

Sinclair, M. (2002) *City Pack: Chicago*, Windsor: AA Publishing.

Smith, V.L. (ed.) (1978) *Hosts and Guests. The Anthropology of Tourism*, Oxford: Blackwell.

Soja, E. (1996) *Thirdspace: Journeys to Los Angeles and Other Real-and-Imagined Places*, Cambridge, MA: Blackwell.

Stephenson, M. (2002) 'Travelling to the ancestral homelands: the aspirations and experiences of a UK Caribbean community', *Current Issues in Tourism* 5(5): 378–425.

Stephenson, M.L. and Hughes, H.L. (1995) 'Holidays and the UK Afro-Caribbean Community', *Tourism Management* 16(6): 429–35.

Tambini, D. (2001) 'Post-national citizenship', *Ethnic and Racial Studies* 24(2): 195–217.

Thanopoulos, J. and Walle, A.H. (1988) 'Ethnicity and its relevance to marketing: the case of tourism', *Journal of Travel Research* 26: 11–14.

Thrift, N. (2000) 'It's the little things', in K. Dodds and D. Atkinson (eds) *Geopolitical Traditions: A Century of Geopolitical Thought*, London: Routledge.

Timothy, D.J. (1997) 'Tourism and the personal heritage experience', *Annals of Tourism Research* 24(3): 751–4.

Timothy, D.J. (2002a) 'Diaspora and genealogy: the experience of personal heritage tourism', paper presented at Tourism Research 2002 – An Interdisciplinary Conference in Wales, Cardiff.

Timothy, D.J. (2002b) 'Tourism and the growth of urban ethnic islands', in C.M. Hall and A.M. Williams (eds) *Tourism and Migration: New Relationships between Production and Consumption*, Dordrecht: Kluwer.

Tölölyan, K. (1991) 'The nation-state and its Others: in lieu of a preface', *Diaspora* 1(1): 3–7.

Tomljenovic, R. and Faulkner, B. (2000) 'Tourism and older residents in a sunbelt resort', *Annals of Tourism Research* 27(1): 93–114.

Urry, J. (1990) *The Tourist Gaze. Leisure and Travel in Contemporary Societies*, London: Sage.

Urry, J. (2000) *Sociology Beyond Societies: Mobilities for the Twenty-First Century*, London: Routledge.

van der Berghe, P.L. (1980) 'Tourism as ethnic relations: a case study of Cuzco, Peru', *Ethnic and Racial Studies* 3(4): 372–92.

van der Berghe, P.L. (1995) 'Marketing Mayas. Ethnic tourism promotion in Mexico', *Annals of Tourism Research* 22(3): 568–88.

van Hear, N. (1998) *New Diasporas. The Mass Exodus, Dispersal and Regrouping of Migrant Communities*, London: Routledge.

Vertovec, S. (1999) 'Conceiving and researching transnationalism', *Ethnic and Racial Studies* 22(2): 447–62.

Vukonić, B. (1996) *Tourism and Religion*, New York: Elsevier.

Wall, G. (1999) 'Partnerships involving indigenous peoples in the management of heritage sites', in M. Robinson and P. Boniface (eds) (1999) *Tourism and Cultural Conflicts*, Wallingford: CAB International.

Werbner, P. (2002) 'The place which is diaspora: citizenship, religion and gender in the making of chaordic transnationalism', *Journal of Ethnic and Migration Studies* 28(1): 119–31.

Williams, A.M. and Hall, C.H. (2000a) 'Tourism and migration', *Tourism Geographies* 2(1): 2–4.

Williams, A.M. and Hall, C.H. (2000b) 'Tourism and migration: new relationships between production and consumption', *Tourism Geographies* 2(1): 5–27.

Williams, A.M., King, R. and Warnes, A. (1997) 'A place in the sun: international retirement migration from northern to southern Europe', *European Urban and Regional Studies* 4(2): 115–34.

Wood, R.E. (1998) 'Tourism ethnicity: a brief itinerary', *Ethnic and Racial Studies* 21(2): 218–41.

Zelinsky, W. (2001) *The Enigma of Ethnicity. Another American Dilemma*, Iowa City: University of Iowa Press.

Zeppel, H. (1998) 'Come share our culture. Marketing Aboriginal tourism in Australia', *Pacific Tourism Review* 2(1): 67–82.

# Part I

# Diasporic experiences of tourism

# 2 Tourism and third space populations

## The restless motion of diaspora peoples

*Keith Hollinshead*

## Introduction: the mixed-up postcolonial world

The juggernauts of internationalization and globalization are mixing up the world in all sorts of new ways. Nation-states are under pressure, either dissolving in the face of these global conformities or changing their form and function while they have to adapt to these new international and transnational coercions (Cohen 1997: 156). Despite the pressure for all places to become globalized and 'cosmopolitan', all kinds of new contestatory 'spaces' have emerged where clearly-observable counter-nationalist, counter-ethnicist, counter-racist, counter-sexist and religious fundamentalist move-ments have begun to flower or re-flower. Scores of subdued populations now use the fissures within the postcolonial mood of the globalizing moment to forge a return (to varying degrees!) to what was/is local and what was/is familiar to them (Hall 1997: 35–6). They reach out for old/new groundings in response to the changing ethical, cultural and spiritual demands which stem from the tensions brought about by the leviathan of new global order.

To comprehend what is going on under the predicaments of the mixed up, fast-internationalizing, postcolonial moment, scholars of nationalism, migration and ethnic relations find they need a whole new legend of conceptual terms to be able to decipher the new identificatory maps of the admixed world. To some – mainly on the left, politically – the pressures induced by globalization are but another form of dominion, whereby the capitalist world economy merely replaces the normalizations of the old imperial order (Wallerstein 1984). To others – often under some postmodernist hue or other – the tensions of globalization do not just provide opportunity for a shift in dominance; rather, they provide opportunities for a re-calculation in things where various sorts of new identifications can voluntarily blossom, gradually (or even, suddenly) procuring a new social space or a new global voice for themselves (Featherstone 1994; Robertson 1994). While these profound disagreements run ever onwards, some suggest a unified global culture is evolving in many respects, while others maintain that we now have not so much a world of removed local cultures, but a more closely connected realm of global cultures, in the *plural*.

It is plain, therefore, that all sorts of new interrogations of national being and new interpretations of cultural association are in ferment. This chapter enters that bubbling froth by exploring the role and place of 'tourism' in the re-mixed, globalizing or glocalizing world. As learned commentators in so-called 'traditional', 'lead' disciplines fast re-address the shifting shapes and textures of 'the national' and 'the social' under the postcolonial/postindustrial moment (Crook 2001: 319), the chapter seeks to explore not so much what scholars in tourism studies are addressing *in vacuo*, but how some of the emergent understandings from the broader humanities – especially from cultural studies (King 1997a) – can help tourism studies researchers plot the agency of tourism as a player or catalyst in the major cultural transformations of the contemporary moment. Hence, the chapter seeks to ask that, if nations are indeed changing (King 1997a: 3), how is, or can, tourism help 'nationally defined societies' change? Otherwise, how is, or can, tourism help populations and sub-populations re-imagine themselves, especially where nations may be in decline?

## Crisis in representation: the gradual acceptance of multiple and transgressive affiliations

### The rise of communitas vis-à-vis societas

In recent decades, those who have sought to depict the play of cultural difference (within matters of national and societal meaning) have endured what Marcus and Fischer (1986) have called a 'crisis of representation'. In so many regions and places it has become harder to maintain a coherent image or sustain a continuous, uninterrupted memory for, or about, populations living there. Consensual traditions of and about, 'organic' ethnic communities are increasingly being challenged and have to be redefined (Featherstone 1995: 10). People within a nation tend to be less confident that they live within a state which has continually 'progressed' since its founding time. Formerly revered symbolic hierarchies are increasingly under attack (Featherstone 1995: 95). Cultural differences, which used to be presumed to lie *between* societies, are now increasingly being found to exist *within* them (Hall 1992). Those representing the nation have increasingly realized that the national image they are charged with projecting not only has an external face to it, but also an internal lineament. After all, cultures are plural in their origins and creole in terms of constituency (Friedman 1988).

Accordingly, in the broader humanities the view has emerged that many orthodox schemes for modelling social and national life are outdated, unable as they are to embrace the risen fluidity of contemporary life. Too many conventional models have been predicated upon universal categories and upon unified identifications (Featherstone 1995: 126). These do not acknowledge the complexities of our received collective identities and community traditions (Simmel 1971). Deleuze and Guattari (1987) have persuaded us to celebrate correctively the qualities of disorder, syncretism and hybridity. They have extolled a new 'nomadological' understanding, which views the globe as a world of runaway individuals and 'moveabout' groups, rather than as an unchanging realm of stable and sedentary populations. For them, such a change in conceptualization would allow us to decently measure the increasing flows of people around the world as walkers, as refugees, as sojourners, as whatever (cf. Featherstone 1995: 128). Furthermore, for Appadurai (1990) it would help us comprehend the very intensity of these flows and the dynamic interchange they so regularly inspire. In these ways, our new models could be based beneficially on metaphors of movement and marginality, for all of our endeavours and experiences are indeed enjoyed in a lifespace that exists 'without frontiers and boundaries' (Gabriel 1990: 396). Moreover, no traditions can ever be totally contained within a boundary; all customs, practices and cultural pursuits eventually cross their containing frontiers. Thus, all cultural activities journey.

These fresh and more flexible classifications for life tend not only to undermine fixed and conventional identities. They freshly privilege transgressive, avant-garde and 'Bohemian' impulses which build new links with supposedly 'other' elements, or with supposedly 'alien' structures. Consonantly, 'the locality' loses its importance as the primary stimulus for our experiences (Morley 1991), as we variously subsume ourselves in the wider global sphere and as we take on board multiple affiliations which undermine the taken-for-granted identifications we and others have held about ourselves. In these enlightened and interdependent ways, the conceivable downturn of exclusivist feeling for 'the nation' has not only allowed us to renew old 'tribal' kinship and ethnic ties of territory, but has catalysed our involvement in all sorts of new or transitory 'neo-tribes', which can be inspired by an infinity of personal, cultural or special interest imperatives (Maybury-Lewis 1992), without rich regard to concerns of spatial territory.

In recent years, therefore, a growing tide of theorists in the humanities have concluded that orthodox classifications of being and bonding have over-celebrated the significance of 'the contained society' and 'the bounded nation' (Tenbruck 1994). More recent classifications of identity and affinity have attempted to account for the conflicts and the hybridities which exist within societies (or within nations) and which can come to terms not only with that given population's interdependencies with so-called 'outsiders', but with the 'leakages of spirit' by and through which the so-called 'insiders' connect with the so-called 'outsiders'. Hence the principal concern could be said to have shifted from *societas* to *communitas* (cf. Featherstone 1995: 130). Such an intensifying regard for communitas has facilitated a more sensitive reading of the ease by which new communities of being and fellowship may be formed (Meyrowitz 1985). 'Psychological neighbourhoods' and 'personal communities' in which individuals desire to immerse themselves are more simultaneously *the excluded others*; that is, groups of customarily other, imaginatively different, psychically dissimilar people who seek to sustain their identity and affinity via non-mainstream or subjugated cultural forms have been exposed. In these ways, 'the other' has not only been seen to be s/he who is dislocated 'from us', but who is so frequently 'amongst us' and that proximal other is more frequently nowadays delineated 'positively' and in a 'non-hierarchical' fashion (Featherstone 1995: 96). Radical new pluralisms abound which are much more responsive to and discriminating about, the 'third cultures' which outsiders from afar initiate, develop and inhabit as they live within mainstream society (King 1990; Featherstone 1995: 91).

### Old imprints and new inscriptions: Gilroy on diasporic identities

Considerable effort is expended by populations in either the attempt to reconfirm precious old identities for themselves, or otherwise in the attempt to gain legitimacy for some newly valued particularity (Lemert 2001: 303–6). Indeed, much of contemporary world politics is today concerned with the ongoing efforts of governments at all levels to regulate the cultural, ethnic and associative powerplays of populations (Barkin and Cronin 1994). This is evident within the identity politics of diaspora. On the one hand, diasporic groups tend to find themselves condemned as 'pariah-people' who tenaciously seek to continue to engage with (perhaps) longstanding lands or with cherished cosmologies, or who are otherwise dismissed as 'abortive-civilizations' (Cohen 1997: 101–2). Thus, they are blocked by other 'native' or 'national' groups from celebrating some seemingly 'new' and 'anomalous' form of cultural expression. In these ways, diasporic populations often stand as either an apparently fossilized people displaced from their traditional heartland, or they are an apparently curious migratory people who are unable to secure a firm, sustainable cultural foothold in the new scattered locales in which they abode. Diasporic populations commonly inhabit difficult psychological spaces both outside of and inside of specific national societies. They are commonly external peoples with unshakeable loyalties reaching outside their host population; but they often are located in large numbers within their new locale, possessive of economic acumen, industrial skills, or other sophisticated practices which position them as a threat to those who are locally mainstream.

In recent years, such diasporic groups have been studied in depth. In examining such collective, relocated aspirations, investigators have focussed not just on what the revered identities of diasporic social movements are, but also how they have (or not) been mobilized *and* how they have (or not) been institutionalized (Delanty 2001: 478). What investigators, especially constructivists (Delanty 2001: 473), have concluded is that worldviews of diasporic populations are not so much as primordial narratives which that group seeks to uphold at turn, but they are resonating discourses. The latter may revolve around core foundational truths of some sort, but are always open to new arrangement and refabricative styling. Diasporic populations may initially meet upon 'essentialist grounds', so to speak, but they readily learn how to mobilize themselves via particular sorts of inventive reformulations of those underlying doxa.

Of the commentators to have worked incisively on the essential and non-essentialist approaches to diasporic identity, Gilroy has focussed upon both the social and the political aspects of identity in an endeavour to explore how diasporic representations tend to speak both to 'sameness' and 'difference' within the single signification. Thus, Gilroy is not an analyst who is comfortable with dualist approaches and he tends to challenge binary understandings which rigidly keep 'essentialist' views well and truly distinct from 'non-essentialist' views. To Gilroy, diasporic identities tend to be complex and multi-faceted; they are inclined to merge received and orthodox foundational-beliefs with new and fertile promissory expressions.

Box 2.1 embodies some of Gilroy's main insights into the coterminous 'stable power' and 'creative authority' of diasporic worldviews and it delineates Gilroy's thinking on what Leroi Jones (1967) termed 'the changing same'. The exhibit clarifies Gilroy's view that, while there is always bedrock which initially helps constitute diasporic worldviews, these should not be deemed to be absolute in their effects or unchanging in their reach (Gilroy 1997: 313). Totalized views of diasporic identity tend to deny important inventive processes of self-making, which may have been involved in their nurturing. Identities are not exclusively predicated upon ancient outlooks on kinship and are not 'fixed'. They do indeed move! Thus, in many nuanced ways diasporic worldviews serve as an alternative view of reality to the stern patterning of primordial inheritance (Gilroy 1997: 328). Diasporic worldviews are in many senses divorced from the disciplined purity of ancient kinship. Each stands as the compounded accretion of a metamorphic identity, where the resultant transcultural mixture has taken considerable advantage of the labile properties of language, culture and contemporary

**Box 2.1 The Protean character of diasporic self-making: ten major insights from Paul Gilroy on contemporary inscriptions of diasporic identity**

Diasporic outlooks on self and society tend to be:

**1 Corrective**
Diasporic inscriptions of identity frequently seek to call up old and previously longstanding notions of people-hood from which a given population has been forcefully ruptured.

**2 Anti-national**
Diasporic inscriptions of identity are generally identifications which the territorial order of (and within) nation-states is inclined to sanction, and diasporas themselves are explicitly anti-national groupings of people who adhere collectively – in part in opposition – to the coercive unanimity of 'the nation'.

**3 Transgressive**
Diasporic inscriptions of identity may be fruitfully understood to be the particularities of dissident outsiders; that is, of those who are comfortable with 'building block' models of being and with 'conflictual' negotiations of bonding.

**4 Difficult to read**
Diasporic inscriptions of identity are, as a rule, accretive compounds, constituting a transcultural mix of 'being' which has become divorced from the purity of any special affiliation or allure.

**5 Emergent**
Diasporic inscriptions of identity may be initially platformed on long-standing cultural, ethnic and other ties, but in the face of the vicissitudes of contemporary globalizing life they tend to be emanative and incomplete rather than fully-formed.

**6 Gelling**
Diasporic inscriptions of identity habitually involve ongoing processes of self-making where a population may initially come together in accordance with long-standing or long-illustrious bonds of being, yet also where that population consciously and actively seeks new and refreshing forms of social interactivity to further its own possibilities of economic or spiritual life.

**7 Negotiated**
Diasporic inscriptions of identity – particularly for individuals caught up in diasporic cross-currents – are continually being changed, re-shaped and re-defined.

**8 Transcultural**
Diasporic inscriptions of identity regularly involve the projection of global networks which have an entreaty that reaches beyond limited and traditional identification with 'roots' and 'common biology'.

**9 Imaginative**
Diasporic inscriptions of identity are not only inclined to draw on distinctive and cherished icons of yester-year and yester-century, but also tend (in

some circumstances) to be productive in terms of the creation of imaginary icons and ancestors.

**10 Promissory**
Diasporic inscriptions of identity ordinarily stand as an invocation to ancient ritual and myth, but they are just as much promissory vocalizations as primordial ones. It is seldom that diasporic outlooks suddenly become not only highly expressive acts of commemoration, but highly articulated acts of affiance or pledged undertaking.

Source: abridged and inspired by Gilroy (1997, esp. pp. 306–41).

life in its gelling. To Gilroy (2001: 323), diasporic identities are invariably protean, 'pointing towards a more refined and wordly sense of culture than [essentializing] characteristic notions of soil, landscape and rootedness [tend to exemplify]' (Gilroy 2001: 328). In his view they must be comprehended not in terms of notions of shared territory or ancestral lineage alone, but rather in terms of the complex dynamics of intercultural living and transcultural inhalation.

## Recognizing and interpreting 'public' and 'counter-public' spheres

### *The role of tourism in diasporic discourses of memory and speculation*

In learning from Gilroy that so many citizens of the world live in difficult psychological spaces, nowadays we recognize that the spatial dialectic which these individuals inhabit is a complex one; they co-terminously exist within various sorts of 'real' or 'imaginary' communities. People who are caught up in this difficult diasporic terrain often have to eke out an uncertain psychic existence sometimes able to celebrate the richness of the life of 'a transnational oversoul' (after Wilson and Dissanayake 1996: 9), yet at other times stumbling along as out-of-kilter 'half-souls' in the nation or region they find themselves. Such are the psychic perturbations of the exile's life caught in the predicaments between failing memories of old hearths and the emergent but speculative hopes of a new cultural homeland. It is these sorts of difficult discon-nectivities and new situated connectivities which are now being tracked by the analysts of 'hybridity'. And it is these sorts of psychic equivocacy located somewhere between 'predicament' and 'promise' which we must explore more rigorously within tourism studies.

Diasporic spaces – particularly the cosmopolitan diasporic spaces – warrant painstaking interpretive and critical investigation. Slowly, we are learning how to undertake close cognitive mapping forays into these difficult 'third spaces' of being and knowing to track the international motions of identity and to trace the local manipulations of affinity. And increasingly we find – as Gilroy intimates – that the diasporic populations of the world are slowly becoming astute in the support of new patriotic expressions of being and belonging as they seek to realign old traditional allegiances, or otherwise to contend cleverly against what Appadurai (1993) deemed the 'trojan nationalisms' of modernity. Such are the necessary neo-tribal and self-consolidating projections as they are caught up somewhere between the conventional canons of 'orthodox', 'traditional' communities and the fledgling formulations of neo-communities and neo-situations.

In this world of hybridizing identities, all manner of diasporic populations face tricky ordeals versioning their re-invigorated, their revised, or their newly adopted visions of 'self' and of 'tribe'. In such difficult claustrophobic configurations of nationality and identity, 'being' can fast become a schizophrenic activity, where those diasporic oversouls/half-souls are trapped in a kind of 'schizo-space' (after Keith and Pile 1993: 2), continually rubbing against the customary interstices of communal and national life. And yet, as evident in Box 2.1, it is insufficient to regard identificatory confrontations as mere binary collisions of contrapuntal imperialisms or nationalisms (after Said 1993). The neo-politics of global being and of local belongings alike are far more complex (Wilson and Dissanayake 1996: 2).

In tourism studies, the requirement of the moment has been primarily to map and monitor the function of tourism as a tool of self-consciousness as particular diasporic populations seek to journey back to old cultural stomping groups, or otherwise as they seek to project new articulations of selfhood on the new terrain they find themselves located (see Chs 12 and 15). In these ways, it is incumbent upon tourism studies researchers to inspect what effect projections of place have in either calling dispersed populations back to old hearthlands, or in otherwise helping those very dispersed populations to re-engineer themselves in new locale(s). Hence, it is incumbent upon scholars to gauge the embeddedness and the power of tourism (and travel) as industrial mouthpieces or as cultural-producers of specific discourses of memory and/or singular discourses of speculation. In the recovery of cultural identity and in the advocacy of new cultural identity, the role of tourism is as yet rather unfigured (Lanfant 1995: 4). Put simply, more informed scholarship is also required on the dialectical movements and the counter-movements of people as they travel between their dual worlds or 'locales' (Shenhav-Keller 1995: 151). And we require much more percipient scholarship of tourism as a 'resource of hope' (cf. Wilson and Dissanayake 1996: 4) for consolidating populations where tourism production and tourism display both play central parts in processes involved in the conscious construction of group selfhood (cf. Friedman 1990: 321). All of this amounts to a far more robust call for acute interpretive projects examining the role of tourism as a coding machine for populations (Horne 1992; see especially Hollinshead 1999). Such is the retooling that is demanded of the tourism studies academy if it is to make significant investigative forays into the difficult psychic situations which diasporic populations face and into the changing cultural and political-economic circumstances, which variously threaten or bolster each and every transnational imagination.

## An introduction to the work of Bhabha in deconstructing diasporic discourse

So far, we have recognized that diasporas are not such neat and discrete phenomena as is generally assumed. For instance, after Gilroy, diasporic positions are best seen not so much as highly specific situations of territorial dislocation where singular invocations of ethnic identity and/or cultural nationalism are clearly discernible.

Rather, they are difficult-to-read states of 'in-between-ness' where many sorts of cultural mutation and restless discontinuity/continuity transpire.

Fortunately, Bhabha's (1994) *The Location of Culture* has equipped us with a much more versatile vocabulary to be able to recognize and deconstruct the uncertain kinds of cultural mutation that Gilroy has signified. The rest of this chapter will be devoted to an advocatory critique of the value of Bhabha's acute commentary on the representation of 'difficult' sites of cultural identity for studies of diaspora in general and for studies of diaspora in tourism, in particular.

*The Location of Culture* is a landmark text about cultural divergence and it stands for a revision of longstanding views about social difference, the 'Self' and the 'Other'. It appears that we have long thought axiomatically only in terms of pristine, intact and fully bounded cultures. Bhabha's scrutiny of the world's emergent ethnicities and halfway locations of culture has been a principal entreaty to think more circumspectly about the apparent soundness, the completeness and the unity of cultures, as well as about the seeming shape and integrity of any subsidiary population (for example, any diaspora) which has been disparately removed or excised from some mainstream society or other. Thus, following Bhabha (1994), increasingly social scientists nowadays reject the view of culture as something concrete. Instead, they understand culture as a rather more supple realm of communal thought. Hence, culture is not regarded as a given 'system' or as a solid 'entity', but rather as a kind of more contextual, fictile arena where specific deeds and events have significances, albeit interpretable, always inherently ambiguous. Hence, after Bhabha the culture of a population is never anything more than an amalgam of contestable codes and interpreted representations. Thus, all cultural forms (like diaspora) are constructed or invented phenomena which are never constant, rather continually renewed and revised.

As such, Bhabha understands culture to be a fundamentally 'manufactured' and 'heavily iconic' activity. Hence, 'culture' is an imaginative 'process' rather than a palpable 'body'. No culture need ever be seen to have a definitive geography or a pervasive socio-historical context. Cultures are not only 'lived' in by the people that think within them, but cultures themselves 'live'. They are dynamic through place, through time and through inter-subjective (i.e. inter-personal) settings. In Bhabha's view, then, it is unhelpful and unwise to contain cultures within tight boundaries of identity, affiliation and/or as being.

Bhabha's critique of culture may be readily translated to diaspora studies. We have learnt how fragile are the ways in which people imagine themselves within their cultures; that is, within their various and situational *en groupe* collectives. Thus, we should also recognize the ways in which we have even hardened up our diasporic boundaries and we have confined people even within their in-between, third space, diasporic settings. We learn from Bhabha to be much more informed about the ways in which we might readily stereotype diasporic peoples, denying them their own 'fantasmatic' actualities and dreams. All subject-positions inside or outside of diaspora are elusive things to capture. We can work far too fast with forms of classificatory logic which not only calcify, but commit acts of 'violent logic' upon that essentialized population (Bhabha 1994: 9). Hence, meanings about diaspora (i.e. meanings about cultural being and belonging) are ultimately incalculable from an ontological point of view (Bhabha 1994: 330). Such matters of being and belonging will always be unstable and therefore should only be interpreted on the kind of ambivalent 'almost . . . *but not quite*' (Bhabha 1994: 86) basis which admits the possibility of other current, or not-yet-read perspectives and the immediate likelihood of the arrival of new fast-changing social knowledges. Subject-positions in culture are not only routinely 'negotiated', but they are routinely 'fictional' (Gandhi 1998: 163). In terms of diaspora *per se*, such moments of in-between-ness might appear to be savage dislocations from past experience for that identified people, but that identification (or that self-interpretation) can only be offered for a short moment in the dynamics of political and lived identity (Bhabha 1994: 25). It is not only people that travel: diasporic affinities and cultured identities also have mobile lives and enjoy effervescent and runaway existences.

## Advanced thoughts on Bhabha's 'New Sense': insights on fractious diasporic positions

Bhabha's commentaries on and into the nightmare of diasporic identifications is of prodigious importance to those who work in or investigate the cultural geography of diasporic 'space'. His critical reflections on emergent/interstitial/halfway/restless locations of culture comprise refreshing examinations of the old-new narratives of selfhood to reveal how various sorts of scattered populations these days live within difficult psychic states of ambiguity, carrying all sorts of historical contingency from our respective mixed pasts. Bhabha's longitudinal and latitudinal explorations of the coterminous terror of displaced 'being' and the new wonders of imagined putative futures is a stunning contradiction of older, limited conceptualizations of 'culture' and 'identity' in the past. *The Location of Culture* stands as an acute condemnation of not only the logocentrisms of nationalist and colonialist authorities who have sometimes advertently (but more damagingly, *inadvertently*) contained diasporic populations within restrictive old-sense classifications of personhood, but also of the logocentrisms with which theorists in the humanities have approached the same hybrid or interstitial spaces. It serves as a brilliant denunciation of certain orthodox binaristic under-standings about rights, citizenship and inter-group relationships and an exhilarating conceptual voyage into the dense and scarcely chartered mise-en-scène of bipolar, even multipolar living.

In many ways, Bhabha's commentaries extend Said's writings on 'Orientalism' (specifically) and on 'Othering' (generally) by revealing the very richness of the Orient's/the Other's inherent and changeable fantasies of Self (Quayson 2000: 62) and of the boundless variety of ambiguous and contesting ways in which those affirmations of being may be interpreted. Drawing from Fanon, Derrida, Foucault and Lacan, Bhabha's interdisciplinary critiques translate well to all manner of emergent cultural settings and unfolding positions of affinity and hope. Equally clearly, however, Bhabha's work is conceptually corpulent and it is deemed to be 'contrived' by many other cultural studies specialists (Quayson 2000: 43). Its obfuscatory style appears to some observers deliberately proliferative, intentionally and unnecessarily resisting decent holistic coherence (Dirlik 1994; Parry 1994). To Sangren (1988), the very postmodernistic style of contemporary critiques (like Bhabha's) is self-indulgent, needlessly dissolving established canons of thought without ever replacing them with any substantial alternative insight on matters of association, belonging and selfness.

Yet, there are scores of other commentators who praise Bhabha's work on the hybridity and ambiguity of postcolonial/diasporic identities for how his commentaries alert us to the differential possibilities of 'being' at or within any moment in time without having to fix those outlooks as being adamantine in the force of their social antagonism (Quayson 2000: 45). Bhabha's work might indeed be dense of substance, but much of that is excusable meticulousness to reveal how the world's emerging identities are indeed highly fractious (Mbembe 1992) and also multifarious, almost to the point of being daily inspected dialectically (Adorno 1973). Examinations of hybrid and emergent identities have to be dissolved very carefully and deconstructed most painstakingly if they are to be meaningfully (re-)aggregated. Just as there is nothing so pure and pristine as 'an unadulterated native position' for indigenous populations (Quayson 2000: 74), so there is nothing so singular and stark as the discursive modality of a singular diasporic or subaltern position (Spivak 1987). The 'soft structures' of heterogenous perspectives – especially in dynamic diasporic settings leave only light vapour trails; Bhabha's new dialectical radar which thankfully directs us towards new alphabets of understanding towards the capture of both fast-changing kinesics of the global gestures of diaspora populations and the postcolonial pantomimes of 'being', within which the latter are caught up.

## Discussion: Bhabha's 'New Sense' insights applied to the hybrid/diasporic spaces of tourism

Bhabha's dialectical inspections of interstitial culture constitute not so much a singular theory on being and becoming; rather, they form a related set of critical accounts of cultural hybridity. His examinations of distantiation and displacement inform us that what counts is not only the negation of many orthodox assessments of the 'Other' which are embedded within dominant representations of diaspora, but the close negotiation and renegotiation of spaces and the temporalities which exist between various 'Others'. Box 2.2 clarifies how Bhabha's ellipsian (i.e. his painstakingly deciphered) constructions (see Hollinshead 1998b: 70–1) of cultural hydridity can help to advance our mapping and monitoring of particular diasporas. More crucially, when teased out, Bhabha's concept of cultural hybridity also raises a series of inter-related, yet distinct questions which should add fundamentally to researchers' understanding of the connectivities between tourism and diaspora (see Box 2.3).

The central tenet of cultural hybridity is that it is composed of those transnational and transitional encounters and negotiations (over differential meaning and value) in 'colonial' contexts where new, ambivalent and indeterminate locations of culture are (fast) guaranteed. However, this inherent nature of cultural hybridity serves to frustrate knowing the concept as the new celebration of identity consists largely of problematic forms of signification which resist discursive closure (Bhabha 1994: 173). A number of postulates attend this meta-idea and Box 2.2 sets out to outline ten major concepts derived from a careful reading of *The Location of Culture*. These are not entirely distinct from one another; there is overlap between some of the statements. However, each has a dual function. It not only exemplifies the importance of cultural hybridity as a concept for understanding diaspora(s) *per se*, but rather it also has great resonance for unpacking the relationships between tourism and diaspora, as the questions in Box 2.3 demonstrate.

In the interests of space, only two questions have been attached to each Bhabhan concept, although more may in fact be precipitated (see Hollinshead 1998a,b for a

**Box 2.2 The emergent and ambivalent locations of culture: ten concepts based on Bhabha's ideas on cultural hybridity and ambiguity revealed in diasporic settings**

Cultural hybridity is composed of those transnational and transitional encounters and negotiations over differential meaning and value in 'colonial' contexts where new ambivalent and indeterminate locations of culture are generated, but where that new celebration of identity consists largely of problematic forms of signification which resist discursive closure.

1 **The emergent and multiple identities of populations**

   Many diasporic populations nowadays live under durable or multiple identifications of being where individuals within that population may claim to belong to competing affinities simultaneously.

2 **The reliance on new vocabularies of identification**

   Many of the new subject-positions of diasporic identity are highly difficult to exist within and a whole new vocabulary of specialist terms is required to cover the vocality, viewing positions, iconology and gestural moments within Third Space locations.

3 **The everyday production of people and places**

   Emergent representations of ethnic identity, racial differences and cultural belongings of diasporic populations are not just derived 'archaeologically' or axiomatically from the past, but are also produced in the here and now.

4 **The paradoxes of plural cultural identification**

   Many diasporic locations of culture constitute crossover zones of cultural identification where the population travelling through 'that space' might respond to several different geographic articulations and cultural imperatives simultaneously and with seeming contrariety to (for example) competing neo- and/or post- and/or colonialist injunctions at the same time in the same space.

5 **The emergent and partial identities of populations**

   At contested diasporic locations of place and identity, it is common for individuals to feel only partially connected to emergent celebrations of identity, since these 'celebrations' and 'projections' are often composed of a mix of as yet incommensurable elements, which have not yet been reconciled to one another.

6 **Gains and losses of syncretism**

   Many delineations of hybridity and ambiguity in the contemporary cultural politics of diaspora not only empathetically herald the new emergent syncretisms which are 'happening' in the world 'out there', but also they are substantively dis-proclaiming the old *a priori* racisms, ethnicities and cultural hierarchies of the world of modernity or modernistic enlightenment.

7 **Cultural politics of resistance**

   Emergent discourses of diasporic racialization, ethnicity and cultural affinity frequently invoke or yield iconic identities, which are mutually alienating and which place the diasporic group/individuals in cross-grained states of physic consciousness where it is not at all easy for it/them to feel 'integrated', 'whole' or 'secure'.

## 8  New performative power of enunciation

Many of the emergent locations of diasporic identity are sufficiently free or liberated to be able to generate ontological states of selfhood which from some outlooks are highly creative and dramatically effective in the manner in which they performatively invent or inscribe fantastic new possibilities of living, yet which from other outlooks, are still restricted in their creativity and effectivity by tough and persevering identifications from yester-year and yester-place.

## 9  New manumissive locations of culture

Many new interstitial locations are manumissive for diasporic peoples inhabiting or passing through such spaces, thereby enabling them to articulate various sorts of 'previously subjugated' or 'new experiential' identifications for themselves.

## 10  New political geography of space

Under new diasporic cultural politics of resistance – and under the new permissiveness of contemporary local/special-interest politics – all sorts of admixed or mesitzo 'populations' will emerge, able to articulate new, mongrel, throbbing articulations of spatial-by-temporal *en groupe* sentiments.

Source: assembled from Hollinshead (1998a: 126–8 and 132)

---

**Box 2.3  Ten Bhabhan diasporic concerns and potential future research questions connected to diaspora tourism which they inspire**

### 1  Emergent peoples: emergent and multiple identities of populations

- To what degree do diasporic tourists seeking to reinforce their identities through diasporic travel feel 'on trial' as cultural members of the place where they are now resident?
- To what degree do diasporic tourists consider themselves to be members of a secure and consolidated diasporic community in the place they are now resident?

### 2  Ethnic group maintenance: the reliance on new vocabularies of identification

- To what degree do diasporic tourists feel 'on trial' as members of the diasporic community in the distant territory or homeland to which they have travelled?
- How do diasporic tourists consider that the diasporic community, of which they are part, maintains that (connected) population as a vibrant entity?

### 3  Chronotype culture: the everyday production of people and place

- How does the diasporic tourist react to the ways in which his or her diasporic community is routinely portrayed (by insiders and outsiders) where s/he lives?

- Does the diasporic tourist intend to return and live in the diasporic territory s/he is visiting on the current trip in question?

### 4 Acts of becoming: the paradoxes of plural cultural identification

- Has the diasporic tourist recently engaged in any trips as public acts of becoming *intentionally* to reinforce visions of selfhood as part of the scattered diasporic community?
- On the trip 'home', were there any particular spaces which were found to serve implicitly, but unexpectedly as unintentional monuments of diasporic memory or association?

### 5 New sense: the emergent and partial identities of populations

- How has the travel journey to the homeland affected the given diasporic tourist's ability to distinguish 'non-sense' (sic) projections of identity from 'new sense' trajectories?
- What role does 'landscape' play in the personal or cultural visions of a given diasporic homeland?

### 6 Agonistics: the gains and losses of syncretism

- To what degree does the given diasporic traveller (who journeys regularly to the homeland) also travel regularly to 'places of pilgrimage' within the place where s/he is actually resident?
- To what degree is the given diasporic traveller's 'vacation' (to the diasporic homeland) a tourist trip and/or a pilgrimage?

### 7 Counter-representations: the cultural politics of resistance

After the first trip to the diasporic homeland, in what ways has that place been misrepresented over time in
- mainstream external representations?
- internal projections of that scattered diasporic population?

### 8 Fantasmatics: the new performative power of enunciation

Has the recently completed trip to the diasporic tourist's homeland confirmed or reinforced any previously held personal fantasmatic
- sacred visions about that territory?
- secular visions about that territory?

### 9 Disseminated storylines: the new manumissive locations of culture

- Which are the most critical sites of or to diasporic communal memory that *must be* visited within the territory of the diasporic homeland?
- Which are the most critical shrines of or to the diasporic self that must be visited within the territory of the diasporic homeland?

### 10 Discontinuous historical realities: the new political geography of space

- What special source-of-identity material objects, if any, do diasporic tourists bring back from trips to the homeland or diaspora-related spaces beyond the homeland?
- After completion of a visit back to the homeland, in what ways does the diasporic tourist feel even further displaced, dislocated, or dispossessed while living away from the homeland?

Source: questions are abridged by distillation from Basu's (2001) ambivalent search for the Scottish motherland.

more intricate discussion of Bhabhan constructs for tourism studies). These are intended to be central issues which researchers in the field may be stimulated to address in their own, new, related, well-grounded and durable investigations into the discursive spaces of identity-making and identity-projection within 'diasporic tourism'.

Perhaps out of all the diasporic concerns the matter of 'enunciation' (Hollinshead 1998b: 71) matters most in inspections of the significance of 'tourism' where tourism sites and storylines can serve as an important launching pad at or from which new/emergent/consolidating diasporic communities can freshly or correctively announce themselves to the world (see Chs 12, 15 and 17). In this respect, Bhabha enlarges the important Foucauldian term *énoncé* (meaning 'a' or 'the' statement) as a vital rhetorical and political instrument by which restless or halfway populations can freshly declare or correctively re-declare themselves to others (and of course, to themselves). For both Foucault and Bhabha, 'an enunciated statement' is not merely a piece of text or language, however, it is a cardinal and corporeal event (Young 2001: 401–2); that is, a material eruption which productively changes not only the way things are perceived but the way things 'are'.

Potentially, the crucial value of tourism to specific emergent/restless/diasporic populations lies within the enunciative 'voltage' (i.e. the communicative reach and articulated effects) of the tourism industry. As nations and states cease to be 'purely endogenous' (Robins 2002: 29) and as ever more large and small diasporas appear around the globe, the twenty-first century will undoubtedly be a period of considerable enunciative activity in and through the articulative power of tourism (Hollinshead 2002). Tourism will not only regularly re-make and de-make old diasporas, it will quite frequently be one of, or *the* principal communicative vehicle by which all sorts of new diasporas and neo-tribal groups seek to reveal and legitimize themselves. Those who work in tourism will therefore continue to be agents of all sorts of empowering new sense, admixed amongst all kinds of plain old non-sense. The trouble is, those who work in both tourism academies and as tourism practitioners may not always be readily equipped to spot the subtle, yet critical differences (Crick 1989; Hall 1994; Tribe 1997; Meethan 2001).

## Conclusion: 'a thousand plateaus'?

So, the first steps have been taken down the routes towards teasing apart the intricate connectivities between tourism and diaspora. As chapters elsewhere in this collection demonstrate, many of the themes articulated by and emergent questions precipitated from, the intense hypothecations of social and cultural theorists of diaspora have formed the basis for current diaspora discourse in tourism. Enunciation, performativitiy, the constant re- and de-making of identity and the negotiation and contestation of multiple diasporic identities, to name but a few themes raised by the seminal thinkers in diaspora studies, have come under the microscope of tourism research. Base camp has been left a long time ago; it is not quite time to push for the summit, but tourism discourse on diaspora must depart from the current plateau it has reached. A more nuanced, theoretically inferred treatment of diasporas and tourism is required based not least on the thoughts of Gilroy and Bhabha, but also those of such luminaries as Appadurai (1994) and Deleuze and Guattari (1987) (a bastardization of whose title inspired the current metaphor). Their work may have inspired existing discourse, their ideas may even be implicit in extant treaties, but now is the time for a more direct engagement with the theoretical footholds of potential enlightenment that their readings of diaspora offer tourism academics. Progress on this path is not going to be easy. Conceptual crevasses must be traversed and we require more individuals in tourism studies who are geared up to be deeply sensitive to the very fluidity and permeability of cultural groupings. More practising individuals in tourism studies must be equipped properly to gaze on the world of exiles, migrants and expatriates; that is, to observe in appropriate 'stereoscopic vision' (Rushdie, cited in Robins 2002: 28) the realms and 'half-realms' of those who are both 'sometime-insiders' and 'sometime-outsiders', rather than continually relying on 'whole sight vision' to make such cultural assessments. And finally, individuals trained in the field of tourism studies must be encouraged to be deeply perceptive about tourism's own population-making, place-making, culture-making importances wherever new realities emerge as endogenous imperatives that admix uncertainly with exogenous impulses.

## References

Adorno, T. (1993) *Negative Dialectics*, New York: Seabury Press.

Appadurai. A. (1990) 'Disjunction and difference in the global cultural economy', *Theory Culture and Society* 1(2–3): 1–24.

Appadurai. A. (1993) 'Patriotism and its futures', *Public Culture* 5: 411–29.

Barkin, S. and Cronin, B. (1994) 'The state and the nation: changing norms and the rules of sovereignty in international relations', *International Organization* 48: 107–13.

Basu, P. (2001) 'Hunting down home: reflections on homeland and the search for identity in the Scottish diaspora', in B. Bender and M. Winner (eds) *Contested Landscapes: Movement, Exile and Place*, Oxford: Berg.

Bhabha, H. (1994) *The Location of Culture*, London: Routledge.

Cohen, R. (1997) *Global Diasporas: An Introduction*, Seattle, WA: University of Washington Press.

Crick, M. (1989) 'Representations of international tourism in the social sciences: sun, sex, sights, savings and servility', *Annual Review of Anthropology* 18: 307–44.

Crook, S. (2001) 'Social theory and the postmodern', in G. Ritzer and B. Smart (eds) *Handbook of Social Theory*, London: Sage.

Delanty, G. (2001) 'Nationalism: between nation and state', in G. Ritzer and B. Smart (eds) *Handbook of Social Theory*, London: Sage.

Deleuze, G. and Guattari, F. (1987) *A Thousand Plateaus: Capitalism and Schizophrenia*, Minneapolis, MN: University of Minnesota Press.

Dirlik, A. (1994) 'The postcolonial aura: Third World criticism in the age of global capitalism', *Critical Inquiry* 20(2): 328–56.

Featherstone, M. (ed.) (1994) *Global Culture: Nationalism, Globalization and Modernity*, London: Sage.

Featherstone, M. (1995) *Undoing Culture: Globalization, Postmodernism and Identity*, London: Sage.

Friedman, J. (1988) 'Cultural logics of the global system', *Theory, Culture and Society* 5(2–3).

Friedman, J. (1990) 'Being in the world: globalization and localization', in M. Featherstone (ed.) *Global Culture: Nationalism, Globalization and Modernity*, London: Sage.

Gabriel, T.H. (1990) 'Thoughts on nomadic aesthetics and the Black independent cinema: traces of a journey', in R. Ferguson, M. Gever, T.T. Minh-ha and C. West (eds) *Out There: Marginalization and Contemporary Culture*, Cambridge, MA: MIT Press.

Gandhi, L. (1998) *Postcolonial Theory: A Critical Introduction*, St Leonards, NSW, Australia: Allen and Unwin.

Geertz, C. (1973) *The Interpretation of Cultures*, New York: Basic Books.

Gilroy, P. (1993) *Small Acts: Thoughts on the Politics of Black Culture*, London: Serpent's Tail.

Gilroy, P. (1997) 'Diaspora and the detours of identity', in K. Woodward (ed.) *Identity and Difference*, London: Sage.

Hall, C.M. (1994) *Tourism and Politics: Policy, Power and Place*, Chichester: John Wiley.

Hall, S. (1992) 'The rest and the west: discourse and power', in S. Hall and B. Gieben (eds) *Formation of Modernity*, Cambridge: Polity Press.

Hall, S. (1997) 'The local and the global: globalization and ethnicity', in A. King (ed.) *Culture, Globalization and the World System: Contemporary Conditions for the Representation of Identity*, Minneapolis, MN: University of Minnesota Press.

Hobsbawm, E. (1990) *Nations and Nationalism Since 1780*, Cambridge: Cambridge University Press.

Hollinshead, K. (1998a) 'Tourism, hybridity and ambiguity: the relevance of Bhabha's "Third Space" cultures', *Journal of Leisure Research* 30(1): 21–156.

Hollinshead, K. (1998b) 'Tourism and the restless peoples: a dialectical inspection of Bhabha's halfway populations', *Tourism, Culture and Communication* 1(1): 49–77.

Hollinshead, K. (1999) 'Tourism as public culture: Horne's ideological critique of tourism', *International Journal of Tourism Research* 1(4): 267–92.

Hollinshead, K. (2002) 'Tourism and the making of the world: the dynamics of our contemporary tribal lives', *Excellence Lectures: Occasional Paper* 1(2), Miami, FL: The Honors College, Florida International University.

Jones, C.O. (1977) *An Introduction to the Study of Public Policy*, North Scituate, MA: Duxbury Press.

Jones, L. (1967) *Black Music*, Santa Barbara, CA: Quill.

Keith, M. and Pile, S. (eds) (1993) *Place and the Politics of Identity*, London: Routledge.

King, A.D. (1990) *Global Cities*, London: Routledge.

King, A.D. (1997a) 'Introduction: spaces of culture, spaces of knowledge', in A.D. King (ed.) *Culture, Globalization and the World-System*, Minneapolis, MN: University of Minnesota Press.

King, A.D. (1997b) 'The global, the urban and the world', in A.D. King (ed.) *Culture, Globalization and the World-System*, Minneapolis, MN: University of Minnesota Press.

Lanfant, M.-F. (1995) 'Introduction', in M.-F. Lanfant, J.B. Allcock and E.M. Bruner (eds) *International Tourism: Identity and Change*, London: Sage.

Lemert, C. (2001) 'Multiculturalism', in G. Ritzer and B. Smart (eds) *Handbook of Social Theory*, London: Sage.

Marcus, G. and Fischer, M.M.J. (1986) *Anthropology as Cultural Critique*, Chicago: University of Chicago Press.

Maybury-Lewis, D. (1992) *Millennium: Tribal Wisdom and the Modern World*, New York: Viking-Penguin.

Mbembe, A. (1992) 'The banality of power and the aesthetics of vulgarity in the postcolony', *Public Culture* 4(2): 1–30.

Meethan, K. (2001) *Tourism in Global Society: Place, Culture and Consumption*, Basingstoke: Palgrave.

Meyrowitz, J. (1985) *No Sense of Place*, Oxford: Oxford University Press.

Morley, D. (1991) 'Where the global meets the local: notes from the sitting room', *Screen* 32(1): 1–14.

Nisbet, R. (1967) *The Sociological Tradition*, London: Heinemann.

Parry, B. (1994) 'Signs of our times', *Third Text* 28: 5–45.

Patterson, O. (1982) *Slavery and Social Death: A Comparative Study*, Cambridge, MA: Harvard University Press.

Quayson, A. (2000) *Postcolonialism: Theory, Practice or Process*, Cambridge: Polity Press.

Robertson, R. (1994) *Globalization: Social Theory and Global Culture*, London: Sage.

Robins, K. (2002) 'Tradition and translation: national culture in its global context', in D. Boswell and J. Evans (eds) *Representing the Nation: A Reader. C Histories, Heritage and Museums*, London: Routledge (in association with The Open University).

Said, E.W. (1993) *Culture and Imperialism*, New York: Alfred Knopf.

Sangren, P.S. (1988) 'Rhetoric and authority of ethnography: "postmodernism" and the social reproduction of texts', *Current Anthropology* 29: 405–35.

Shenhav-Keller, S. (1995) 'The Jewish pilgrim and the purchase of a souvenir in Israel', in M.-F. Lanfant, J.B. Allcock and E.M. Bruner (eds) *International Tourism: Identity and Change*, London: Sage.

Simmel, G. (1971) 'The metropolis and mental life', in D. Levine (ed.) *George Simmel on Individuality and Social Forms*, Chicago, IL: University of Chicago Press.

Smith, A. (1986) *The Ethnic Origins of Nations*, Oxford: Blackwell.

Spivak, G. (1987) *In Other Worlds: Essays in Cultural Politics*, New York: Methuen.

Tenbruck, F. (1994) 'International history of society or universal history', *Theory, Culture and Society* 11(1): 75–93.

Tribe, J. (1997) 'The indiscipline of tourism', *Annals of Tourism Research* 24(3): 638–57.

Wallerstein, I. (1984) *The Politics of the World Economy*, Cambridge: Cambridge University Press.

Wilson, R. and Dissanayake, W. (1996) 'Introduction: tracking the global-local', in R. Wilson and W. Dissanayake (eds) *Global–Local: Cultural-Production and the Transnational Imaginary*, Durham, NC: Duke University Press.

Wolff, E.R. (1990) 'Facing power – cold insights, new questions', *American Anthropologist* 92(3): 586–96.

Young, R.J.C. (2001) *Postcolonialism: An Historical Introduction*, Oxford: Blackwell.

# 3 Conceptualizing return visits

## A transnational perspective

*David Timothy Duval*

Had a brief conversation downtown [Toronto] at the XYZ bank with a teller who had moved here from eastern Jamaica about 30 years ago. She described her experiences in Canada since her arrival and concluded that while she has lived here for such a long time, she did not feel comfortable. She didn't consider it to be her 'home'. Interestingly, while she had made occasional trips back to Jamaica to see family and friends, she did not regard Jamaica as 'home' and did not feel entirely comfortable visiting. She actually described herself as 'homeless'.

(Diary entry, author's fieldwork)

## Introduction

The vignette captured in the diary entry above raises a number of salient issues involving migrant populations, individual migrants and diasporas. The first of these is the recognition of broad relationships, interconnections and circulations (Tsing 2000) incorporating multiple social identities and spaces (Faist 2000a,b). In many respects, the mobility of individuals has all but replaced the more common conceptualization of 'community' or 'communities' (Lash and Urry 1994) and the resultant formation of 'sociospheres', to borrow from Albrow (1997), might best capture the temporal and spatial variability in social arrangements that are augmented at the level of the individual based on dynamic interpretations of home, lifestyle, affiliation and a sense of belonging.

What is represented in the diary entry above, on the one hand, is the diaspora in a generic sense, although more particularly a Caribbean or Jamaican diaspora. On the other hand, also featured, however implicitly, is the bank teller's external homeland. In the 'global ecumene' (Hannerz 1996), relationships between the external homeland and the diaspora incorporate and revolve around multiple identity structures that are prevalent in many diasporic communities, where members may socially align themselves to more than one ethnic group, nationality, or social consciousness.

Not surprisingly, several critical readings of 'culture', especially in the social sciences, attempt to resist the 'assumed isomorphism' (Gupta and Ferguson 1997: 34) of culture by, alternatively, recognizing the existence of individuals across borders, both conceptually and physically. Thus, the second issue to emerge from this vignette is precisely how periodic trips between the diaspora and external homeland (which is, quite often, the natal home) might function as both an adaptive mechanism and a tool for the negotiation of identities in both localities. Thus, of interest here is the use of such trips to facilitate cultural continuity and social arrangements that span multiple localities.

The broad purpose of this chapter is to outline the meaning and role of the return visit within diasporic identity structures. In doing so, the arguments forwarded suggest that the return visit – elsewhere characterized by Duval (2002) as a form or type of VFR travel but here discussed as a discrete form – can be viewed through the lens of transnationalism in order to fully capture and understand the link(s) between diaspora, migration, place and identity. Thus, return visits are shown to reinforce, reiterate and solidify social fields, such that, as a *transnational exercise*, identities and social relationships between the diaspora and the external homeland are, in effect, propagated by tourism episodes. Whereas the return visit permits the recognition of multiple social spaces at the level of the individual (as opposed to the community or ethnic group), transnationalism is shown to be a conceptual framework through which multi-stranded identities are rationalized. For the study of tourism, transnationalism provides an insight into patterns of touristic movement, meanings and the linkages within and between broad social networks, but it also considers notions of the local and the global beyond static notions of host and guest.

The chapter begins with a broad exploration of the concept of the return visit. Following this, the connection between the return visit and transnationalism is offered and examples from the existing migration and cultural studies literature are briefly discussed. Finally, findings from the author's ethnographic fieldwork among Commonwealth Eastern Caribbean migrants living in Toronto are provided.

## The return visit

Return visits may be defined as periodic, but temporary, sojourns made by members of diasporic communities to either their external homeland or another location in which strong social ties have been forged. To some degree, such visits originate in diasporic environments, which themselves are shaped through migration. The return visit, then, bridges the cultural and the social among various social 'nodes and hubs' (Vertovec 1999: 449) through which cultural information passes. Taken further, the return visit incorporates three key characteristics (Duval 2002). In the first instance, the return visit makes the assumption that the returning visitor has past non-tourist experience at a particular destination. In other words, the returning visitor has intimate social and cultural knowledge of the destination that can only come from first-hand experience. It can be argued, therefore, that they have extensive social and cultural foundations at a destination.

Second, the returning visitor, as opposed to the tourist who visits friends or relatives (or perhaps both), may be characterized as having extensive familial and social ties at the particular destination to which he or she is visiting. It is theorized, therefore, that temporary contact in the form of the return visit functions as a means to renew, reiterate and solidify familial and social networks. Such networks are perceived to be transnational in nature, the significance of which is examined below. Third and closely following the idea that the return visit allows for the reinvigoration of social capital, the return visit involves individuals who are often part of a larger and necessarily self-ascribed social unit that is associated with diasporic communities formed as a result of past voluntary migratory episodes. Diasporic communities and social units are ultimately based within transnational frameworks largely because many individuals or migrants may retain those cultural and social patterns salient to their country or location of origin. The traditional view of the nation-state, then, is under challenge, as Schmidtke (2001: 7) points out: 'Through social interactions and communication beyond national borders, new forms of communal ties can be formed that do not follow the pattern of the nation-state'. The return visit can ultimately be positioned as one vehicle through which transnational identity structures between diasporic communities and homelands are maintained.

That returning visitors claim extensive social and familial ties within a particular destination might lead some to claim similarities with the more common category or classification of VFR tourism. The return visit, however, should perhaps been seen more as a social function than an activity. Without question, numerous other means are available to migrants and members of diasporic communities that facilitate the solidification of social ties. The Internet, inexpensive telephone calls and remittances certainly allow for 'the maintenance of old cultural ties among modern immigrants' (Schmidtke 2001: 7), but the return visit is often overlooked with respect to how (and why) such exercises are used to solidify social contexts.

Broadly speaking, the association of the return visit with cultural diasporas ultimately highlights the transnational component of such trips; that is, the return visit can be seen to exemplify elements inherent with current (although often contested) under-standings of transnationalism. For Glick Schiller (1997: 155), transnational cultural studies examine the 'growth of global communications, media, consumerism and public cultures that transcend borders', while attention to transnational migration has sought to explore 'the actual social interactions that migrants maintain and construct across borders . . .'. Thus, it can be argued that return visits enable such social interactions, perhaps almost as easily as technological advances in media and communications, but with the added burden of travel costs and time.

Further linkages between the return visit and transnationalism can be gleaned from Vertovec's (1999) description of six detailed theoretical premises from which studies of transnationalism have traditionally been approached. Of particular interest, however, is Vertovec's characterization of transnationalism as a form of 'social morphology' in that it gives rise to communal identities, diasporas and social networks. What can be drawn from this is how the return visit functions within multiple social spaces, or perhaps more accurately, how it serves to fuse, strategically, various social localities and spaces across which transnational identities have been shaped and maintained. What can be suggested is that the return visit serves to balance these identities and networks through physical contact and linking. As an extension of Vertovec's social morphology characterization, the visit is a means through which individuals position themselves socially in more than one locality or place and can even be seen to act as a centrifuge in Sheffer's (1986) triadic relationship of scattered 'self-identified' ethnic groups, their current geopolitical associations; and their 'homeland' states or territories.

## Transnationalism, diasporas and tourism

Proposing a conceptual link between return visits and transnationalism from a diasporic perspective requires further exploration of how diasporic communities retain ties to their external homelands. Rather than attempt to fit such explorations within the broad notion of diasporas (recognizing the presence of political, labour and imperial diasporas, for example), Cohen's (1997: 128) description of the 'cultural diaspora' best captures, in his words, 'the lineaments of many migration experiences in the late modern world'. Although the notion of the cultural diaspora is approached here in the context of transnationalism – that is, the existence of the cultural diaspora is presumed to incorporate multiple transnational relationships and, by extension, incorporate numerous personal and corporate social relationships spanning the globe (Faist 2000b) – it should be pointed out that much of the scholarly literature on transnationalism suggests that the interconnected nature of lived experience(s) and identity does not necessarily require membership of communal networks or formal and informal migrant institutions.

Glick Schiller and Fouron (2001: 23) argue that the use of the term diaspora to describe transborder affiliation is problematic because it does not address the process of 'nation-state building' by migrants. For them, diaspora is more closely associated with 'dispersed populations that share an ideology of common descent and a history of dispersal, racialization and oppression' (Glick Schiller and Fouron 2001: 23). However, there is a pre-supposition in this argument that the migrant indeed has an overt interest in nation-state building. In other words, some migrants wish to be part of what Glick Schiller and Fouron refer to as the 'transnational nation-state'; that is, a state and its affiliated governmental authorities that are increasingly transcending geopolitical borders. While this may be a useful means by which Haitian transnational fields are organized and conceptually juxtaposed, such positioning may not always be applicable. Many migrants, for instance, may not declare allegiances to such transnational nation-states and, alternatively, may elect to affiliate themselves socially to a region or area (or place) that is based more on social and cultural compositions than those which are inherently political. Further, many may situate themselves socio-politically within a diaspora or place of residence by overtly claiming affiliation with both and effectively alternating accordingly, a nation-state and a region.

By extension, it can be suggested that migrants do not necessarily need to belong to a diasporic group in order to subscribe to the social customs and values native to their external homeland. The return visit could well work for everyone. Individuals themselves are becoming deterritorialized, as people 'now feel they belong to various communities despite the fact that they do not share a common territory with all the other members' (Papastergiadis 2000: 115). We have, therefore, individuals engaged in what Pries (2001: 8) calls 'pluri-local social practices'. This argument, combined with the above suggestion that the interconnected nature of transnational personal experience and affiliation does not necessarily require membership within closed-corporate diasporic communities, would seem to suggest that the notion of the diaspora is conceptually fluid (largely because its membership reflects this), almost to the point of being meaningless. Even Iyer's (2000) 'global traveller' is perhaps a more appropriate metaphor than assigning collectively oriented associations to individuals who exhibit varying degrees of meaningful social affiliation and association.

The question remains, however, as to how an understanding of transnationalism will help shed light on the use of the return visit as a means to bridge social networks among members of diasporas. Much of the literature on transnationalism speaks directly to the processual nature of multi-stranded identities and diasporas. For example, Spoonley (2000: 4) notes that:

transnationalism is the existence of links between a community in its current place of residence and its place of origin, however distant and between the various communities of a diaspora. . . . Transnationalism signals that significant networks exist and are maintained across borders and, by virtue of their intensity and importance, these actually challenge the very nature of nation-states. . . .

We can expand on Spoonley's comment regarding 'links' between a community and its external homeland by suggesting that the return visit, as a form of tourism, is afforded a position in the process of maintaining transnational ties. In other words, cross-border movements and diasporic networks are ultimately facilitated through the return visit. Another key aspect of transnationalism that is particularly relevant to this discussion is the suggestion, as outlined by Basch *et al.* (1994: 7), that it is not so much a salient and tangible phenomenon but more a process through which the multiple strands of social relationships (which could be read as multiple 'locals' acting globally) link societies in both the homeland and the locality in which settlements have been established. Again, it is this linkage that the return visit perpetuates. Tourism in the form of the return visit might be said to act as the means by which transnational relationships are strengthened.

As much as the study of transnationalism, most notably definition issues and conceptual arguments, receives considerable attention in the scholarly literature (see Ch. 1), comparatively little addresses the actual processes involved in the social linkage between two localities, outside of, perhaps, monetary remittances and technological advances that continue to facilitate global communication (or at least local to local communication). The return visit, as outlined above, suggests that individuals often travel for the purpose of seeking renewed contact with friends and family. Seen this way, the return visit represents the physical connection between, in some cases, the diaspora and the external homeland, while transnationalism as a conceptual framework can be used as an explanatory framework that highlights such connections as socially meaningful exercises. A significant body of literature exists that has given the nature of diasporic identities in the context of post-colonial, often urbanized environments in developed countries. Yet, it is imperative that we understand how such identities are maintained after the migration event and, spatially and temporally, they are connected to other places and localities. In other words, what is happening in the diaspora post-migration (or, more precisely, what is it about the social organization within the diaspora) that facilitates how identities within the diaspora are negotiated, re-affirmed and even globalized?

Of particular interest here is how transnationalism, return visits and diasporic communities (and individuals) can be conceptually tied together. The theoretical arguments behind the convergence of these concepts have been outlined. Several examples from the literature on transnationalism and migration serve to illustrate this potential conceptual blending and are discussed briefly below. Following these, an overview of the transnational meaning of return visits among Commonwealth Eastern Caribbean migrants is offered.

Among migrants to New York City from Mexico, Smith (2001) found that second-generation youth regularly use their parents' hometown in Mexico (Ticuani) as a place to visit on a holiday. Moreover, Smith (2001: 44) provides evidence of what he calls 'transnational participation':

Many Ticuanenses and the US-born children return regularly to Ticuani and communicate regularly by phone with relatives and friends there. The most common time for return is during the Feast of the patron saint, Padre Jesus, in January, but many also return during summer vacation during the Feast of the patron saint of the neighbouring town. These trips serve as vacations, but are often structured around the rituals associated with the town's religious cargo system.

Highlighted here is the fact that multiple localities spawn movements that are culturally influenced. Such movements involve individuals located within a broader diasporic identity (Mexico/Mexicans) who embark upon movement (in the form of return visits) to a specific locality that represents an alternative 'home' (Ticuani) as a social construction and for the purpose of maintaining (transnationally) social threads between communities. In this particular case, such social threads are strong enough to render social organization and kinship obligations to transcend borders and nation-state definitions (see Itzigsohn *et al.* 1999).

In a study of Dominican transnational migration and return migration, Guarnizo (1997) examined the broad transnational social processes of movement and mobility. In seeking to extrapolate the borderless nature of identities and cultural patterns, Guarnizo (1997) purposely excluded those whom he called 'the most obvious transnational migrants'; that is, 'those who shuttle back and forth for short working periods abroad and *visiting migrants temporarily in the country*' (Guarnizo 1997: 290, emphasis added). On the other hand, Levitt's (1998) study of social remittances, or the 'ideas, behaviours, identities and social capital that flow from receiving – to sending-country communities', between Miraflores in the Dominican Republic and Jamaica Plain, Boston, revealed that various normative structures are exchanged. Of particular interest here is how those normative structures are transmitted. For example, in the words of one migrant living in Boston:

> When I go home, or speak to my family on the phone, I tell them everything about my life in the United States. What the rules and law are like. What is prohibited here. I personally would like people in the Dominican Republic to behave the way people behave here. . . . When I'm in Miraflores, when I see people throwing garbage on the ground, I don't go and pick it up because that would be too much, but I get up and throw my own garbage away and everyone sees me do it. These things and many more, the good habits I've acquired here, I want to show people at home.
>
> (Levitt 1998: 933)

Layton-Henry (2002: 18–19) found that African-Caribbeans in Birmingham also maintained links to their respective home countries. One respondent in Layton-Henry's study (2002: 18) noted:

> It is important to have links as I have said before. They keep you informed of what's going on over there. Also, for holidays it's great. Even though I haven't been to the Caribbean, most of my family are over there. So I know when the time comes it will be easier to stay with my family.

In their study of *Viet kieu* attitudes toward travelling back to Vietnam from Australia, Nguyen and King (1998) found that many migrants who travel to Vietnam do in fact stay with family and friends. Nguyen and King (1998: 359) note, however, that the extent to which such travel 'consists of members of families returning to reaffirm their family membership' and whether or not such travel ensures that 'they and their children remain as recognized members of the family group' remains to be fully investigated.

## Commonwealth Eastern Caribbean nationals in Toronto

Toronto's Commonwealth Eastern Caribbean community refers here to various (and often overlapping) social networks, often collectively identified as a diaspora (Henry 1994), arising from the settlement of individuals from the Commonwealth Eastern Caribbean. Anthropological, geographical and sociological studies of the broader Caribbean diaspora have focused on socio-economic patterns (such as community studies, issues of employment and coping and adjustment mechanisms), issues of racism and patterns of migration and return migration (e.g. Anderson 1985, 1993; Anderson and Grant 1975; Chamberlain 1998; Clarke 1984; Foner 2001; Henry 1994; Peach 1984). Various works of fiction have also sought to capture the migration of experience of 'Caribbeans' living abroad in a variety of countries, a notable example being Selvon's (1985) *The Lonely Londoners*.

Throughout the 1990s, the total number of immigrants from the Caribbean region to Canada has been variable, from a high of just over 16,000 in 1993 to only 6,700 in 1998 (Statistics Canada 2001a). In the 1996 census of the Metropolitan Toronto region (also known as the Greater Toronto Area), over 155,000 people identified the Caribbean (including Bermuda) as their country of birth (Statistics Canada 2001c) and over 166,000 indicated the Caribbean region as their 'ethnic origin' (Statistics Canada 2001b). Richmond (1989) notes that the majority of Caribbean migrants to Canada have settled (and continue to settle) in large metropolitan cities such as Toronto and Montreal. Of those individuals identifying a Commonwealth Eastern Caribbean background, slightly more than 64,000 live in the Greater Toronto Area, with the majority stating Trinidad and Tobago as their country of origin. The wider Caribbean community in the Greater Toronto Area is serviced by various newspapers ('Caribbean Camera' and 'Share', both of which are distributed free throughout Metropolitan Toronto), local shops offering imported fresh and canned food from the region, many of which serve as a focal point for meeting others and catching up on news and gossip and various nation-based organizations (e.g. St Vincent and the Grenadines Association, or the Grenadian Association).

Not unlike other studies discussed above, ethnographic research conducted by the author among social networks within the Commonwealth Eastern Caribbean diaspora in Toronto (see Duval 2002) revealed that return visits were seen to satisfy, in the first instance, the maintenance of specific social ties, such that return visits by members of this particular network within the Commonwealth Eastern Caribbean community were taken in order to link social fields and communities. For example, one migrant felt that his first trip was necessary to further affirm his social space in his external homeland:

> It was something I had to do, to go back and keep in touch with my family and to keep in touch with myself. It would make my family seem more real, more reality,

going back to seeing the land itself, nature, remembering every little corner, little stone, the symbols of childhood. So that was my first time.

For many, the return visit is used to 'keep the network alive and active'. Much like Smith's (2001) discussion of Mexicans in New York City, such network connections often extend to subsequent generations. Many migrants will often make trips so extended family members living in the Eastern Caribbean can see the children. One migrant interviewed by the author in Toronto felt quite strongly about exposing his children to another locality:

> For me, I feel a sense of responsibility to take my children to SVG [St Vincent and the Grenadines], to make them understand the SVG. But they enjoy it because they have been brought up that way, so they'll go home and be quite comfortable in SVG. I have two homes so they have to know about both.

On occasion, trips were made voluntarily, but often were seen as obligatory. More importantly, the return visit is used as a strategy to maintain visibility within a former home community. Basch *et al.* (1994: 84), in a discussion on the mobilization of resources and support systems, found that:

> Vacations also become important pegs in the transnational social field, contributing to its viability and continuity. . . Migrants' relatives living in St Vincent and Grenada and especially those with more economic resources, also vacation in New York, some fairly often, where they stay with relatives.

For Eastern Caribbeans living in Toronto, the external homeland represents an alternative reality and locality. It is balanced, ultimately, between the spatial locality of the present (Toronto) with that of the past (the external homeland), yet the latter is reified through return visits and the distinction between past and present is effectively blurred. Olwig (1993: 180) was able to rationalize the establishment of similar connections between 'home' and diaspora in her study of migration involving the island-state of Nevis:

> Though most Nevisian migrants experience a lessening of social and economic ties to their home island, as they settle abroad and their parents on Nevis die, Nevis nevertheless has remained an important source of cultural identification for most of them. Indeed it seems that as the migrants lost their 'natural' ties to Nevis, they try to strengthen the cultural ties.

Return visits are also made to measure change and transformation. Migrants would compare the homeland as remembered to what is seen and experienced today. Baldassar (2001) found similar sentiments among first-generation migrants from Italy in Perth. To this end and from the author's fieldwork in Toronto, what can be suggested is that the return visit is an exercise through which the returning visitor relates himself or herself to the environment (which is broadly defined here as the physical and social elements of the external homeland). In other words, the return visit is perhaps a unique process of identity negotiation incorporating several social fields separated only geographically. It is an additional 'sociosphere' (Albrow 1997) that is incorporated into the lives of migrants.

By way of context, emigration in the Caribbean is often seen as socially desirable, but remaining visible and socially 'in contact' is of equal importance. Interestingly, in the Commonwealth Eastern Caribbean community in Toronto, maintaining such visibility facilitates the potential for return migration, whether planned or not (Duval 2002). Furthermore, maintaining social and cultural ties does not always require movement to one's external homeland. In effect, many individuals would regularly make trips to the UK where family and friends now live and who themselves once lived in the Eastern Caribbean. The return visit, as a transnational exercise binding multiple ties, need not, therefore, include the external homeland. As such, we should perhaps be conscious of multiple localities requiring multiple mobilities. Recalling Vertovec's (1999) notion of nodes and hubs in transnational relationships, it can be suggested that while the connection to the 'homeland' is significant, transnational networks need not necessarily include 'homeland' nodes, but rather incorporate global nodes and hubs of other transnational communities. One might even argue that 'diasporas' as units of study are becoming irrelevant as networks between migrants and homelands become sustained through advances in air transportation and technology such as the Internet.

In light of this example from the Commonwealth Eastern Caribbean diaspora in Toronto, there is at least some room for the consideration of the *reason* return visits are transnational exercises. This reason is found in the work of Robotham (1998; see also Mintz 1998), who offers some exceptional, albeit tentative, insights into drawing this connection between two social spaces that invariably represent, at different time, one's 'home'. Robotham (1998) historicizes the transnational Caribbean experience into four broad periods. For each period, he demonstrates how the Caribbean, as a region, has been connected to the larger global stage of issues and adjustments. Robotham suggests that post-colonial national identities began to develop in the Caribbean immediately following the Second World War. These nationalist identities were largely the result of the state-sponsored economic policy that effectively discouraged manufacturing in the Caribbean in an effort to maintain the 'classic colonial model of the local society as a raw materials producer importing its manufactured goods from the mother country' (Robotham 1998: 310).

The significance of this period and the period that immediately followed, from Robotham's perspective, is that the maintenance of social and cultural ties through return visits in the present may genuinely reflect the dominant ideology of Caribbean at the time when these individuals initially emigrated from the Caribbean. In other words, it can be suggested that perhaps the social meaning behind return visits extends even further beyond the maintenance of social and cultural ties. Perhaps the return visit is, in many respects, an effort to continually recognize the nationalism and nationalistic ideology that emerged from the Caribbean during this period. Likewise, return migration may also be a by-product or a representation of nationalist (or even regionalist) ideologies. The return visit is more of a mechanism in the maintenance of social and communal identities that, in the twenty-first century, now more than ever exist irrespective of modern geo-political borders. It is these transnational identities that ultimately warrant a closer examination with respect to the meaning of the return visit within diasporic communities.

## Conclusion

The purpose of this chapter has been to forward the suggestion that return visits are closely linked with transnational identities. As such, it has been argued that investigating the mobility, from a touristic sense, of migrants living within diasporic communities ultimately benefits from a transnational perspective. It can be concluded, therefore, that members of diasporic groups who maintain identity structures in multiple localities may utilize the return visit as a transnational exercise such that temporary and periodic movement between a diaspora and the external homeland renews, reiterates and solidifies both social networks and the cultural values and norms carried by individuals. In this sense, the return visit closely resembles Portes' (1999: 464) notion of the transnational activity, which takes place 'on a recurrent basis across national borders and that require a regular and significant commitment of time by participants'.

While transnationalism, as a cultural concept, allows for an understanding of how multiple identities are manipulated, this chapter has shown how transnationalism is manifested in the context of cultural diasporas. As Portes *et al.* (1999: 217) point out, transnationalism is characterized as a field that is 'composed of a growing number of persons who live dual lives: speaking two languages, having homes in two countries and making a living through continuous regular contact across national borders'. Similarly, for Basch *et al.* (1994) transnationalism represents 'multiple ties and interactions linking people or institutions across the borders of nation-states'. Thus, it is this contact that is of interest to tourism scholars. However defined, studies of the cultural meaning of tourist flows would do well to consider a broader transnational framework.

Given these conclusions, at this point it is perhaps not enough to suggest that some diasporic groups are necessarily transnational due to the obvious social, economic and political linkages established and routinely fostered (Levitt 2001). The challenge for tourism scholars engaged in attempting to understand 'mobility' as a broader construct of tourism is to interrogate the meaning behind the return visit by members of diasporic groups who utilize such movement to maintain involvement and social relevance in multiple localities. A further challenge is to interrogate the migration literature in the hope of bridging useful concepts relating to mobility and temporary mobility. Overall, the potential certainly exists for future studies of temporary human movement that would be framed within the context of transnational identities. Such research would ideally be centred upon the diaspora as a meaningful node embedded within a complex network of social relations.

## References

Albrow, M. (1997) 'Travelling beyond local cultures: socioscapes in a global city', in J. Eade (ed.) *Living the Global City: Globalization as a Local Process*, London: Routledge.

Anderson, W.W. (1985) *Caribbean Orientations: A Bibliography of Resource Material on Caribbean Experience in Canada*, Toronto: William Wallace.

Anderson, W.W. (1993) *Caribbean Immigrants: A Socio-demographic Profile*, Toronto: Canadian Scholars Press.

Anderson, A. and Grant, R. (1975) *The Newcomers: Problems and Adjustments of West Indian Immigrant Children in Metro Toronto Schools*, Toronto: York University.

Baldassar, L. (2001) *Visits Home: Migration Experiences between Italy and Australia*, Melbourne: Melbourne University Press.

Basch, L., Glick Schiller, N. and Szanton Blanc, C. (1994) *Nations Unbound: Transnational Projects and the Deterritorialized Nation-state*, New York: Gordon and Breach.

Chamberlain, M. (1998) 'Introduction', in M. Chamberlain (ed.) *Caribbean Migration: Globalised Identities*, London: Routledge.

Clarke, C. (1984) 'Pluralism and plural societies: Caribbean perspectives', in C. Clarke, L. Ley and C. Peach (eds) *Geography and Ethnic Pluralism*, London: George Allen and Unwin.

Cohen, R. (1997) *Global Diasporas: An Introduction*, Seattle: University of Washington Press.

Duval, D.T. (2002) 'The return visit-return migration connection', in C.M. Hall and A. Williams (eds) *Tourism and Migration: New Relationships between Production and Consumption,* Dordrecht: Kluwer.

Faist, T. (2000a) 'Transnationalization in international migration: implications for the study of citizenship and culture', *Ethnic and Racial Studies* 23(2): 189–222.

Faist, T. (2000b) *The Volume and Dynamics of International Migration and Transnational Social Spaces*, Oxford: Clarendon Press.

Foner, N. (ed.) (2001) *Islands in the City: West Indian Migration to New York*, Berkeley: University of California Press.

Glick Schiller, N. (1997) 'The situation of Transnational Studies', *Identities* 4(2): 155–66.

Glick Schiller, N. and Fouron, G. (2001) *Georges Woke Up Laughing: Long-distance Nationalism and the Search for Home*, Durham, NC: Duke University Press.

Guarnizo, L.E. (1997) 'The emergence of a transnational social formation and the mirage of return migration among Dominican transmigrants', *Identities* 4(2): 281–322.

Gupta, A. and Ferguson, J. (1997) 'Beyond 'culture': space, identity and the politics of difference', in A. Gupta and J. Ferguson (eds) *Culture, Power, Place: Explorations in Critical Anthropology*, Durham, NC: Duke University Press.

Hannerz, U. (1996) *Transnational Connection: Culture, People, Places*, London: Routledge.

Henry, F. (1994) *The Caribbean Diaspora in Toronto: Learning to Live with Racism*, Toronto: University of Toronto Press.

Itzigsohn, J., Cabral, C.D., Hernandez Medina, E. and Vazquez, O. (1999) 'Mapping Dominican transnationalism: narrow and broad transnational practices', *Ethnic and Racial Studies* 22(2): 316–39.

Iyer, P. (2000) *The Global Soul: Jet-lag, Shopping Malls and the Search for Home*, London: Bloomsbury.

Lash, S and Urry, J. (1994) *Economies of Signs and Space*, London: Sage.

Layton-Henry, Z. (2002) 'Transnational communities, citizenship and African-Caribbeans in Birmingham', working paper, ESRC Transnational Communities Programme (WPTC-02–07).

Levitt, P. (1998) 'Social remittances: migration driven local-level forms of cultural diffusion', *International Migration Review* 32(4): 926–48.

Levitt, P. (2001) 'Transnational migration: taking stock and future directions', *Global Networks* 1(3): 195–216.

Mintz, S. (1998) 'The localization of anthropological practice', *Critique of Anthropology* 18(2): 117–33.

Nguyen, T.H. and King, B.E.M. (1998) 'Migrant homecomings: Viet kieu attitudes towards travelling back to Vietnam', *Pacific Tourism Review* 1: 349–61.

Olwig, K.F. (1993) *Global Culture, Island Identity: Continuity and Change in the Afro-Caribbean Community of Nevis*, London: Routledge.

Papastergiadis, N. (2000) *The Turbulence of Migration: Globalization, Deterritorialization and Hybridity*, Oxford: Polity Press.

Peach, C. (1984) 'The force of West Indian island identity in Britain', in C. Clarke, L. Ley and C. Peach (eds) *Geography and Ethnic Pluralism*, London: George Allen and Unwin.

Portes, A. (1999) 'Conclusion: towards a new world – the origins and effects of transnational activities', *Ethnic and Racial Studies* 22(2): 463–77.

Portes, A., Guarnizo, L.E. and Landolt, P. (1999) 'The study of transnationalism: pitfalls and promise of an emergent research field', *Ethnic and Racial Studies* 22(2): 217–37.

Pries, L. (2001) 'The approach of transnational social spaces: responding to new configurations of the social and the spatial', in L. Pries (ed.) *New Transnational Social Spaces: International Migration and Transnational Companies in the Early Twenty-First Century*, London: Routledge.

Richmond, A.H. (1989) *Caribbean Immigrants: A Demographic Economic Analysis*, Ottawa: Statistics Canada.

Robotham, D. (1998) 'Transnationalism in the Caribbean: formal and informal', *American Ethnologist* 25(2): 307–21.

Schmidtke, O. (2001) 'Transnational migration: a challenge to European citizenship regimes', *World Affairs* 164(1): 3–16.

Selvon, S. (1985) *The Lonely Londoners*, Harlow: Longman.

Sheffer, G. (1986) 'A new field of study: Modern diasporas in international politics', in G. Sheffer (ed.) *Modern Diasporas in International Politics*, London: Croom Helm.

Smith, R.C. (2001) 'Comparing local level Swedish and Mexican transnational life: an essay in historical retrieval', in L. Pries (ed.) *New Transnational Social Spaces: International Migration and Transnational Companies in the Early Twenty-First Century*, London: Routledge.

Spoonley, P. (2000) 'Reinventing Polynesia: the cultural politics of transnational communities', working paper (WPTC-2K-14), Transnational Communities Research Programme, Institute of Social and Cultural Anthropology, University of Oxford, UK.

Statistics Canada (2001a) *People and Population*. Online. Available: www.statcan.ca/english/Pgdb/People/Population/ demo35f.htm (accessed 26 January 2001).

Statistics Canada (2001b) *People and Population*. Online. Available: http://www.statcan.ca/english/Pgdb/People/Population/demo28f.htm (accessed 26 January 2001).

Statistics Canada (2001c) *CANSIM Series*, Matrix 2: D39, Ottawa: Statistics Canada.

Tsing, A. (2000) 'The global situation', *Cultural Anthropology* 15(3): 327–60.

Vertovec, S. (1999) 'Conceiving and researching transnationalism', *Ethnic and Racial Studies* 22(2): 447–62.

# 4 Tourism, racism and the UK Afro-Caribbean diaspora

*Marcus L. Stephenson*

## Introduction: the UK Caribbean diaspora

The term 'Afro-Caribbean' commonly refers to those individuals who are of African origin but also of recent Caribbean descent (Phoenix 1988). The UK Afro-Caribbean diaspora is intergenerational, consisting mainly of first-, second- and third-generation populations. In this chapter, the term 'black' is often applied synonymously with the term 'Afro-Caribbean', as opposed to other groups which have been conventionally labelled black (e.g. 'Asians'). In addition to physical appearance or skin colour, this term represents both a state of political consciousness and a 'positive source of identity' (Pilkington 2003: 37). It is a more appropriate form of expression than 'black British', because individuals are not able to fully adopt a British identity given the distinctive ethnic and cultural characteristics of the wider Caribbean community (Stephenson 2002).

The UK Afro-Caribbean diaspora is culturally diverse, consisting of migrants and their descendants from various Caribbean islands of the former British colonies. Although this chapter deals with the Afro-Caribbean diaspora in total, there are a number of sub-diasporic groups living in the UK (e.g. 'Jamaicans', 'Barbadians' and 'Vincentians'). The 'Afro-Caribbean'/'black' diaspora is the second largest of all groups of non-European origin living in the UK, totalling over 600,000 in number (Ballard and Kalra 1994) and living predominantly in inner city areas (Skellington 1992). Caribbean migrants sought residency in the UK after the Second World War period and were employed mainly in manufacturing and services, especially tourism, catering and transport (Fryer 1984). Residence was largely differentiated by place of origin: Dominicans resided in Preston; Nevisians migrated to Leicester; and Jamaicans settled in various urban centres including Birmingham, Derby, London and Manchester (Peach 1984; Byron 1994).

For Cohen and Kennedy (2000: 32), one key conceptual characteristic of a 'diasporic group' relates to which degree its members still 'evince a common concern for their homeland and come to share a common fate with their own people'. Importantly, Stephenson's (2002) ethnographic study of the Afro-Caribbean community of Moss Side (Manchester) revealed that first- and second-generation Afro-Caribbeans overwhelmingly focus their tourism aspirations on the 'ancestral homelands', although this group has resided abroad for the past five decades. Aspirations to travel to the homeland relate to the need to pursue voluntary and personal quests, as well as the need to fulfil social and family commitments. These aspirations are largely determined by 'mental images and retained cultural knowledge, reconstructed and transmitted within metropolitan societies' (Stephenson 2002: 416). The experience of travelling to particular ancestral (Caribbean) islands predominantly inspired individuals to maintain social ties with family members and to consummate long-term ambitions to become (re-) acquainted with their cultural heritage and ethnic roots.

Symbolic association with the homeland may not only be influenced by cultural and ethnic determinants. Aspects relating to social marginalization and racial alienation among disenfranchised communities may offer further explanations as to why inter-subjective attachments to the homeland often prevail in contemporary societies. Although it is not the intention here to focus on how far racial alienation influences people's desire to travel to the homeland, the following suggests there are limited aspirations to travel to destinations that are not culturally sanctioned by attributes of ethnicity. This relates to the racialized realities of destinations dominated by white populations; that is where the anticipation and/or experience of racial encounters impacts upon other people's ability to participate in tourism freely.

Nonetheless, socio-cultural boundaries that pre-exist between different ethnic communities are currently in the process of disintegrating, particularly as individuals are increasingly becoming interested in others' lives (Urry 1990, 1995, 2002; Rojek 1993a,b). This assertion points to a range of factors behind the transculturalization of societies, cultures and locales, including the globalization of the travel and tourism industry; the multiculturalization of cosmopolitan societies; and the international role of the media. From this perspective, it could be assumed that ethnic groups such as the Afro-Caribbean community are increasingly in a position to extend their aspirations beyond ethnic-based choices in an attempt to experiment with other different cultures and societies. In contrast, this chapter contends that members of the Afro-Caribbean diaspora experience a range of socially alienating encounters during various travellings. Any significant attempt to travel to an unfamiliar environment runs that risk of over-exposing individuals to racialized circumstances and situations. The racial problems illustrated here relate to nationalistic gestures, verbal insults, suspicious glances, fearful encounters and acts of racial surveillance. Such concerns will be assessed in the context of exploring the social ramifications of visiting UK countryside destinations and other destinations in Europe.

If relationships of racial difference between black visitors and white hosts are believed to exist within various tourism environments and contexts, then interactions and interrelationships between individuals and groups may not always be perceived as being interchangeable or flexible. Accordingly, this chapter accounts for the ways in which racialized divisions potentially inhibit people's capacity to engage in mutual exchanges with those cultures and communities that exist outside of their own familiar environments, territorial boundaries and diasporic settings. Nevertheless, there is a concerted attempt to identify ways in which travel can encourage members of the Afro-Caribbean diaspora to think seriously about their identities and diasporic associations. This concern is evident through the recognition that cosmopolitan forms of travel can encourage individuals to experience shared encounters and constructive interactions with other members of the wider black (transatlantic) diaspora (i.e. those individuals who share similar ethnic origins). Consequently, suggestions are raised with regard to how it may be possible for individuals to achieve positive experiences within domains that lie outside of the ethnic and cultural boundaries of the homeland.

## Racialism and racism

'Racism' is traditionally defined as a process by which individuals and groups popularly categorize and stereotype those whom they feel are inferior (Yeboah 1988). These racial categories are formed on the basis of intrinsic properties such as colour and national origin. Arguably, racism relates to the construction of cultural and religious stereotypes, with the effect of producing the 'stigmata of otherness' (Balibar 1991: 18); that is, a social disposition with implications beyond biological determinism. This process may more appropriately be conceived of as 'racialism', a social construct that accounts for people's fear and intolerance of others (Anthias and Yuval-Davis 1993). Racialism is frequently characterized by irrational conduct, such as racial violence prevalent in the UK from the early 1950s and is responsible for creating social discomfort within black communities (Dummett and Dummett 1987). Racialism, or personal racism, or prejudice, transmits to racism once those who exercise institutional power and cultural authority have sanctioned discriminatory practices.

Critical discourses have overwhelmingly focussed on why members of the UK Afro-Caribbean community frequently experience difficulties in attaining equal opportunities and benefits of British citizenship (Fryer 1984; Gilroy 1987; Pilkington 2003). Accounts of race relations have profiled the inhumanities faced by black communities, including racial surveillance (Fryer 1984); racial violence (Keith 1995); and economic exploitation (Rex and Tomlinson 1979). Evidence suggests that racial inequality is prevalent within institutional settings such as housing (Peach and Byron 1993) and education (Pilkington 2003). Unfortunately, concerns relating to the prevalence of racial inequalities within tourism institutions and spaces have been overlooked within sociological debate on race and ethnicity.

Racism primarily functions on the assumption that it is not considered natural for people from minority cultural and ethnic backgrounds to be part of a 'bounded community', or a 'nation' (Barker 1981: 21). For Barker (1981), racism relates to the territorial claims of the dominant group to restrict the political and social rights of those classified as 'outsiders'. Nationalistic discourses often construct definitions of the 'British Nation' on the basis of socio-biological beliefs, appropriated by the dominant ethnic group in an attempt to preserve its status and position in society.

However, it is too simplistic to understand racism within the boundaries of a nation or a national culture. Racism manifests itself in the practices and ideologies of countries and societies that have mutual interests in dealing with ethnic minority groups. Therefore, it is necessary to acknowledge forms of racism that transcend national boundaries to represent the collective practices and ideological agendas of countries that share similar orientations towards minority groups. Accordingly, it is crucial to deal with aspects of 'pan-European racism' (Sivanandan 1990: 153; Jenkins 1987). As Jenkins (1987: 3) explains, 'European racism is a unique manifestation of ethnicity, historically formed by slavery, colonial expansion, nineteenth-century evolution and twentieth-century labour migration'.

The rights and interests of ethnic minority groups and minority citizens living in the European Union (EU) have become increasingly marginalized (Shore and Black 1992; Shore 1993). The socio-political and economic movement towards a more integrated Union has reinforced the values and objectives of dominant ethnic groups. For instance, various centralized policies (such as the Social Charter) have not comprehensively established clear objectives for the establishment of racial equality (Gabriel 1994; Kennett 1994; Mitchell and Russell 1994). Moreover, pan-European racism manifests itself in EU practices and strategies to 'culturally integrate' the main populations and cultural nationalities of member states (Shore 1993). This rationalization, 'Europeanization of cultures and societies', is augmented through the proliferation of shared cultural initiatives and events (e.g. sport, travel, heritage and cinema), with the effect of celebrating a 'European identity' (Shore and Black 1992; Shore 1993). Further challenges to the notion of a 'socially integrated Europe' have stemmed from racial hostilities towards minority groups living in countries such as Belgium, France, Germany and the UK (Keith 1995; Willems 1995; Witte 1995).

The following focuses on how perceptions and experiences of racial hostility and ethnic dominance affect ways in which members of a (non-white) diasporic community encounter other communities and societies. Anecdotal evidence is presented on people's aspirations and experiences of travelling to English countryside domains and other European domains.

## Visiting the English countryside

### *Racialized boundaries and representations*

Several writers have tentatively acknowledged that non-white minorities, visiting or considering visiting rural environments, often anticipate or encounter racial prejudice (Agyeman 1989; Malik 1992; Agyeman and Spooner 1997). Racial prejudice arguably disenfranchises minorities from actively participating in rural tourism activities. Malik (1992), who compared Asian and non-Asian attitudes of the British countryside, found that the former group overwhelmingly expected to be confronted by racial abuse.

Ingrid Pollard, a photographer with an interest in rural landscapes, explains her perception of countryside environments:

> It's as if the black experience is only lived in an urban environment. I thought I liked the Lake District, where I wandered lonely as a black face in a sea of white. But a visit to the countryside is always accompanied by a feeling of unease, dread . . . feeling that I don't belong.
>
> (Cited in Taylor 1993: 265)

Julian Agyeman, former Chair of the Black Environment Network (BEN), describes his perception of countryside communities:

> As a black man, I can walk into a country pub and it's like the Wild West – the piano stops and people stare expectantly – waiting for you to swing from the rafters.
>
> (Cited in Derounian 1993: 71)

The Commission for Racial Equality (CRE) produced two reports highlighting the problems and implications of rural racialism. The first profiled racial difficulties that ethnic minorities experienced while living in rural areas in south-west England (Jay 1992). The second report covered similar experiences of minority groups living in rural Norfolk, east England (Derbyshire 1994). Partly based on personal accounts, experiences and encounters of racialism, this report emphasized that local reactions are often detrimental to racial harmony. The two studies conclude that members of host societies operate a criterion of exclusion based on attempts to sustain the tradition of rural domains for white communities.

Media representations and popular perceptions of 'rurality' indirectly help to safeguard the racialized boundaries of countryside communities, creating an impression of the countryside as a domain exclusive to the interests and value systems of (white) English populations. Several commentators concur that representations of the 'countryside environment' are often constructed through authentic notions of 'Englishness' (Lowenthal 1991; Bunce 1994; Daniels 1994). Lowenthal (1991: 213), for instance, noted that mythical and symbolic images of the countryside, popularly transmitted through countryside magazines, brochures, television programmes, poetry and novels, powerfully promote the 'quintessential national virtues' of English culture. Bunce (1994: 29–34) discusses how representations of the English countryside embody romantic myths such as 'rural nostalgia', 'traditionalism', 'agrarian simplicity' and 'green and pleasant lands' – as portrayed in the poetic works of William Blake, John Keats and William Wordsworth. Nationalistic sentiments and perceptions of the English countryside are dismissive of those individuals considered as a 'threat to its sanctity' (Sibley 1997: 219). For Sibley (1997: 228),

> Just as idealised and romanticised representations of English rural landscapes have no place for chicken factories, gravel pits, electricity pylons and council houses, so the representations of rural society articulated by its protectors have no room for Travellers, factory workers or ethnic minorities.

Nonetheless, perceptions and encounters of racial hostility have a direct role in terms of making the countryside less accessible to Afro-Caribbeans, especially by constructing social and physical boundaries between visitors and local residents. As black visitors are essentially more conspicuous than others, they are potentially subjected to covert and/or overt ridicule. In this context, colour is a primary indicator influencing how black visitors are socially received in rural communities. Yet conceiving race simply in terms of biological differences does not necessarily explain how and why various disparities and differences between groups evolve and develop (Miles and Phizacklea 1984). Boundaries and divisions between black visitors and white locals are not entirely naturally determined but formed as a consequence of the ideologies of 'ethnic identification' and 'folk cosmology' (Jenkins 1997: 50). Racial ideologies are evident in the production of spurious images of black communities such as popular myths of black criminality (Hall *et al.* 1978; Keith 1989). These stereotypes potentially impact the way in which 'black others' are generally viewed in public places and spaces.

'Black others' may be commonly perceived as a source of 'social and cultural pollution' in rural areas, thereby impinging upon daily life and endangering local cultural systems and institutions. Although constructs of 'pollution' and 'danger' are popularly addressed in anthropological enquiries on the cultural manifestations of the human body (Douglas 1984), their application to the study of inter-ethnic relations and host/guest environments could help explain the reasons for the occurrence of racialized encounters and exchanges. With reference to this particular case study, these concepts indicate why symbolic frames of expression (e.g. physical provocation and verbal insult) are potentially directed towards those considered to be ethnically and racially different from the expected norm. Consequently, the restriction and control of black people's access to white spaces signifies how 'blackness' is perceived to pollute 'whiteness' and how there are subsequent attempts by the dominant group to 'purify' the primary space.

As Afro-Caribbeans are often perceived as belonging to urban domains, the perception of 'blackness' as a source of pollution ought to be contextualized through the socio-spatial representations of black communities. Conceptions of urban domains are often guided by popular images of social decay and environmental contamination (Herbert and Smith 1989). The fact that particular inner city areas (e.g. Moss Side, Manchester) have not been significant sites for 'visual consumption' illustrates a perception of such areas as 'socially polluting' (Urry 1992: 19). According to Agyeman and Spooner (1997: 199), 'In the white imagination people of colour are confined to town and cities, representing an urban, "alien" environment and the white landscape of rurality is aligned with "nativeness" and the absence of evil or danger'. Nonetheless, it should not be assumed that members of the UK Caribbean diaspora do not have a genuine empathy with, and social attachment to, rural environments. Migration from rural areas in the Caribbean implies that perceptions of rurality are not completely devoid of nostalgic sentiment and symbolic association.

### Landscape connections and symbolic links to the homeland

Several enquiries have identified the social significance of rural landscapes for a number of ethnic minority groups living in the UK. Eaton (1994: 5) reported on regular weekly visits to an Essex country park by large groups of Turkish minorities from London. The park 'apparently reminds them of Turkey, the open spaces and peaceful surroundings'. As these events embody a variety of ethnic signs and cultural symbols (e.g. Turkish music, barbecues, kebabs, square carpets and traditional dress), they illustrate how it is possible for an ethnic minority group to transform a section of the English countryside into a culturally defined space.

Coster (1991) discussed the experiences of a group of Asian women travelling by coach through the Welsh mountains. The group generally commented on how these landscapes were similar to the mountainous areas in Pakistan, Kashmir and Mirpur. Malik (1992) relates how members of the Indian community preferred to visit Snowdonia and the Lake District as reminders of their own rural heritage, especially the mountainous holiday retreats (i.e. the 'hill stations'). Stephenson's (1997) ethnographic observations of countryside visits by first – and second-generation Jamaicans highlighted how individuals formed landscape connections between the British countryside and the 'old country' (i.e. Jamaica). People's sensory perceptions focussed on a range of familiar sights (e.g. green fields, pothole roads and milestones), smells (e.g. fresh air and pollen) and sounds (e.g. farm animals and birds).

The social process of gazing at rural landscapes enables ethnic minorities to reminisce about life in the 'ancestral homeland'. The personal relevance of tangible landscape links presents an alternative way of understanding rural representations beyond popular views concerning the countryside as a signifier of national identity and sentiment (Lowenthal 1991; Bunce 1994; Daniels 1994). Accordingly, visions of the countryside transcend national boundaries to represent identities that are not holistically British or English, but Indian, Pakistani or Jamaican. Thus, rural landscapes are imbued with a multi-ethnic and multi-cultural appeal.

Fundamentally, however, the symbolic links that exist between two physically separate and topographically distinct landscapes illustrate how it may be possible for diasporic communities to confront their spatial displacements constructively. This situation would encourage individuals to positively address their social detachments from their ancestral homeland. Although the term 'diaspora' may essentially be defined on the basis of the physical and social dispersal of individuals from their place of origin, the conceptual significance of a diasporic group 'embodies a notion of a centre, a locus, a "home" from where the dispersal occurs' (Brah 1996: 181). Therefore, landscapes positively encourage members of a diasporic group or community to inscribe a 'homing desire' (Brah 1996: 180). Moreover, the process of gazing at the British countryside delivers a sense of personal and collective ownership of such environments, which may otherwise be assumed unlikely given perceptions and experiences of racism.

Although rural areas may possibly attract the attention of diverse social groups with the outcome of creating positive experiences and (culturally) sustainable encounters, racialized perceptions of 'black others' must be challenged. This is indeed an arduous task. However, the presence of ethnic minorities within rural domains has recently become a politicized concern. Organizations like the CRE and BEN have actively addressed the rights of minorities to have equality of access to rural places and spaces in the UK.

## Visiting places and destinations in Europe

### *Racialized experiences and encounters*

The prevalence of racial tensions, xenophobia and ethnic nationalisms and the re-emergence of patriotic movements and neo-Nazi groups throughout various European countries illustrates the level of social opposition to 'dominant outsiders' in Europe (Gabriel 1994; Kennett 1994; Witte 1995). In Germany racial violence against foreigners has intensified particularly as a consequence of reunification. Minority Jewish, Vietnamese, Mozambiquan and Turkish communities have all been targets of racial attacks (Willems 1995; Witte 1995). Sections of the UK black community have been concerned with the increase of racial violence towards black travellers in Europe. Reports in the 'black press' have often covered racialized incidents occurring in various tourism destinations. Headlines have read: 'Spanish Bouncers Leave Holiday Pals Injured' (*Voice* 1996a: 11) and 'African in German Skinhead Attack' (*Weekly Journal* 1994a: 7).

Arguably, covert forms of racialism also have a personal impact on how individuals encounter particular European destinations. One concerned individual described her encounters in Italy:

> Having recently returned from Europe I can only say that I will never be returning there again as the whole experience has given me serious misgivings . . . My

family and I went to Milan, which is supposed to be a cosmopolitan city. Judging from the reaction we got from the locals anyone would think we had just touched down from Mars. Everywhere we went we received stares and comments. My Italian is not wonderful but you do not need to be an expert to know when you are not wanted . . . I must point out that I did not go there totally naïve and did expect some hostility, but never on such a scale. Unfortunately, the future looks nothing but bleak for black people both travelling and living in Europe.

(Weekly Journal 1994b: 17)

This illustrates ways in which the 'white gaze' contributes to a range of negative experiences. It also indicates how the maltreatment of travellers potentially encourages individuals not to return to places perceived to engender acts of racialism. Caryl Phillips (1993: 83) recounts his experiences of being a black visitor to Munich. He records how,

After eighteen hours I wanted to escape. The cold Germanic faces snapped round in the street to look at me. They gazed as though I had just committed an awful crime, or was about to cannibalize a small child. I began to stare back and conduct imaginary arguments. 'My skin was not burned in Europe', I murmured silently.

(Phillips 1993: 83)

Michael Lomotey (2001: 4), an outreach worker for *Tourism Concern*, describes his experience of visiting France:

My visits to France are marked by constant abuse and harassment. I am snubbed by taxi drivers and shop keepers who pretend not to see me or not to understand me. In rural France people visibly freak out when they see my black face. I remember one woman crossing the street clutching her bags when she saw me. I don't go to France any more – it has little appeal.

A post-modern reading views tourism experiences as having a significant role in dissolving the socio-cultural boundaries that pre-exist between local and foreign cultures. For Urry (1995: 166–7),

International tourism produces international familiarization/normalization so that those from other countries are no longer seen as particularly dangerous and threatening – just different and this seems to have happened on a large scale in Europe in recent years.

However, in instances where racial boundaries between visitors and guests are socially and ideologically constructed, opportunities for the development of normative exchanges and productive encounters are seriously limited. Racial differences decelerate the formation of mutual affiliations and congenial social relations between hosts and guests. The Afro-Caribbean visitor can simply become an 'element of spectacle' in white communities and destinations; that is, viewed and censored by the prevailing power of the 'white gaze'.

People's aspirations and perceptions do not fully conform to postmodern theorizations of tourists, in particular the need to explore isolated destinations. This aspiration apparently belongs to the 'post-tourist', or someone prepared to take chances and experience challenging situations (Feifer 1985: 259; Urry 1990, 2002; Munt 1994). Increasing opportunity to indulge in chance encounters and experimental experiences is read as a function of living in (post-industrial) 'risk societies' where individualized endeavours, fraught situations and personalized uncertainties are ever significant (Beck 1992). For Afro-Caribbeans, however, the desire or need to take risks may be a lower priority than the need for secure and safe experiences. Social and personal risks may be heightened in regions of Europe that do not have a significant black/Afro-Caribbean presence. The novelist Mike Phillips (2001: 197) explains how visits to East European countries may embody 'peculiar and dislocating experience(s)'. One of his trips to Prague was clouded by feelings of despair, such that,

> What I was thinking about was the fact that I hadn't seen another black person since I'd left Heathrow. In these circumstances you get a sense of isolation. You wonder a little about what would happen if you were attacked by a mob of racists, or even by a solitary nutter. Sometimes you feel you are the centre of attraction – everyone in the street must know you're there. Sometimes you feel completely anonymous – no-one knows who you are . . . What was certain was the fact that, unlike Western Europe, the countries of Eastern Europe had practically no contact with Africa or Asia.
>
> (Phillips 2001: 195–6)

Consequently, individuals may prefer to travel to destinations that limit their exposure to racialism. Although the ancestral homeland is one obvious destination where Afro-Caribbean visitors are likely to feel comfortable, especially as racialized encounters are limited and opportunities for ethnic and cultural familiarization prevail (Stephenson 2002), travel may be less threatening in destinations where there is a significant presence of 'black others'.

## Diasporic identifications and bilateral associations

As members of the Caribbean diaspora are in a better position to empathize and identify with other ethnic minority groups, for instance in sharing similar experiences of racism and inequality (James 1989), the feeling of being a 'complete outsider' or 'stranger' in Europe may be challenged. Travel encounters with other black minorities may encourage individuals to engage in a sense of collective identity with people of similar ethnic origins and compatible cultural and political histories. Within various multi-cultural destinations, such as Amsterdam with a significant Caribbean population, black visitors and inhabitants signify the geographical and social reach of the transatlantic diaspora.

Bonsu, a radio reporter, relates his experiences and affinity with other black people he met at Berlin airport and in the city, 'Whether African students, or entrepreneurs from elsewhere in the diaspora, they were happy to speak to me – another foreigner, an oasis of tolerance' (Bonsu 1994: 17). Consequently, cosmopolitan societies in Europe have more of a comforting appeal to Afro-Caribbean visitors than white-dominated destinations, particularly if they are perceived to be tolerant of ethnic and racial differences. As Phillips (1993: 102) recalls,

In a way, I came to Norway to test my own sense of negritude. To see how many of 'their' ideas about me, if any, I subconsciously believed. Under a volley of stares it is only natural to want eventually to recoil and retreat. In a masochistic fashion, I was testing their hostility. True, it is possible to feel this anywhere, but in Paris or New York, in London or Geneva, there is always likely to be another black person around the corner or across the road. Strength through unity in numbers is an essential factor in maintaining a sense of sanity as a black person in Europe.

Jules-Rosette (1994) discusses ways in which various Parisian tourist sites and enclaves reflect the diverse cultures and communities of the African diaspora (e.g. African market places, Afro-Antillian entertainment venues and places associated with famous African-American writers). Here, 'black Paris' represents a range of cultural/ethnic attributes, signs, symbols and products from various regions of black/African diaspora (e.g. 'black-America', sub-Saharan Africa, East Africa and the Caribbean). Importantly, 'a multilayered black tourist experience will surface in Paris, masking the living conditions of ethnic communities with nostalgic and exotic myths' (Jules-Rosette 1994: 696).

Issues concerned with understanding collective identifications in different travel contexts contribute to a clearer perspective on the importance of formulating inter-cultural relationships and shared cultural experiences with other black (African) diasporic cultures. This affinity could help to contextualize the role of travel in forming symbolic exchanges and bilateral alliances with those of similar ethnic origins, or those who share similar racial experiences (cf. Gilroy 1993a,b). Although Gilroy (1993b) was concerned with how particular elements of a shared cultural heritage (e.g. music) can symbolically bring black cultures closer together, travel may also be another link directly connecting black communities to a common socio-cultural reality.

Nevertheless, what is very apparent is the way in which threatening experiences and fearful encounters and the incursive nature of the 'white gaze' are counter-effective to productive experiences. However, racialized situations and events are not only evident within a country's physical boundaries, but at its borders and frontiers.

## Problems encountered at Europe's borders and frontiers

European Union member states have made repeated attempts to strengthen immigration restrictions towards asylum seekers, refugees and migrants from the 'South'. At points of entry, individuals have experienced prolonged periods of detainment or immediate repatriation 'home' (Mitchell and Russell 1994). Some of the problems experienced by non-nationals have extended to those ethnic minorities who have legal status of entry and/or residency – as they too are 'visibly different' from Europe's ethnic majorities (Allen and Macey 1990: 385). Sivanandan (1990: 159–60) has warned that 'free movement' of black citizens from one domain to another would be politically uncertain, stating:

> Citizenship may open Europe's borders to blacks and allow them free movement, but racism which cannot tell one black from another, a citizen from an immigrant, an immigrant from a refugee . . . is going to make such a movement fraught and fancy.

During the 1990s, the black press published complaints from the black community of incidents of racial harassment during entry into and/or exit from countries such as Belgium, France, the Netherlands and the UK (*Voice* 1996b: 6). One particular incident occurred when over 300 Jamaican visitors, travelling to the UK to visit their families for Christmas celebrations, were detained for significant periods by British immigration. At least thirty of the visitors were forced to return to Jamaica (Jones 1994; Oakes 1994; Francis 1995: 3).

The UK Home Office move in January 2003 to introduce a legal requirement for Jamaican nationals to acquire a visa to enter the UK (Travis 2003: 6) exemplifies the extent to which travel patterns are increasingly monitored by the state. Jamaica is one of several black countries of the Commonwealth to suffer restrictions on its citizens' entry into the UK, although Guyanans are the only other Caribbean nationals who require a visa. Members of the wider Jamaican diaspora have emphasized the socio-cultural ramifications of visa requirements. The 'visa regime' will directly impact upon extended family networks in the UK, by restricting and monitoring visits of relatives from Jamaica (Gallimore 2003: 13). Attention also turned to Jamaicans living in the USA and Canada who wish to visit the UK, as they too are likely to face the inconvenience of obtaining a travel visa (Lindsay 2003: 13). This case exemplifies the problems that people experience in moving from one diasporic space to another.

The Joint Council for the Welfare of Immigrants, the Society of Black Lawyers and the Anti-Racist Alliance have acknowledged that repetitive cross-examination actually inhibits racial tolerance and social equality (Butscher 1995: 2; *Weekly Journal* 1995: 3). Problems experienced by non-white minorities relate to national concerns over illegal entry of 'immigrants'/'refugees' and the importation of illegal substances. For HM Customs and Excise, it is imperative to execute extra security measures and searches at points of entry from Jamaica, Nigeria and South America, since illicit drugs originate from such places. Yet, systematic searches and incidents of perceived racial harassment have also been reported by black (i.e. 'British') citizens entering the UK from various European countries (*Weekly Journal* 1994a: 2). Stephenson (1997: 192) reports why Valerie, a second-generation Jamaican living in Manchester, believed black people are often subject to racial harassment by customs officials.

> I know black people have problems at customs . . . even with a British passport! It doesn't make any difference, you have got a black face and that's that . . . it's happened to me before, you know, the harassment thing . . . I think one of the main reasons is that they have stereotypes in their heads. They suspect them of drug dealing. It's the ganja thing! Black people have been made scapegoats . . . So we're not always seen to be travelling for the enjoyment of it all!

Although 'black others' have a tendency to be perceived as 'threatening' or 'dangerous', 'white others' can also be viewed in a similar light. hooks (1992) presents an encounter that she experienced in France as an example of the importance of reading racial situations from a perspective that considers perceptions of 'whiteness':

> I was strip-searched by French officials, who were stopping black people to make sure we were not illegal immigrants and/or terrorists, I think that one fantasy of whiteness is that the threatening Other is always a terrorist. This projection enables

many white people to imagine there is no representation of whiteness as terror, as terrorizing.

(hooks 1992: 174)

The 11 September 2001 incident has further aggravated the difficulties experienced by ethnic minority travellers. Members of Muslim communities have been subjected to hostile treatment from immigration officials, aviation authorities and other passengers (Wazir 2001: 4). Nevertheless, the journeys of minority groups do not always constitute legitimate touristic ventures. This is evident in situations where one's status or identity as a tourist is not fully sanctioned by state authorities. Thus, the purpose of travel is potentially perceived as involving the pursuit of criminal or illegal activities rather than activities of a pleasurable or educational nature. Moreover, surveillance by customs officials and frontier guards represents the distinctive attributes of both the 'nation-state' and the 'European-state'. Aspects of control and containment also reflect a continuation of those coercive and threatening events of the past (Clifford 1992; hooks 1992; Curtis and Pajaczkowska 1994). Travel histories of the Afro-Caribbean diaspora expose the 'horrors of their exile' (James 1993: 244), exemplified by acts of 'enslavement', 'enforced migration', 'immigration' and 'relocation' (hooks 1992: 173). Contemporary travel experiences not only manifest elements of the past but also epitomize the structural dislocations and institutional problems existing within everyday environments. Once more, journeys involve degrees of control, protest and possible mediation.

Despite the possession of British/EU passports, the contested nature of journeys suggest that Afro-Caribbeans are implicitly denied a British and/or European identity. For individuals to acquire such an identity fully, they would have to adopt additional attributes of identity, or even acquire 'specific forms of double consciousness' (Gilroy 1993a: 1). This would be difficult to achieve in situations where self-perceptions and the perceptions of 'white others' have an overwhelming influence on the construction and/or reconstruction of individual and collective identifications. Accordingly, travel encounters and experiences encourage individuals to reflect on their own diasporic existences and identities, particularly the non-British/non-European attributes of ethnic identification.

## Conclusions and research implications

Destinations, which are dominated by white ethnic groups and perceived to be racially motivated, are not always conducive to productive cultural exchanges. Racial reactions overshadow positive experiences and are detrimental to 'self-satisfaction'. Although racial ideologies and incidents manifest perceptions of black travellers/visitors as 'threatening' and/or 'dangerous', the dilemma identified here alludes to how 'white others' are also similarly perceived. Subsequently, travel encounters do not, as Urry (1995: 166) assumed, fully reflect mutual 'familiarization' with other cultures. They do not denote, as Rojek (1993a: 199) maintained, the de-differentiation of boundaries between 'local' and 'foreign' cultures. Afro-Caribbean travellers are often victims of the 'racialized condition' and viewed as a threat to the social order and norms of particular communities. In a similar manner to Bauman's (1991: 56) conceptual description of the 'stranger', black visitors are perceived to 'bring the outside to the inside and poison the comfort of order with suspicion of chaos'.

Unless travel events are well organized, racialized realities may overshadow the experience of visiting other cultures and communities. Furthermore, the view that tourists are forever in search of 'new places to visit and capture' (Urry 1992: 5) does not necessarily apply to an understanding of the travel motivations of members of the UK Afro-Caribbean diaspora. The search for cultural familiarization and identification with others, and the need to feel safe and secure, would perhaps be better indicators for explaining people's aspirations.

Crucially, the problems individuals encounter during particular movements compel them to reflect on their own diasporic roles, statuses and identities. Racialized experiences are likely to influence people's perceptions and impressions of particular destinations, societies and cultures. Given the structural circumstances of racism apparent in everyday life, perceptions are often preconditioned within the (metropolitan) home environment. The anticipation and experience of racialism in 'white spaces' in Europe indicates that Afro-Caribbeans are routinely marginalized from experiencing the possible social benefits of tourism as other national citizens, including: recreational pleasure, educational awareness, self-actualization, social esteem and mutual inter-action. This is despite the fact that they have a legitimate right to British/European citizenship.

Given that this work strongly implies the importance of dealing with the rights of individuals to enjoy the benefits of tourism experiences without prejudice or discrimination, future investigations ought to consider the socio-political rights of individuals to travel to places unrestricted by actions and/or reactions of others. Moreover, the social inequalities and injustices that black diasporic groups experience during their various travellings suggests that there is an immediate need for these concerns to reach the widest audience of academics, policy makers and representatives of the travel and tourism industry. Consequent strategies for achieving mutual exchanges, constructive experiences and positive encounters may then develop to the socio-cultural advantage of diasporic groups.

Finally, attention to the experiences of other UK racialized diasporas, such as the Jewish, Pakistani and Irish communities, would help to develop a wider analysis of diasporic forms of travel. This would further encourage other social dimensions to emerge, particularly with respect to aspects of religion and gender. For the moment, however, members of the UK Afro-Caribbean community have limited opportunities to travel outside of their own cultural and ethnic domains and their own diasporic boundaries and territories.

## References

Agyeman, J. (1989) 'Black people, white landscape', *Town and Country Planning* 58(12): 336–38.

Agyeman, J. and Spooner, R. (1997) 'Ethnicity and the rural environment', in P. Cloke and J. Little (eds) *Contested Countryside Cultures: Otherness, Marginalization and Rurality*, London: Routledge.

Allen, S. and Macey, M. (1990) 'Race and ethnicity in the European Context', *British Journal of Sociology* 41(3): 375–93.

Anthias, F. and Yuval-Davis, N. (1993) *Racialized Boundaries: Race, Nation, Gender, Colour and Class and the Anti-Racist Struggle*, London: Routledge.

Balibar, E. (1991) 'Is there a neo-racism?', in E. Balibar and I. Wallerstein (eds) *Race, Nation, Class: Ambiguous Identities*, London: Verso.

Ballard, R. and Kalra, V.S. (1994) *The Ethnic Dimensions of the 1991 Census: A Preliminary Report*, Census Dissemination Unit, Manchester: University of Manchester.

Barker, M. (1981) *The New Racism*, London: Junction Books.

Bauman, Z. (1991) *Modernity and Ambivalence*, Cambridge: Polity.

Beck, U. (1992) *Risk Society: Towards a New Modernity*, London: Sage.

Bonsu, H. (1994) 'The cutting edge of Fortress Europe', *Weekly Journal*, 3 November 1994: 17.

Brah, A. (1996) *Cartographies of Diaspora: Contesting Identities*, London: Routledge.

Bunce, P.M. (1994) *The Countryside Ideal: Anglo-American Images of Landscape*, London: Routledge.

Butscher, M. (1995) 'MP is "playing the race card"', *Voice*, 21 February 1995: 2.

Byron, M. (1994) *Post-War Caribbean Migration to Britain: The Unfinished Cycle*, Aldershot: Avebury.

Clifford, J. (1992) 'Travelling cultures', in L. Grossberg, C. Nelson and P.A. Treichler (eds) *Cultural Studies*, London: Routledge.

Cohen, R. and Kennedy, P. (2000) *Global Sociology*, London: Macmillan Press.

Coster, G. (1991) 'Another country', *The Guardian*, 1 June 1991: 4–6.

Curtis, B. and Pajaczkowska, C. (1994) 'Getting there: travel, time and narrative', in G. Robertson, M. Mash, L. Tickner, J. Bird, B. Curtis and T. Putnam (eds) *Travellers' Tales: Narratives of Home and Displacement*, London: Routledge.

Daniels, S. (1994) *Fields of Vision: Landscape Imagery and National Identity in England and the United States*, Cambridge: Polity.

Derbyshire, H. (1994) *Not in Norfolk: Tackling the Invisibility of Racism*, Norwich: Norfolk and Norwich Racial Equality Council.

Derounian, J.G. (1993) *Another Country: Real Life Beyond Rose Cottage*, London: NCVO Publications.

Douglas, M. (1984) *Purity and Danger: An Analysis of the Concepts of Pollution and Taboo*, 2nd edn, London: Routledge and Kegan Paul.

Dummett, M. and Dummett, A. (1987) 'The role of local government in Britain's racial crisis', in C. Husband (ed.) *'Race' in Britain: Continuity and Change*, 2nd edn, London: Hutchinson.

Eaton, L. (1994) 'Turks delight in an Essex Sunday', *The Independent*, 19 August 1994: 5.

Feifer, M. (1985) *Going Places: The Ways of the Tourist from Imperial Rome to the Present Day*, London: Macmillan.

Francis, V. (1995) 'Airport officials "irrational"', *Weekly Journal*, 13 April 1995: 3.

Fryer, P. (1984) *Staying Power: The History of Black People in Britain*, London: Pluto Press.

Gabriel, J. (1994) *Racism, Culture, Markets*, London: Routledge.

Gallimore, P.B. (2003) 'The case of a few bad apples', *Weekly Gleaner*, 22–28 January 2003: 13.

Gilroy, P. (1987) *There Ain't No Black in the Union Jack: The Cultural Politics of Race and Nation*, London: Unwin Hyman.

Gilroy, P. (1993a) *The Black Atlantic: Modernity and Double Consciousness*, London: Verso.

Gilroy, P. (1993b) *Small Acts: Thoughts on the Politics of Black Cultures*, London: Serpent's Tail.

Hall, S., Critcher, C., Jefferson, T., Clarke, J. and Roberts, B. (1978) *Policing the Crisis: Mugging, the State and Law and Order*, London: Macmillan.

Herbert, D.T. and Smith, D.M. (eds) (1989) *Social Problems and the City: New Perspectives*, Oxford: Oxford University Press.

hooks, b. (1992) *Black Looks: Race and Representation*, Boston, MA: South End Press.

James, W. (1989) 'The making of black identities', in R. Samuel (ed.) *Patriotism: The Making and Unmaking of British National Identity*, London: Routledge.

James, W. (1993) 'Migration, racism and identity formation: the Caribbean experience in Britain', in W. James and C. Harris (eds) *Inside Babylon: The Caribbean Diaspora in Britain*, London: Verso.

Jay, E. (1992) *Keep them in Birmingham*, London: Commission for Racial Equality.

Jenkins, R. (1987) 'Countering racial prejudice: anthropological and otherwise', *Anthropology Today* 3(2): 3–4.

Jenkins, R. (1997) *Rethinking Ethnicity: Arguments and Explorations*, London: Sage.

Jones, L. (1994) 'We haven't a clue about Xmas flight JAs', *Voice*, 28 June 1994: 7.

Jules-Rosette, B (1994) 'Black Paris: touristic simulations', *Annals of Tourism Research* 21(4): 679–700.

Keith, M. (1989) 'Riots as a social problem in British cities', in D.T. Herbert and D.M. Smith (eds) *Social Problems and the City: New Perspectives*, Oxford: Oxford University Press.

Keith, M. (1995) 'Making the street visible: placing racial violence in context', *New Community* 21(4): 551–65.

Kennett, P. (1994) 'Exclusion, post-Fordism and the "New Europe"', in P. Brown and R. Crompton (eds) *A New Europe? Economic Restructuring and Social Exclusion*, London: University College London Press.

Lindsay, A. (2003), 'Visa regime strains ties to "Mother Country"', *Weekly Gleaner*, 22–28 January 2003: 13.

Lomotey, M. (2001) 'No holiday from racism', *In Focus* 40: 4–5.

Lowenthal, D. (1991) 'British national identity and the English countryside', *Rural History* 2(2): 205–30.

Malik, S. (1992) 'Colours of the countryside: a whiter shade of pale', *ECOS* 13(4): 33–40.

Miles, R. and Phizacklea, A. (1984) *White Man's Country: Racism in British Politics*, London: Pluto.

Mitchell, M. and Russell, D. (1994) 'Race, citizenship and "Fortress Europe"', in P. Brown and R. Crompton (eds) *A New Europe? Economic Restructuring and Social Exclusion*, London: University College London Press.

Munt, I. (1994) 'The "other" postmodern tourism: culture, travel and the new middle classes', *Theory, Culture and Society* 11(3): 101–23.

Oakes, J. (1994) 'No hitches! First Christmas charter flight lands', *Gleaner*, 13 December 1994: 1.

Peach, C. (1984) 'The force of West Indian island identity in Britain', in C. Clark, D. Ley and C. Peach (eds) *Geography and Ethnic Pluralism*, London: Allen and Unwin.

Peach, C. and Byron, M. (1993) 'Caribbean tenants in council housing: "race", class and gender', *New Community* 19(3): 407–23.

Phillips, C. (1993) *The European Tribe*, London: Pan Books.

Phillips, M. (2001) *London Crossings: A Biography of Black Britain*, London: Continuum.

Phoenix, A. (1988) 'The Afro-Caribbean myth', *New Society* 83(1314): 10–13.

Pilkington, A. (2003) *Racial Disadvantage and Ethnic Diversity in Britain*, Basingstoke: Palgrave Macmillan.

Rex, J. and Tomlinson, S. (1979) *Colonial Immigrants in a British City: A Class Analysis*, London: Routledge and Kegan Paul.

Rojek, C. (1993a) *Ways of Escape: Modern Transformations in Leisure and Travel*, Basingstoke: Macmillan.

Rojek, C. (1993b) 'After popular culture: hyperreality and leisure', *Leisure Studies* 12(4): 277–89.

Shore, C. (1993) 'Inventing the "People's Europe": Critical approaches to EC "cultural policy"', *Man* 28(4): 779–800.

Shore, C. and Black, A. (1992) 'The European communities and the construction of Europe', *Anthropology Today* 8(3): 10–11.

Sibley, D. (1997) 'Endangering the sacred: nomads, youth cultures and the English countryside', in P. Cloke and J. Little (eds) *Contested Countryside Cultures: Otherness, Marginalization and Rurality*, Routledge: London.

Sivanandan, A. (1990) *Communities of Resistance, Writings on Black Struggles for Socialism*, London: Verso.

Skellington, R. (with Morris, P.) (1992) *Race in Britain Today*, London: Open University Press/Sage.

Stephenson, M.L. (1997) 'Tourism, race and ethnicity: the perceptions of Manchester's Afro-Caribbean community concerning tourism access and participation', unpublished PhD Thesis, Manchester: Manchester Metropolitan University.

Stephenson, M.L. (2002) 'Travelling to the ancestral homelands: the aspirations and experiences of a UK Caribbean Community', *Current Issues in Tourism* 5(5): 378–425.

Taylor, D.E. (1993) 'Minority environmental activism in Britain: from Brixton to the Lake District', *Qualitative Sociology* 16(3): 263–95.

Travis, A. (2003) 'Jamaicans dismayed by visa requirement', *Guardian*, 9 January 2003: 6.

Urry, J. (1990) *The Tourist Gaze: Leisure and Travel in Contemporary Societies*, London: Sage.

Urry, J. (1992) 'The tourist gaze and the environment', *Theory Culture and Society* 9(3): 1–26.

Urry, J. (1995) *Consuming Places*, London: Routledge.

Urry, J. (2002) *The Tourist Gaze: Leisure and Travel in Contemporary Societies,* 2nd edn, London: Sage.

*Voice* (1996a) 'Spanish bouncers leave holiday pals injured', 9 June 1996: 11.

*Voice* (1996b) 'EU faces quiz over travel hassle: why are black tourists stopped, MEP demands?', 18 June 1996: 6.

Wazir, B. (2001) 'British Muslims fly into a hostile climate', *Observer*, 21 October 2001: 4.

*Weekly Journal* (1994a) 'African in German skinhead attack', 29 September 1994: 7.

*Weekly Journal* (1994b) 'Letters to the editor: European doubt', 10 November 1994: 17.

*Weekly Journal* (1994c) 'Smart dogs patrol green channel', 11 August 1994: 2.

*Weekly Journal* (1995) '"Criminal" slur haunts travellers', 6 April 1995: 3.

Willems, H. (1995) 'Right-wing extremism, racism or youth violence? Explaining violence against foreigners in Germany', *New Community* 21(4): 501–23.

Witte, R. (1995) 'Racist violence in Western Europe', *New Community* 21(4): 489–500.

Yeboah, S.K. (1988) *The Ideology of Racism*, London: Hansib Publishing Limited.

# 5   Linking diasporas and tourism

## Transnational mobilities of Pacific Islanders resident in New Zealand

*C. Michael Hall and David Timothy Duval*

## Introduction

Migration and mobility are inherent features of the peoples of the South Pacific. From Polynesian migration throughout the islands of the South Pacific (including New Zealand) during the pre-European historical period to the labour migrations in comparatively recent decades, mobility remains a central component of the lives of many Pacific Islanders (Curson 1973; Ahlburg 1996; Bedford 1997). Yet, while islander populations are highly mobile in terms of employment and education, substantial bonds of kinship and relationships to village and land remain. As Hau'ofa (1993: 11) observes, the networks and linkages that characterize diasporic mobilities are integral to the Pacific peoples:

> so much of the welfare of ordinary people of Oceania depends on informal movement along ancient routes drawn in bloodlines invisible to the enforcers of the laws of confinement and regulated mobility . . . [Pacific peoples] are once again enlarging their world, establishing new resource bases and expanding networks for circulation.

The formal movement of transnational travel and tourism is an important, though more recent, component of such contemporary circulation. Recognition of its significance, however, has not yet been fully acknowledged in either the literature on tourism in the Pacific or by the tourism industry itself. It can be suggested, however, that such movement can be important for political and socio-economic reasons as well as providing an almost secure market segment for those destinations that are particularly reliant upon international tourist markets.

This chapter has two goals. The first is to position a Pacific Islander diaspora in the context of multiple identities and social contexts within New Zealand. The second goal is to show how these multi-local, multi-social and pluri-local networks – to lend from Pries' (2001) observation – can have an impact on understanding post-migration movement between diaspora and the external homeland. It is argued that understanding the social relationship that migrants have with their post-migration home leads to a better understanding of post-migration travel patterns and the relationship of those patterns to wider issues within Pacific island tourism. Similarly, recognizing the movement of diasporic populations to external homelands following the migration event can potentially allow for a better understanding of their broader social relationships and

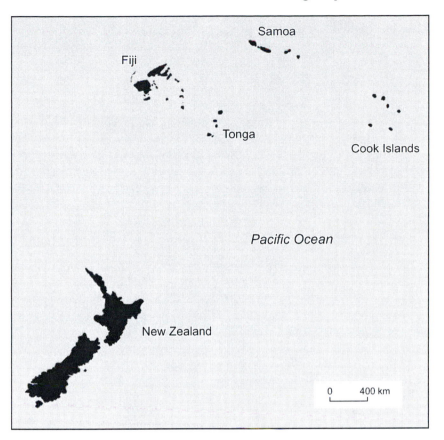

*Figure 5.1* New Zealand and the South Pacific Islands.

identity negotiation strategies, especially those which are, in a transnational manner, cross-border by nature.

## Pacific Islanders in New Zealand

In the 2001 New Zealand Census (Statistics New Zealand 2002b), a total of 231,801 people identified their ethnicity as one of Pacific origins. It is important to note, however, that this is a declared response and does not take into consideration place of birth. While this figure represents an increase of 39 per cent over the 1991 Census, the increase may, in part, be attributed to more individuals electing to declare a Pacific Island (or relevant nation-state) ethnic background. In fact, from the 2001 census, nearly six in ten people who identify themselves as Pacific Islander were born in New Zealand. In the census of 1991, only 50 per cent were born in New Zealand. The largest Pacific Islander subgroup in New Zealand are Samoans (50 per cent), followed by Cook Island Maori (23 per cent), Tongan (18 per cent), Niuean (9 per cent), Fijian (3 per cent), Tokelauan (3 per cent) and Tuvalu Islander (1 per cent) (Statistics New Zealand 2002b).

The substantial rate of out-migration from the Pacific Islands has often been regarded as problematic, particularly within those countries in the region, such as Australia and New Zealand, which are often the destination of migrants (Hooper 1961a,b; Curson 1970, 1972; Graves and Graves 1976; Ahlburg 1996). Not unlike the Caribbean region (Gmelch 1992), some of the smaller nation-states can claim that more of their nationals live abroad than currently within the country. In fact, when the ethnic composition of New Zealand's population, with special reference to the Pacific Islands, is compared to the estimated population size of the associated nation-states which Pacific Islanders theoretically identify as their external homeland, several observations can be made (Table 5.1). First, the balance between the two populations shows that the proportional size of New Zealand's Pacific Island population is not entirely different from that of the external homelands. In some cases, especially the Cook Islands and Tuvalu, the size of these ethnic groups in New Zealand exceeds the populations in the nation-state itself (Bedford 1997). Second, these figures would suggest that the Pacific Islands region is home to societies which are very much migration-oriented and thus not unlike other regions comprised of small-islands states (Philpott 1973). The proportionately high number of Cook Island Maori and Tokelauans in New Zealand, when compared to the actual population figures from these states, is in large part, a consequence of the automatic dual citizenship (Cook Islands/New Zealand and Tokelau/New Zealand) held by these nationals.

The overall increase in Pacific Islander migration reinforces Hau'ofa's (1993) suggestion that Pacific Islanders' world view encompasses more than their external homeland. Further, although a significant number of individuals who live in New Zealand and who identify themselves as Pacific Islander were born in New Zealand, this nonetheless represents a strong cultural affinity to an external homeland, a situation that provides a firm foundation for travel between 'homelands'.

## Migration

The magnitude and importance of migration from the Pacific Islands has strong historical roots. The interface created between expanding colonial powers and indigenous peoples in the nineteenth century helped to broaden the base of movement in the region. Between the 1950s and 1970s, the rate of migration from the Pacific Islands, particularly Samoa, to New Zealand was substantial. Macpherson (1997) notes that the size of the Samoan population in New Zealand rose from 6,481 in 1961 to 27,950 in 1976. The primary reasons for migration were employment and family.

*Table 5.1* New Zealand Census counts in comparison with island population estimates

| Ethnicity | New Zealand census count | Island population estimate |
|---|---|---|
| Samoan | 115,017 | 170,900 |
| Cook Island Maori | 52,569 | 19,300 |
| Tongan | 40,716 | 99,400 |
| Niuean | 20,148 | 1,900 |
| Tokelauan | 6,204 | 1,500 |
| Tuvalu Islander | 1,965 | 10,000 |

Source: Statistics New Zealand 2002b.

Labour shortages in the New Zealand economy that existed until the mid-1970s provided for both short- and long-term migration opportunities (Bres and Campbell 1975). In the 1950s, the main point of migration was to seek financial returns for the home community, with single females often being selected by communities to become migrants because they were regarded as being more likely to remit their wages. By the 1960s migration began to have a greater family focus with reunification becoming an important factor (Macpherson 1997).

While post-war economic growth within the New Zealand economy has, overall, been strong, many Pacific Island countries have experienced considerable fluctuations in their own economies. Many Pacific nation-states, for example, struggled under the recession of the early 1980s, which was largely a result of the reliance on relatively few economic activities in the region (agriculture and fisheries). As a consequence, while the region was, as Connell (1987: 381) points out, on the 'extreme periphery', movement to fringe nations which were more developed was facilitated. As more migrants elected to emigrate, family members would often join their kin in their new homes.

Policies in the receiving countries with respect to migration, particularly the restriction of numbers, have also played an important role. Many restrictions, as Connell (1987) notes, operate within the larger sphere of international economic conditions and thus dictate the rate and flow of migrants. As well, and as mentioned above, nationals from the Cook Islands, Niue and Tokelau are New Zealand citizens, so their movement is legally and politically unrestricted. Furthermore, political unrest has also influenced some migration flows in the region, particularly within the Fijian Indian community (Bedford 1989).

## Ethnicity and identity: dialectics and dualism

To speak of a Pacific diaspora in New Zealand clouds the diversity with which identities and affiliations are formed. Following decades of migration to New Zealand, from the 1950s through to the present day, Macpherson (2001: 71) noted that, for Samoans, many 'saw themselves as expatriate members of families and villages rather than members of a coherent migrant community'. In New Zealand, however, the broader cultural and ethnic label of Pacific Islander was applied, which 'made little political or practical sense for a generation who were brought up in, and identified with, partic-ular islands' (Macpherson 2001: 71). Such affiliations run throughout the local community but also incorporate ties to external homelands, other diasporas and even other social spheres. The essence of Pacific peoples' lives in New Zealand is not, however, solely associated with the ties that are formed between New Zealand and the external homeland. Pacific peoples have clearly become part of New Zealand society, yet have afforded themselves, in many cases, customs within their new homeland. At the risk of trivializing the ability of the migrant to function in a new country by stating that Pacific peoples have 'integrated' well (Macpherson 1997: 95), individual-based affiliations and social spheres have indeed been created and maintained. Many of these mirrored world views are held in the external homeland. In fact, in a discussion of the Samoan diaspora, Macpherson (1997: 95) poses perhaps the most pertinent question and one that is certainly applicable to other Pacific island groups in New Zealand: 'why [did] the migrant Samoans . . . [choose] to re-establish significant parts of a kin-based, religious, rural, village-based world view and lifestyle, in a secular, urban, industrial society?'

While the relative size of the Pacific Island population in New Zealand has grown in the past three to ten years (5 per cent of the total resident population of New Zealand in 1991, versus 6.5 per cent in 2001), the actual composition of this group raises a number of conceptual questions with respect to identity and how that identity relates to the meaning behind other localities and homelands. Without question, both migration and self-identification play a role in determining the ethnic identity of Pacific Islander peoples in New Zealand. In other words, the number of individuals identifying a Pacific Island identity is the result of both direct migration from the numerous Pacific Island nations as well as the rearing of children by these migrants to respect and recognize their parents' (and ultimately their own) social and cultural roots. However, a definitive 'Pacific Island' identity is problematic, both in an academic sense and practically. As Macpherson (2001: 67) notes, multiple identities are effectively 'nested within an emerging Pacific identity which embodies certain common experiences'. Inadequate recognition of this tends to reinforce an underlying assumption that a universal, singular categorization of culture and social values that can be attributed to Pacific Islanders actually exists.

In fact, distinct social and cultural differences exist between and among Pacific Islander groupings in New Zealand. Further, the question remains as to how Pacific Islander populations align themselves socially and culturally, above and beyond what it means to be a 'Pacific Islander'. We are left, then, with several questions that relate a broad 'Pacific Islander experience' in the context of 'mobile migrants' (see Ch. 3): do Pacific Islanders maintain a level of connection and attachment to an external place of locality? How might this connection be manifested, both 'on the ground' and in terms of actual movement and mobilities? Can it be suggested that Pacific identities and social structures within New Zealand are mirrored through the connection, both physical and ideological, between places, where one place represents an idealized home?

Built into this discussion and also in terms of understanding social allegiances and ascription, is how ethnicity is classified. There are, of course, various sources for identifying Pacific Islander ethnicity in New Zealand. Substantive research on the meaning of being 'Pacific Islander' has shown that such broad cultural monikers are problematic (e.g. Linnekin and Poyer 1990). Individuals may occupy several strands of identity as is the case among Haitians in New York (Glick Schiller and Fouron 2001). To this end, Albrow's (1997) 'sociospheres', where individuals variably harvest relationships at different times and points within their life, based in large part on their position and need for recognition within a social realm or sphere, serves as a useful context within which identity structures are considered. One might even suggest that the wider networks with which Pacific Islander identities are fostered and nurtured leads to the creation of such sociospheres. Following Hannerz's (1996) suggestion that social networks provide the basis for which linkages and meanings can be traced, we are left with the tentative proposition that Pacific Islanders in New Zealand have at their disposal broad networks of social and political meaning that they can call upon to define themselves. Such networks can be inherently social or artificially created based on political realities. In other words, governments and official bodies find it easy to administer social programmes to those who identity themselves as Pacific Islander.

Added to this is the issue of how ethnicity is measured. Measuring the ethnicity of Pacific Islanders in New Zealand is, politically, charged to the New Zealand census, which, like many developed countries, occurs every five years. However, there has been some concern and analysis regarding how ethnic groups are identified in an official government census (Bedford and Didham 2001; Macpherson *et al.* 2001). Statistics New Zealand, for example, envisions ethnicity as the ethnic group or groups (note the allowance for plurality) that an individual 'feels they belong to'. A Statistics New Zealand publication in 2002 (Statistics New Zealand 2002a) aptly engages the issue of understanding the fluid nature of ethnicity, suggesting that ethnic groups:

- share a sense of common origins
- claim a common and distinctive history and destiny
- possess one or more dimensions of collective cultural individuality and
- feel a sense of unique collective solidarity

Such characterizations do little to address the reality that multi-stranded identities and connections exist. The social context in which one individual is asked to assign themselves may be external to that which another individual reports.

The point of this discussion and the reason for introducing the problematic nature of defining ethnic identities in New Zealand, is to raise the implications for how diasporas are characterized and how tourism statistics are collected. Because identities are fluid and often transparent, the common understanding of the diaspora, at least in this context, may be inappropriate. In other cases, geographic clustering and patterns of residence reflect a micro-locality that is definitively socially aligned. Hence, in many large urban areas, neighbourhoods come to be known by ethnic nicknames, such as 'Little Italy' or 'Chinatown' (Timothy 2002). As such, is it the diaspora that links the migrant to the external homeland or is it the transnational nature of identities and social meanings that allows for such connections to be manifested? Moreover, how does this express itself in the various mobilities between 'homelands'?

## Movement and mobilities

To address some of the issues raised above, a structuralist approach was adopted to understand ethnic identities and diasporic populations. That is, identities and the wider diasporas from which these identities are situated are theorized as representing a world view or hierarchy of social meaning. The argument hinges on the basis of the trans-national identity structures among Pacific Islanders in New Zealand which lead to an understanding of how aspects of mobility and movement are, first, at the forefront of negotiations of such identity structures in New Zealand (i.e. through mechanisms of adaptation) and, second, how such movement facilitates the broader notion of a Pacific Islander identity (i.e. one that is pan-Pacific and not necessarily routed in one particular locality).

Using available data from the New Zealand Census and Statistics New Zealand (Statistics New Zealand 2001), a rough picture of the extent of movement of Pacific Islanders from New Zealand to their external homeland emerges. Because of substantial limitations in the secondary data available, only the nation-states of Samoa, the Cook Islands, Tonga and Fiji are discussed here. For Tonga and Fiji, while fewer data are available, it is possible at least to make some generalizations based on the inferences made for Samoa and the Cook Islands. These data are then contextualized with data outlining the visitation from these nation-states to New Zealand. Unfortunately, it is generally not possible to sub-analyse travel to and from New Zealand on the basis of ethnicity. This is a limitation in the departure and arrival cards that all travellers are required to complete upon arrival to or departure from New Zealand and on other statistical analyses of tourism in the region. As a result, it is only possible to make a tenuous connection between visitor arrivals and departures based on the destination and origin countries (see Tables 5.2 and 5.3).

### Samoa

Travel to Samoa by New Zealand residents has fluctuated considerably since the early 1980s. During the period 1985–1988 inclusive, there were annual average increases in departures of 1,400 individuals. This increase of temporary travel from New Zealand may be due to the fact that many migrants at the time were on strict work permits that required them to have a job if they were to remain in New Zealand. As New Zealand's economy was undergoing rather dramatic structural re-adjustments at that time, employment was not certain for many migrants and many lost their jobs. On the other hand, for the years 1989, 1990 and 1991, annual departures of New Zealand residents to Samoa fell by 3.5 per cent per annum. This may be explained by the tenuous economic conditions in both New Zealand and Samoa and indeed the generally bleak economic outlook around the world at that time. Throughout the 1990s, however, departures increased annually by 5 per cent (Statistics New Zealand 2002b). In the absence of ethnically based data one of the most important *de facto* measures of Samoans resident in New Zealand travelling to Samoa are the visiting friends and relatives (VFR) data (Table 5.4). Significantly, those countries with the highest populations of Samoan residents (American Samoa, New Zealand, Australia and the USA) are major sources of VFR arrivals (Tourism Council of the South Pacific 1998). In the year 2000, 42 per cent of New Zealand resident departures to Samoa were for the purposes of VFR.

In statistical terms, VFR traffic is therefore a very significant contributor to inbound travel in Samoa. Approximately 40–45 per cent of New Zealand visitors to Samoa are VFR, with an average spend of SAT$1,078 in 1997 (Tourism Council of the South Pacific 1998). Interestingly, while there is a substantial literature on the importance of remittance money to the economy of Samoa and other Pacific islands (e.g. Ahlburg 1991; Brown 1995, 1998; Brown and Walker 1995), there is little discussion of the value of the expenditure of migrants when they return as visitors. Indeed, in the case of some Samoans living overseas, there is room for consideration that, as their level of economic well-being increases, some of their remittance expenditure becomes transformed into travel expenditure that is ultimately 'remitted' in person. Indeed, one of the most remarkable facets of tourism planning and marketing in Samoa is the extent to which the visiting Samoan who is resident overseas is ignored in official documentation (Tourism Council of the South Pacific 1998) with the focus consistently being on the leisure-oriented holiday-maker. Such a perspective is somewhat ironic given not only the size of the market but also its behavioural characteristics in terms of repeat visitation, the relative lack of stress it places on existing scarce capital in

Table 5.2 Tourist arrivals in South Pacific Tourism Organization countries, 1992–1999 (by island destination)

| Country | 1992 | 1993 | 1994 | 1995 | 1996 | 1997 | 1998 | 1999 |
|---|---|---|---|---|---|---|---|---|
| Cook Islands | 50,009 | 52,868 | 57,331 | 48,500 | 48,819 | 49,964 | 48,628 | 55,599 |
| Fiji | 278,534 | 287,462 | 318,874 | 318,495 | 339,560 | 359,441 | 371,342 | 409,955 |
| French Polynesia | 123,619 | 147,847 | 166,086 | 172,129 | 163,774 | 180,440 | 188,933 | 210,800 |
| Kiribati | 3,747 | 4,225 | 3,888 | 2,697 | 3,406 | 8,200 | 14,211 | 6,107 |
| New Caledonia | 78,264 | 80,754 | 85,103 | 86,256 | 91,121 | 105,137 | 103,835 | 83,016 |
| Niue | 1,668 | 3,358 | 2,802 | 2,161 | 1,522 | 2,041 | 1,736 | 1,778 |
| Papua New Guinea | 42,816 | 33,552 | 39,420 | 32,578 | 61,215 | 66,143 | 67,465 | 67,368 |
| Samoa | 37,531 | 47,073 | 67,089 | 68,392, | 73,155 | 67,960 | 77,926 | 85,124 |
| Solomon Islands | 12,446 | 11,570 | 11,918 | 11,795 | 11,217 | 15,894 | 13,229 | 2,474 |
| Tonga | 23,020 | 25,513 | 28,408 | 29,520 | 26,642 | 26,162 | 27,132 | 30,883 |
| Tuvalu | 862 | 929 | 1,224 | 922 | 1,039 | 1,029 | 1,077 | 770 |
| Vanuatu | 42,673 | 44,562 | 42,140 | 43,554 | 46,123 | 49,605 | 52,100 | 50,746 |
| Total | 695,189 | 739,713 | 824,283 | 816,999 | 867,593 | 932,016 | 967,614 | 1,004,620 |

Source: South Pacific Tourism Organization (2000).
Note: Figures for American Samoa not available.

*Table 5.3* Tourist arrivals in South Pacific Tourism Organization countries, 1992–1999 (by source market)

| Country | 1992 | 1993 | 1994 | 1995 | 1996 | 1997 | 1998 | 1999 |
|---|---|---|---|---|---|---|---|---|
| Australia | 173,758 | 160,123 | 170,478 | 157,846 | 179,189 | 193,531 | 215,734 | 228,418 |
| New Zealand | 92,969 | 96,285 | 120,362 | 124,445 | 133,705 | 142,307 | 143,509 | 149,720 |
| Pacific Island | 59,396 | 66,241 | 77,279 | 77,445 | 79,592 | 82,852 | 93,012 | 85,590 |
| USA | 96,740 | 116,798 | 121,953 | 114,903 | 109,600 | 114,677 | 131,288 | 156,782 |
| Canada | 21,328 | 21,075 | 20,119 | 17,103 | 17,893 | 20,942 | 20,956 | 23,106 |
| UK | 26,438 | 29,039 | 34,489 | 34,942 | 39,014 | 45,261 | 50,667 | 51,459 |
| Germany | 29,532 | 30,067 | 32,490 | 28,645 | 24,028 | 23,761 | 21,199 | 19,466 |
| France | 35,838 | 49,724 | 65,223 | 70,586 | 78,421 | 86,799 | 87,547 | 76,667 |
| Other European | 49,196 | 53,607 | 59,402 | 61,725 | 60,603 | 67,931 | 67,737 | 68,034 |
| Japan | 81,425 | 86,756 | 90,212 | 96,116 | 88,154 | 100,373 | 90,832 | 88,877 |
| Other Asian | 16,896 | 17,500 | 19,122 | 20,672 | 35,144 | 33,942 | 24,484 | 23,819 |
| Other countries | 11,673 | 12,498 | 13,154 | 12,571 | 22,250 | 19,640 | 20,649 | 32,683 |
| Total | 695,189 | 739,713 | 824,283 | 816,999 | 867,593 | 932,016 | 967,614 | 1,004,620 |

Source: South Pacific Tourism Organization (2000).

Table 5.4 Annual VFR arrival statistics for Samoa, 1991–1999

| Year | 1991 | 1992 | 1993 | 1994 | 1995 | 1996 | 1997 | 1998 | 1999 |
|---|---|---|---|---|---|---|---|---|---|
| Total arrivals | 39,414 | 37,531 | 47,073 | 67,089 | 58,392 | 73,155 | 67,960 | 77,926 | 84,124 |
| Purpose: VFR | | | | | | | | | |
| American Samoa | 2,009 | 3,443 | 9,543 | 9,383 | 12,254 | 11,482 | 10,409 | 15,323 | 15,039 |
| New Zealand | 2,578 | 2,888 | 5,772 | 8,284 | 9,118 | 8,835 | 8,683 | 9,585 | 10,532 |
| Australia | 926 | 835 | 2,679 | 1,862 | 2,029 | 2,089 | 2,304 | 2,704 | 3,228 |
| USA | 730 | 895 | 3,132 | 2,189 | 2,236 | 2,458 | 2,212 | 2,804 | 2,477 |
| Other Pacific Islands | 664 | 544 | 2,315 | 931 | 1,018 | 708 | 746 | 1,053 | 1,007 |
| Other European countries | 96 | 83 | 508 | 206 | 283 | 132 | 99 | 144 | 179 |
| Germany | 57 | 100 | 1,027 | 100 | 150 | 88 | 66 | 78 | 91 |
| Other countries | 221 | 151 | 505 | 408 | 230 | 153 | 139 | 161 | 378 |
| UK | 43 | 65 | 395 | 94 | 82 | 91 | 79 | 95 | 122 |
| Japan | 59 | 96 | 336 | 93 | 72 | 83 | 83 | 103 | 90 |
| Canada | 29 | 32 | 155 | 71 | 41 | 95 | 62 | 44 | 61 |
| NZ VFR (percentage of all arrivals) | 6.50 | 7.70 | 12.30 | 12.35 | 15.60 | 12.10 | 12.80 | 12.30 | 12.50 |
| Total VFR arrivals | 7,410 | 9,132 | 26,367 | 23,621 | 27,513 | 26,214 | 24,882 | 32,094 | 33,204 |
| Percentage of all arrivals | 18.8 | 24.3 | 56.0 | 35.2 | 41.0 | 35.8 | 36.6 | 41.3 | 39.5 |

Source: Samoa Visitors Bureau 2000.

terms of infrastructure, its contribution to the local economy and the lack of conflict with traditional Samoan values.

## Tonga

Like Samoa, travel to Tonga by New Zealand residents has been characterized by a similar degree of fluctuation since the early 1980s. Increases in visitation in the years 1985–1988 inclusive (an average increase of 23 per cent) can be attributed to structural adjustments in the national economy of New Zealand. Arrivals from Tonga seem to have peaked at 9,400 in 1997, but fell in the following two years. The 2000 arrival figures, however, show an increase of 16 per cent over 1999 figures. Of the approximately 8,500 arrivals in the year 2000, one in two stated their purpose of visit as visiting friends and relatives. Like Samoa, the magnitude of VFR trips would seem to suggest the presence of substantial social (and obviously familial) ties between Tonga and New Zealand.

## Cook Islands

Arrivals from the Cook Islands to New Zealand have been steady since the early 1980s. Within the past 10 years, average annual growth in the number of visitors from the Cook Islands has been 5 per cent. Not surprisingly, the bulk of these visitors declare their intended trip purpose as VFR. For 2000, 46 per cent of the 6,641 trips make by Cook Islanders to New Zealand were for this reason. Departures by New Zealand residents far surpass arrivals by Cook Island nationals. However, these departure figures also include New Zealand Pakeha (Europeans). As well, the popular vacation destination of Raratonga is part of the Cook Islands. The net migration of Cook Islanders to New Zealand over the past century has remained relatively flat in terms of growth. Between 1995 and 2000, the net growth in migration has been approximately 200 individuals per annum.

Unlike other destinations in the South Pacific, limited data is available on the arrival of Cook Islanders resident overseas. Table 5.5 provides details of visitor arrivals to the Cook Islands by country of residence 1994–2000. Of considerable interest is the long-term contribution of Cook Islanders resident overseas visitor market (Table 5.6). Accounting for as high as 11.15 per cent of visitor arrivals in 1988 the figure has subsequently dropped to below 5 per cent at the end of the 1990s, most likely because of the drop in the relative value of the New Zealand dollar from 1997–2000 as a result of the Asian crisis and the subsequent increase in the relative cost of international travel.

An alternative explanation, however, may relate to the relative increase in size and extent of dispersal of the transnational Cook Islanders and the pressures this may place for VFR travel away from the Cooks. Table 5.7 clearly indicates the significance of VFR travel at certain times of the year, particularly Christmas and, to a lesser extent, Easter, both of which are periods at which church and home exert an extremely strong influence on Cook Islanders lives. However, these peak demand periods have created some stresses with the formal tourism sector because of competing demands for seat availability (Buck and Hall 1996). Nevertheless, despite such frictions, the resident overseas market makes a substantial steady contribution to visitor arrivals and again highlights the significance of transnational relationships for travel patterns.

*Table 5.5* Visitor arrivals to the Cook Islands by country of residence, 1994–2000

| Market | 1994* | 1995* | 1996 | 1997 | 1998 | 1999 | 2000 |
|---|---|---|---|---|---|---|---|
| New Zealand | 15,312 | 14,161 | 11,942 | 12,700 | 12,239 | 15,448 | 19,564 |
| Australia | 5,186 | 4,361 | 3,647 | 3,681 | 3,680 | 6,347 | 11,194 |
| USA | 7,839 | 5,270 | 6,088 | 6,417 | 5,365 | 5,853 | 6,734 |
| Europe/UK | 20,380 | 18,600 | 18,024 | 19,896 | 19,290 | 18,382 | 23,638 |
| Canada | 3,962 | 2,667 | 2,757 | 3,074 | 3,622 | 5,230 | 5,992 |
| Tahiti | 3,282 | 1,620 | 1,845 | 1,055 | 1,072 | 742 | 756 |
| Asia | 627 | 646 | 447 | 548 | 450 | 443 | 654 |
| Cook Islanders living overseas | n.a. | n.a. | 3,024 | 1,593 | 2,252 | 2,281 | 3,390 |
| Other | 705 | 574 | 580 | 902 | 659 | 873 | 1,027 |
| Total | 57,293 | 47,899 | 48,354 | 49,866 | 48,629 | 55,599 | 72,994 |

Source: Cook Islands Tourism Corporation.

* 1994 and 1995 figures included Cook Islanders living overseas within market figures of other countries and regions.

## Fiji

Fiji has long been a vacation destination for New Zealanders. Until 2000, when political disruptions were covered extensively in the press in New Zealand, growth in the number of arrivals had been steady. In 1994, for example, some nearly 54,000 New Zealanders travelled to Fiji, while in 2000 the number was just over 72,000 (Fiji Islands Bureau of Statistics 2002). Consequently, a sizeable majority of travel from New Zealand to Fiji is for purposes of taking a holiday (70 per cent in 2000). Trips to visit friends and relatives accounted for only 14 per cent of all travel to Fiji in 2000. Travel from Fiji to New Zealand, on the other hand, is predominantly for the purposes of visiting friends and relatives (nearly 50 per cent in 2000). Like Tonga, it can be surmised that the nature of return visits to Fiji from New Zealand has much to do with the social linkages maintained between both countries.

## Conclusion: networks of interrelationships

The above discussion is meant to provide an overview of travel both to and from New Zealand in the context of selected Pacific Island nation-states. As indicated, however, limitations in the data prevent an accurate portrayal of the ethnicity of travellers entering and leaving New Zealand. As a result, the relative size of ethnic groups moving in and out of New Zealand for the purposes of visiting friends and relatives in the external homeland is speculative.

Nonetheless, some broad trends are notable. Networks of interrelationships undoubtedly aid in the adjustment of migrants from the Pacific Islands as they settle in New Zealand. Cultural relationships, the solidification of social meaning and the linking of familial structures and ties are likely all facilitated by the regular movement of Pacific Islanders in New Zealand to their external homeland. For the Pacific Island states discussed above, it was shown that many migrants returned temporarily to the islands during the structural reform period of the mid- to late 1980s. In this case, those

*Table 5.6* Cook Islanders living overseas as visitor arrivals in the Cook Islands

| 1983–1994 | 1983 | 1984 | 1985 | 1986 | 1987 | 1988 | 1989 | 1990 | 1991 | 1992 | 1993 | 1994 |
|---|---|---|---|---|---|---|---|---|---|---|---|---|
| Total | 19,799 | 26,488 | 28,782 | 31,245 | 32,110 | 33,886 | 32,907 | 34,218 | 39,984 | 50,009 | 52,688 | 57,321 |
| Cook Islanders in NZ | 1573 | 2,020 | 2,324 | 2,887 | 3,303 | 3,778 | 3,401 | 3,371 | 3569 | 3235 | 3358 | 3386 |
| Percentage of total arrivals | 7.9 | 7.6 | 8.1 | 9.2 | 10.3 | 11.2 | 10.3 | 9.9 | 8.9 | 6.5 | 6.4 | 5.9 |
| Cook Islanders in Australia | – | – | – | – | – | – | 282 | 310 | 370 | 365 | 550 | 473 |
| Percentage of total arrivals | – | – | – | – | – | – | 0.9 | 0.9 | 0.9 | 0.1 | 1.0 | 0.8 |

| 1996–2000 | 1996* | 1997* | 1998 | 1999 | 2000 |
|---|---|---|---|---|---|
| Total visitor arrivals | 48,354 | 49,866 | 48,629 | 55,599 | 72,994 |
| Cook Islanders living overseas | 3,024 | 1,593 | 2,252 | 2,281 | 3,390 |
| Percentage of total arrivals | 6.3 | 3.2 | 4.6 | 4.1 | 4.6 |
| of which: | | | | | |
| Cook Islanders resident in NZ | 3,052 | 1,456 | 1,911 | 1,689 | – |
| Percentage of total arrivals | 6.3 | 2.9 | 3.9 | 3.0 | – |
| Cook Islanders resident in Australia | 497 | 244 | 341 | 592 | – |
| Percentage of total arrivals* | 1.0 | 0.5 | 0.7 | 1.1 | – |

Source: Cook Islands Tourist Authority, Cook Islands Tourism Corporation, various statistical publications.

*Figures for Cook Islanders resident in Australia and New Zealand are recorded as greater than Cook Islanders living overseas for 1996 and 1997. Figures are reported as provided by the Cook Islands Tourism Corporation. It is possible that the figure is higher because of dual citizenship between Cook Islands and New Zealand.

*Table 5.7* Visitor arrivals to the Cook Islands, monthly market shares by country of residence, 1998–2000 (%)

|  | Jan | Feb | Mar | Apr | May | Jun | Jul | Aug | Sep | Oct | Nov | Dec |
|---|---|---|---|---|---|---|---|---|---|---|---|---|
| *Market share Cook Islanders resident in New Zealand* | | | | | | | | | | | | |
| 1998 | 3 | 3 | 5 | 4 | 6 | 7 | 4 | 4 | 1 | 1 | 1 | 9 |
| 1999 | 1 | 2 | 2 | 2 | 1 | 1 | 1 | 1 | 1 | 0 | 3 | 17 |
| 2000 | 5 | 1 | 2 | 4 | 3 | 3 | 2 | 1 | 2 | 1 | 3 | 11 |
| *Market share Cook Islanders resident in Australia* | | | | | | | | | | | | |
| 1998 | 0 | 0 | 0 | 1 | 0 | 1 | 1 | 1 | 0 | 0 | 2 | 2 |
| 1999 | 0 | 0 | 0 | 0 | 0 | 0 | 0 | 0 | 1 | 0 | 0 | 8 |
| 2000 | 1 | 0 | 0 | 1 | 1 | 1 | 0 | 0 | 1 | 0 | 0 | 8 |

Source: Cook Islands Tourist Authority.

networks proved to be invaluable in that they allowed migrants to fall back to their external homeland when economic conditions in their new home were unfavourable.

Spoonley (2000) suggests that the transnational nature of migrant identities among Pacific peoples in New Zealand is demonstrated by the flow of capital and human assets to multiple locations. Technology and the availability of travel have certainly added a degree of speed to this mix (Spoonley 2000). Thus, the modernization of the Pacific region and its affiliated globalizing links to other regions, has meant that an increasing amount of return visits have been facilitated. Using secondary data from Statistics New Zealand, this chapter has demonstrated that some inferences from existing data can be made about the magnitude of such trips. Despite the rhetoric of sustainable tourism, the relative value of such markets has been all but ignored in regional tourism planning (South Pacific Tourism Organization 2002), even though expenditure is still being directed into the economy with only minor demands on local culture and infrastructure. It seems that, from the perspective of government and the tourism industry in the region, unless visitors stay in hotels they do not necessarily count as tourists.

This chapter has attempted to illuminate some of the causal links between understanding ethnic identities and the propensity for travel in post-migration environments. While a cursory picture of movement and mobility has been presented, the social connections between and among diasporic and external homelands holds some significance for the motivation behind travel between these two environments. Ward (1997) has suggested that the degree of interconnectedness throughout Oceania is analogous to the geomorphological system of anastomosis, which refers to patterns and systems of rivers and arteries across a landscape. In effect, the pluri-local behaviour and associations demonstrated by some Pacific Islander migrants in New Zealand leaves room for a more spatially centred analysis of identities, especially in those circumstances where external and internal characterizations of belonging, affiliation and social meanings are incongruent. In many respects, this fits well within the new paradigm (as defined by Itzigsohn and Saucedo 2002) of understanding immigration, which sees immigrants having intentional cognition of multiple localities of social meaning.

Beyond the more academic implications that this chapter has put forward, there are several reasons why it is important that governments and national tourism organizations in the Pacific Rim, as well as worldwide, recognize the importance of cross-border transnational movement of former emigrants. First, the scale of such movement can potentially be increased through directed marketing at diasporic communities. As Williams and Hall (2000) and Duval (2002) have shown, such movement can potentially lead to return migration. Second, tracking migrant mobilities can be important for political and certainly socioeconomic reasons (see also Ch. 16). While the extent to which financial remittances are being replaced with borderless movement is far from certain, the contacts forged with overseas diasporic communities can potentially bring rewards of knowledge and social interaction that may be beyond the capabilities of governments. As such, accurate collection of data relating to migrant mobilities is essential and perhaps should be treated with the same degree of importance afforded to the tracking of international visitor arrivals.

## References

Ahlburg, D.A. (1991) *Remittances and Their Impact: A Study of Tonga and Western Samoa*, Canberra: Research School of Pacific Studies, Australian National University.

Ahlburg, D.A. (1996) 'Demographic and social change in the island nations of the Pacific', *Asia-Pacific Population Research Reports* 7: 1–26.

Albrow, M. (1997) 'Travelling beyond local cultures: socioscapes in a global city', in J. Eade (ed.) *Living the Global City: Globalization as a Local Process*, London: Routledge.

Bedford, R. (1989) 'Out of Fiji: a perspective on migration after the coups', *Pacific Viewpoint* 30(2): 142–53.

Bedford, R. (1997) 'Migration in Oceania: reflections on contemporary theoretical debates', *New Zealand Population Review* 23: 45–64.

Bedford, R. and Didham, R. (2001) 'Who are the "Pacific Peoples"? Ethnic identification and the New Zealand census', in C. Macpherson, P. Spoonley and M. Anae (eds) *Tangata O Te Moana Nui: The Evolving Identities of Pacific Peoples in Aotearoa/New Zealand*, Palmerston North: Dunmore Press.

Bres, J. de and Campbell, R.J. (1975) 'Temporary labour migration between Tonga and New Zealand', *International Labour Review* 112: 445–57.

Brown, R. (1995) 'Hidden foreign exchange flows: estimating unofficial remittances to Tonga and Western Samoa', *Asian and Pacific Migration Journal* 4(1): 35–54.

Brown, R. (1998) 'Do migrants' remittances decline over time? Evidence from Tongans and Western Samoans in Australia', *The Contemporary Pacific* 10(1): 107–51.

Brown, R. and Walker, A. (1995) *Migrants and their Remittances: Results of a Household Survey of Tongans and Western Samoans in Sydney*, Pacific Studies Monograph No. 17, Sydney: Centre for South Pacific Studies, University of New South Wales.

Buck, P. and Hall, C.M. (1996) 'Cook Islands', in C.M. Hall and S. Page (eds) *Tourism in the Pacific: Issues and Cases*, London: International Thomson Publishers.

Connell, J. (1987) 'Paradise left? Pacific Island voyagers in the modern world', in J.T. Fawcett and B.V. Cariño (eds) *Pacific Bridges: The New Immigration from Asia and the Pacific Islands*, New York: Centre for Migration Studies.

Curson, P.H. (1970) 'Polynesians and residential concentration in Auckland', *Journal of the Polynesian Society* 79: 421–32.

Curson, P.H. (1972) 'Population change in the Cook Islands. The 1966 population census', *New Zealand Geographer* 28: 51–65.

Curson, P.H. (1973) 'Birth, death and migration: elements of population change in Rarotonga, 1890–1926', *New Zealand Geographer* 29: 51–65.

Duval, D.T. (2002) 'The return visit–return migration connection', in C.M. Hall and A. Williams (eds) *Tourism and Migration: New Relationships between Production and Consumption,* Dordrecht: Kluwer.

Fiji Islands Bureau of Statistics (2002) *Fiji Statistics.* Online. Available: http:www.statsfiji. gov.fj/s_arrivals.html (accessed 15 November 2002).

Glick Schiller, N. and Fouron, G. (2001) *Georges Woke Up Laughing: Long-distance Nationalism and the Search for Home,* Durham, NC: Duke University Press.

Gmelch, G. (1992) *Double Passage: The Lives of Caribbean Migrants Abroad and Back Home,* Ann Arbor: University of Michigan Press.

Graves, T.D. and Graves, N.B. (1976) 'Demographic changes in the Cook Islands: perceptions and reality: or, where have all the Mapu gone?', *Journal of the Polynesian Society* 85(4): 447–61.

Hannerz, U. (1996) *Transnational Connections: Culture, People, Places,* London: Routledge.

Hau'ofa, E. (1993) *A New Oceania: Rediscovering Our Sea of Islands,* Suva: School of Social and Economic Development, University of the South Pacific.

Hooper, A. (1961a) 'The migration of Cook Islanders to New Zealand', *Journal of the Polynesian Society* 70: 11–17.

Hooper, A. (1961b) 'Cook Islanders in Auckland', *Journal of the Polynesian Society* 70: 147–93.

Itzigsohn, J. and Saucedo, S.G. (2002) 'Immigration incorporation and sociocultural transnationalism', *International Migration Review* 36(3): 766–98.

Linnekin, J. and Poyer, L. (1990) 'Introduction', in J. Linnekin and L. Poyer (eds) *Cultural Identity and Ethnicity in the Pacific,* Honolulu: University of Hawaii Press.

Macpherson, C. (1997) 'The Polynesian diaspora: new communities and new questions', in K. Sudo and S. Yoshida (eds) *Contemporary Migration in Oceania: Diaspora and Network,* Osaka: The Japan Center for Area Studies, National Museum of Ethnology.

Macpherson, C. (2001) 'One trunk sends out many branches: Pacific cultures and cultural identities', in C. Macpherson, P. Spoonley and M. Anae (eds) *Tangata O Te Moana Nui: The Evolving Identities of Pacific Peoples in Aotearoa/New Zealand,* Palmerston North: Dunmore Press.

Macpherson, C., Spoonley, P. and Anae, M. (2001) 'Introduction', in C. Macpherson, P. Spoonley and M. Anae (eds) *Tangata O Te Moana Nui: The Evolving Identities of Pacific Peoples in Aotearoa/New Zealand,* Palmerston North: Dunmore Press.

Philpott, S.B. (1973) *West Indian Migration: The Montserrat Case,* London School of Economics Monographs in Anthropology No. 47, London: Athone Press.

Pries, L. (2001) 'The approach of transnational social spaces: responding to new configurations of the social and the spatial', in L. Pries (ed.) *New Transnational Social Spaces: International Migration and Transnational Companies in the Early Twenty-First Century,* London: Routledge.

South Pacific Tourism Organization (2002) *Tourism Strategy for the South and Central Pacific,* Suva: South Pacific Tourism Organization.

Spoonley, P. (2000) 'Reinventing Polynesia: the cultural politics of transnational communities', working paper (WPTC-2K-14), Transnational Communities Research Programme, Institute of Social and Cultural Anthropology, University of Oxford, UK.

Statistics New Zealand (2001) *Tourism and Migration 2000.* Online. Available: http://www. stats.govt.nz/domino/external/pasfull/pasfull.nsf/web/Reference+Reports+Tourism+and+ Migration+2000 (accessed 10 November 2002).

Statistics New Zealand (2002a) *Measuring Ethnicity in the New Zealand Population Census (February),* Wellington: Statistics New Zealand.

Statistics New Zealand (2002b) *2001 Census, Snapshot 6, Pacific Peoples,* Wellington: Statistics New Zealand.

Timothy, D.J. (2002) 'Tourism and the growth of urban ethnic islands', in C.M. Hall and A.M. Williams (eds) *Tourism and Migration: New Relationships Between Production and Consumption*, Dordrecht: Kluwer.

Tourism Council of the South Pacific (1998) *Tourism Economic Impact Study*, Suva: TCSP.

Ward, R.G. (1997) 'Expanding worlds of Oceania: implications of migration', in K. Sudo and S. Yoshida (eds) *Contemporary Migration in Oceania: Diaspora and Network*, Osaka: The Japan Center for Area Studies, National Museum of Ethnology.

Williams, A.M. and Hall, C.M. (2000) 'Tourism and migration: new relationships between production and consumption', *Tourism Geographies* 2(3): 5–27.

# 6 Jewish past as a 'foreign country'

## The travel experiences of American Jews

*Dimitri Ioannides and Mara Cohen Ioannides*

> Migration is not part of Jewish history, it is Jewish history itself.
>
> (Gartner 1998: 107)

## Introduction

The preceding quote is extremely apt considering the countless instances throughout history when Jews have been forced to move either in response to prejudice and persecution, or in search of improved economic opportunities. Among the most notable instances of Jewish migration are the flight of Sephardic Jews following their expulsion from Spain in 1492 and the mass movement away from Eastern Europe of millions of Ashkenazic Jews during the period 1890–1920 as they sought to escape pogroms and dire economic conditions. No country benefited more from Jewish migration than the USA, which is now home to more Jews than any other state in the world (Goldstein 1987).

Today, in response to increasing secularization and liberalization, more and more Jewish-Americans have begun pursuing ways of fulfilling their desire to strengthen their ties with their Jewish background (Sarna 2002). This phenomenon is especially evident upon examination of Jewish-Americans' travel behaviour since much of that behaviour, expressed as their decision on where they go and what to do when they get there, appears to be determined by a relentless search for their cultural roots (Cohen 1992; Levy 1997). Thus, this chapter aims to investigate the travel patterns of American Jews, arguing that Judaism as cultural background bears a significant influence on destination choice.

The argument is based on Eisen's (1998) contention that Jews of the diaspora possess an inherent need to perform what he regards as nostalgic pilgrimages in search of their past. These are journeys through which contemporary Jews (religious and secular), wherever they are, strive to keep in touch with the Judaism of their ancestors and discover their cultural heritage, albeit from a safe distance (see also Lowenthal 1985). This need to travel to their past, their constant pining for different places at different times, is embodied through 'pilgrimages' to spaces that either directly or indirectly are associated with Judaism (Cohen 1983). These spaces include the countries from where their ancestors came, old Jewish neighbourhoods, homes of famous Jewish personalities, synagogues and graveyards, Jerusalem, the death camps of the Holocaust and museums exhibiting Jewish artefacts (religious and secular) (Epstein and Kheimets 2001; Golden 1996).

## The Jewish diaspora in America

### *Jewish-Americans in the past*

Today, the USA boasts the largest Jewish community in the world, even larger that that of Israel, the sole Jewish state (CIA 2002). Nevertheless, there is no consensus as to how many Jews actually live in the USA. The *CJF 1990 National Jewish Population Survey*, which provides the most recent information about Jews in the USA, indicates there are as many as 8.1 million people who describe themselves as Jews (Council of Jewish Federations n.d.). According to this source, nearly 91 per cent are American-born and 48 per cent classify themselves as Ashkenazic Jews, whose origins are Central or Eastern European.

Other observers, such as Goldstein (1987) and Kosmin *et al.* (1986), provide far more conservative estimates, placing the country's Jewish population at just under 6 million. Goldstein attributes the lack of accurate statistics concerning America's Jewish population to the US Constitution's prohibition on government inquiries relating to creed and, thus, the Decennial Census conducted by the Census Bureau does not include questions relating to religious background (see also Rosenwaike 1989). Also, since the American-Jewish community is not homogeneous – there are many different denominations of Jews, plus those describing themselves as secular – it has been impossible ever to carry out an internal census. Thus, existing numbers concerning the US's Jewish population are derived from various methods including projections and estimates based on historical data and indirect measures, such as recorded birthplace of parents and grandparents and/or mother tongue (Goldstein 1987; Rosenwaike 1989).

An examination of the historical patterns of Jewish migration to the USA is extremely useful for shedding light on American Jews' choice of travel to specific sites, which, regardless of their religious denomination, they view as important to their cultural heritage. As Boyarin and Boyarin (1993) maintain, all persons with Jewish roots, whether they consider themselves religious or not, are bound by the idea of group identity. They explain that the term 'group identity' is based on two definitions, both of which are applicable to Judaism, 'on the one hand as the product of a common genealogical origin and, on the other, as produced by a common geographical origin'. The genealogical origin, which defines one's Jewish identity, was briefly touched upon in the preceding paragraph, whereas the geographical origin is explained in some detail below. The geographical origin approach is used here and helps explain the travel choices made by US-based Jews.

Although undoubtedly the largest period of Jewish migration to the USA occurred at the end of the nineteenth and the beginning of the twentieth centuries, historical records indicate two previous major waves of Jewish migrants (Libo 1989). The first lasted almost 200 years from 1654 to 1830. Initially, Sephardic (Spanish and Portuguese) Jews came to this country in the 1650s from Brazil, to where they had originally fled following the spread of the Inquisition throughout Iberia the century before (Chaliand and Rageau 1995). These early migrants were followed by other Sephardic and Ashkenazi Jews who made their way from various north-western European countries, especially Germany. The original Jewish immigrants in America settled mostly along the East Coast in places like Newport (Rhode Island), New York City, Philadelphia, Charleston (South Carolina) and Savannah (Georgia) (see Figure 6.1). Illustrating this fact, Libo (1989: 107) writes that:

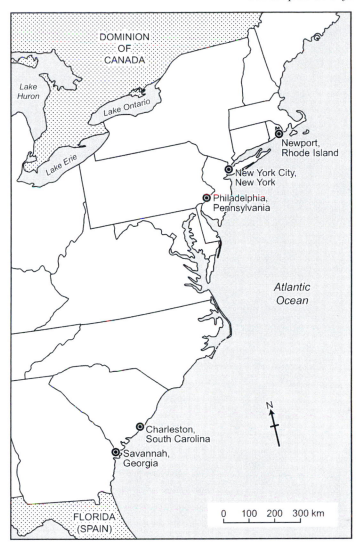

*Figure 6.1* Jewish settlement in the USA before 1830.

In 1733, the directors of London's Bevis Marks Congregation shipped 42 Ashkenazic and Sephardic Jews to Georgia, a colony in the American South established a year earlier by James Oglethorpe. . . . In 1735, a synagogue was established and a few years later a mikveh [ritual baths] was opened for the use of the congregation.

Even though the first wave of Jewish migration to America lasted almost two centuries, in 1800 there were only 2,000 and in 1830 fewer than 10,000 Jews in the country (Chaliand and Rageau 1995). By 1815, an increasing number of German Jews began arriving in the USA and by 1830 the second stage of Jewish migration had started in earnest. The new migrants did not confine themselves to the East Coast like their predecessors; instead, they spread throughout the country, to places such as farming communities of the Ohio Valley and as far as the Pacific and Gulf coasts. According to Ritterband (1997: 195), German Jews often followed their non-Jewish compatriots to areas where they had settled and 'they continued to provide economic services to German gentiles as they had done in their native land'. By 1850, there were 50,000 Jews throughout the USA, and just 30 years later their population had exploded to over 250,000 (Libo 1989).

The 1880s marked the beginning of the final and by far the largest, wave of Jewish migration to the USA, lasting until the beginning of the First World War. These new migrants were primarily Yiddish speakers from Eastern Europe, including the Russian Pale of Settlement and parts of the Austro-Hungarian Empire. A total of approximately 2 million Jews entered the country during this period and by 1917 there were 3.4 million American Jews (Chaliand and Rageau 1995). Most of these migrants settled in the major ports of entry, in particular New York City and to a lesser extent Boston, Philadelphia and Baltimore (Kosmin *et al.* 1986). A smaller number of Jewish immigrants moved into the interior of the country for economic reasons and settled in smaller communities such as railway (e.g. Springfield, Missouri) or mining towns (e.g. Calumet, Michigan) (Weissbach 1997). Often, these immigrants ran small businesses.

The imposition of quotas on new migrants to the USA during the 1920s led to a decline in overall numbers of arrivals, although the policy worked to the advantage of immigrants from Western Europe and Germany in particular (Chaliand and Rageau 1995). By 1930, the Jewish population in the USA was estimated at 4.2 million and the proportion of Jews peaked at 3.7 per cent of the American population in the mid-1930s (Goldstein 1987) (see Figure 6.2). Goldstein argues that after the Second World War, a sharp downturn in immigration to the USA, lower fertility rates and factors such as intermarriage led to a substantial decline in the Jewish growth rate to a point where 'by 1984 Jews constituted only between 2.3 and 2.5 per cent of the total population' (Goldstein 1987: 135). By the 1980s, more than 80 per cent of Jewish-Americans had been born in the USA and approximately half were third- or fourth-generation Americans. This is despite the immigration of 80,000 Soviet Jews and up to 100,000 Israelis during the 1970s (Goldstein 1987).

### Jewish-Americans in the twenty-first century

There are Jewish communities throughout the USA, even though there remains an obvious geographic concentration in the Northeast, especially in the area including New York City, southern New England and northern New Jersey (Kosmin *et al.* 1986) (see Figure 6.3). According to Chaliand and Rageau (1995), there are more than forty persons of Jewish origin per 1,000 inhabitants in this region. Nevertheless, the share of the region's Jewish population has declined gradually over the last three decades as many Jews have dispersed throughout the country and especially to the Sunbelt states such as Florida, California and Arizona. In 1986, Los Angeles had replaced Chicago as the city with the second largest Jewish population after New York City, while the Miami-Ft. Lauderdale region was ranked third with 6.3 per cent of the nation's Jews (Kosmin *et al.* 1986) (see Figure 6.4). Kosmin *et al.* (1986) suggest that the recent pattern of concentration of Jewish population throughout the USA and, certainly, the

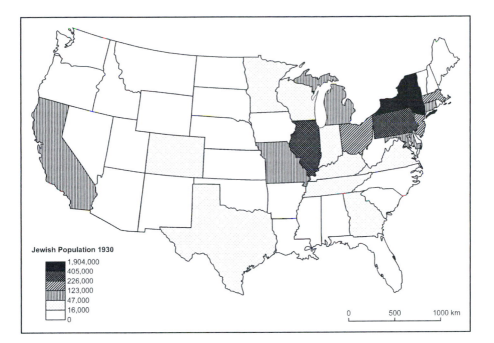

*Figure 6.2* The Jewish population in the USA in 1930.

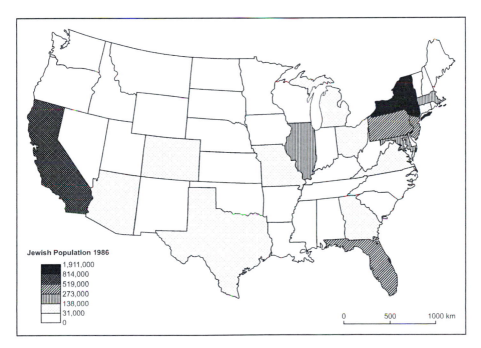

*Figure 6.3* The Jewish population in the USA in 1986.

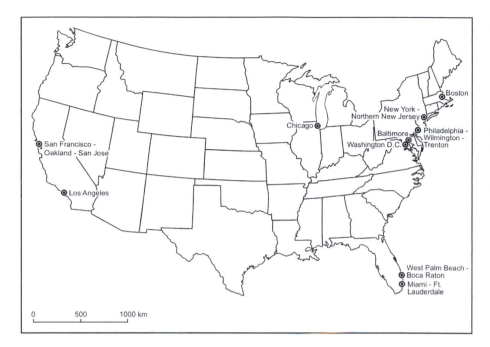

*Figure 6.4* Major Jewish settlements in the USA in 1986.

impressive rate of growth of some communities coincides with regions experiencing rapid economic development: in addition to major metropolitan regions, these include high-tech areas such as North Carolina's Research Triangle, university towns such as Champaign-Urbana, Illinois and places like Huntsville, Alabama, where there is a major NASA facility.

Generally speaking, American Jews are highly mobile, reflecting trends characterizing other well-educated sub-groups of the population who move in the pursuit of economic goals. Approximately 75 per cent of Jews in the USA do not live in the cities where they were born and up to 'a third have moved within the last 5–6 years' (Goldstein 1987: 142). Moreover, even if they choose to remain within the metropolitan areas where they were born, the majority of Jews like most other Americans abandon their original neighbourhoods in favour of the suburbs. Ritterband (1997) suggests that this high mobility has inevitably led to a situation where in the last 40 years American Jews have become increasingly differentiated in social, cultural and religious terms.

Perhaps with the exception of a few ultra-religious groups (e.g. the Hassidim and certain Orthodox Jewish communities), most Jews can now be described as assimilated within the broader US population. Certainly, Jews in America are accepted as:

> first class citizens . . .[and] (with few exceptions) [they] have enjoyed equality of status and have been as free as their Gentile neighbors to vote, hold public office, own property, move freely and earn their livelihoods unimpeded by custom or law.
>
> (Libo 1989: 117)

There are rising concerns that the growing assimilation of American Jews, resulting at least in part from their excessive mobility and high rates of intermarriage, may have substantially weakened their ties to Judaism (Goldstein 1987). Conversely, other scholars believe these trends may have actually spurred a revival of interest on the part of Jewish-Americans towards their religious and, especially, their cultural heritage (Sarna 2002). Sarna contends that in reality there have been numerous occasions in the last 150 years when American Jews have revived their interest in Judaism precisely because of concerns that assimilation would destroy their culture. This revival appears to have been substantially strengthened during the late 1930s when Jews in this country became aware of the annihilation of their people in Europe. Following the Second World War, interest on behalf of American Jews towards Judaism was strengthened even further, exactly at a time when they became increasingly mobile and seemingly more integrated within the mainstream population. Sarna argues that during this period membership in synagogues of all denominations increased, the attendance of Hebrew school rose to unparallel levels and there was a record construction of new synagogues around the nation. In more recent years, even as the Jewish-American population has been further assimilated into the mainstream (Shapiro 1992), there is ample evidence that American Jews are not abandoning Judaism outright. Instead, even though fewer Jews raised in Orthodoxy remain, they choose to join other movements of Judaism (CJF n.d.). The Reform movement now claims the largest number of members of any Jewish denomination, while the Orthodox movement still has the largest number of synagogues (Witham 2002).

One way in which many Jews in America and elsewhere seek to reaffirm and strengthen their affiliation to Judaism is demonstrated through their travel patterns. Many journeys undertaken by Jews have been described by some observers as 'pilgrimages' to the places where their forebears lived (Eisen 1998; Ioannides and Cohen Ioannides 2002; Levy 1997). In the case of American Jews, these trips could be to Jewish neighbourhoods within the USA, where they or their parents grew up or longer journeys abroad to the countries where their ancestors originally came from (e.g. Central and Eastern Europe). In the rest of this chapter, the travel patterns of Jewish-Americans are examined, paying particular attention to how Judaism influences their destination choice. First, though, it is important to summarize briefly the relationship of the Jewish religion to travel and pilgrimage.

## Judaism, pilgrimage and travel

The authors have argued elsewhere that pilgrimage in Judaism does not assume the same level of theological significance that it does in other major religions such as Christianity, Islam or Hinduism, all of which are associated with many sites their followers regard as sacred spaces (Ioannides and Cohen Ioannides 2002). True, the Bible (Deuteronomy 16: 16) pinpoints Jerusalem as an important pilgrimage site for Jews and that city is still regarded as the focal point of the Jewish religion. Moreover, in Jerusalem there are a number of spaces that have traditionally attracted Jewish pilgrims. Perhaps the one best known as a religious site is the *Kotel Ma'aravi*, the Wailing Wall where Moses' tablets are reputedly located (Epstein and Kheimets 2001; Shenhav-Keller 1993; Vukonić 1996).

Nevertheless, unlike religions such as Christianity, which has multiple pilgrimage sites around the world (e.g. Lourdes and Medjugorje), the Jewish faith does not really have major holy spaces outside Jerusalem. An explanation for this is that Judaism is fundamentally a religion that pays heavy attention to historical events and remembrance rather than attaching special meaning to specific holy spaces (Heschel 1955). It should be considered, for instance, that unlike churches, synagogues are not regarded as holy places. It is not as important in Judaism where specific services (such as those for holidays like Yom Kippur or Sukkot) take place as long as there is minyan (traditionally defined a minimum of ten men) present (Glustrom 1988). Indeed, the celebration of the Passover *Seder* traditionally takes place at home and not at the synagogue (Strassfeld 1985).

The absence of holy spaces outside Jerusalem does not mean Jews never travel for religious/cultural purposes. On the contrary, there is enough evidence of Jews undertaking 'pilgrimages' to the homelands of their parents in Eastern Europe or northern Africa (Cohen 1983). 'Organized tour groups of Moroccan-born Israelis have been travelling back to their native land' responding to Jews' search for identity (Levy 1997: 25). Levy describes these tours which attract approximately 2,000 Moroccan-Israelis per year as 'quasi-pilgrimages' (Levy 1997: 28), because in addition to including secular activities like trips to historic cities and *suqs* (Arab markets), they encompass visits to the tombs of the so-called *tsaddiqim*, whom Moroccan Jews regard as holy persons; such a *hagiolatric* practice is unique to this group of Jews and stems from 'indigenous maraboutism (veneration of holy men endowed with supernatural power)' (Levy 1997: 28). Similarly, certain Hasidic Jews based in the USA make a point of visiting the grave of the founder of the Breslov Hasidic Movement in the Ukrainian town of Uman at least once in their lifetime, believing that by doing so they will be spared purgatory (Gershom 2001). When visiting Krakow, Hasidic and other religious Jews come to the Remuh Synagogue to pray and 'venerate the tombs of ancestors and sages, above all the great sixteenth-century Talmudic scholar Moses ben Israel Isserles, known as *Remuh*, who is buried in the Old Jewish Cemetery' (Gruber 2002: 130).

It should be noted that it is not just the Jews belonging to minority groups such as the Hassidim who travel in relation to their religious and cultural identity. Even so-called secular Jews and those affiliated with moderate forms of Judaism (Conservative or Reform) undertake trips relating to their Jewish identity. The *New York Times* reports that 'each year, more than 100,000 Israeli and American Jews visit Poland, viewing plaques that mark significant sites in the wartime ghetto and visiting former Nazi death camps . . .' (Green 2003). These visits are primarily undertaken out of respect to the memory of the millions of Jews who perished during the Holocaust. Most importantly, the trips reflect once again the significance that Judaism attaches to remembrance of people and events; that is, the trip is not undertaken to a specific place (like a concentration camp) because that place is regarded as holy (in the same way as a Hindu pilgrimage site like Varanasi draws worshippers), but rather because the journey allows the Jewish people, whether they are religious or not, to come into touch with their past and commemorate their ancestors (Weiss 1991).

### *'Mitzvahs of nostalgia'*

Eisen (1998: 184) views the trips undertaken by Jews of the diaspora as the way through which they can perform the 'mitzvah (good deed) of nostalgia'. He uses the term in the same way Lowenthal (1985) sees nostalgia as the yearning for a past way of life, albeit from a safe distance. The need 'to know how and why things happened is a compelling motive for witnessing past events' (Lowenthal 1985: 22) and it is such a need that drives Jews to visit the sites where their ancestors once lived (Ashworth 2003).

Although the remembrance of ancestors is a fundamental element of Jewish faith and a vehicle through which Jews have historically expressed their devotion toward God, the mitzvah of nostalgia was not institutionalized until the mid-1880s. In the 1840s, for the first time the 'literature of nostalgia ... *summoned Jews to return and identify with the ways of their ancestors* ... The rabbis, editors and publicists of France and England stressed the obligation to follow in the paths of parents, grandparents and more distant ancestors' (Eisen 1998: 172). During this period Jewish travel guides appeared like Daniel Steuben's (1860) *Scènes de la vie juive en Alsace* (in Eisen) and I.J. Benjamin's (1956) *Three Years in America: 1859–1862*. Benjamin's publication was the first travel diary concerning the historical and cultural importance of Jewish communities of various sizes throughout the USA. Since these publications described the places where Jews had once lived and because they extended an invitation to visit, they enabled Jews of the diaspora to come closer to their forebears' way of life (Eisen 1998).

During the nineteenth century, yet another phenomenon reinforced the Jewish need to travel in order to perform mitzvahs of nostalgia. Specifically, by the late 1800s it became increasingly common in major cities of the western world to host exhibits of Jewish history, culture and religion. This following was written in reference to one such exhibit held in London in honour of Queen Victoria's jubilee:

> No event so captured the [*Jewish*] *Chronicle's* attention in the last four decades of the century, or was covered so generously in its pages, as the *Anglo-Jewish Historical Exhibition* held at the Albert Hall in 1887. . . . The exhibit displayed the 'inner connection of your Past and Present'. English Jews were 'patriots attached to this happy isle' but as such, wanted nonetheless to 'preserve connection and continuity with the long series of generations of Israel'.
>
> (Eisen 1998: 173–4)

At the time this was the largest collection of Jewish religious artefacts ever presented at a single exhibit (Adler 1895). Similar exhibits appeared in the USA. The *New York Evening Post* in 1894 gave a detailed account of one such exhibit by the Smithsonian Institution at which a variety of items used by Jews for religious services were displayed, including a Torah, a *Ketubah* (marriage contract) from 1816 and an eighteenth-century spice box used for the *havdalah* (conclusion service of the Sabbath). Eisen (1998) contends that the importance of these exhibits for visiting Jews was that they enabled them to better understand from a cultural perspective their ancestors' way of life and, thus, encouraged them to demonstrate greater loyalty toward their forebears; that is,

Jews of another time and place who, like Jews in the present, had wandered far from their own homes, altered tradition in keeping with the times, but retained enough of what they inherited to pass something down.

(Eisen 1998: 174)

## Travel patterns of American Jews

### *The early years*

As most Jewish-Americans in major urban areas of the USA saw a general improvement in their living standards during the late nineteenth and early twentieth centuries, they sought opportunities to spend an annual vacation. In response to this growing demand, a number of resort areas became popular destinations for Jews. Among the most popular were the New York Catskills, dubbed the Borscht Belt, and South Haven in Michigan (Brown University 1995; Kraus 1999) (see Figure 6.5).

There were many reasons for the appearance of these Jewish-only destinations. The first is they were a response to the fact that non-Jewish resorts at the time were often anti-Semitic. It was not uncommon, for instance, to see signs in vacation areas specifically prohibiting Jews from renting holiday homes (Kraus 1999). Also, during the early 1900s, most Jews continued to maintain, mostly by choice, their traditional

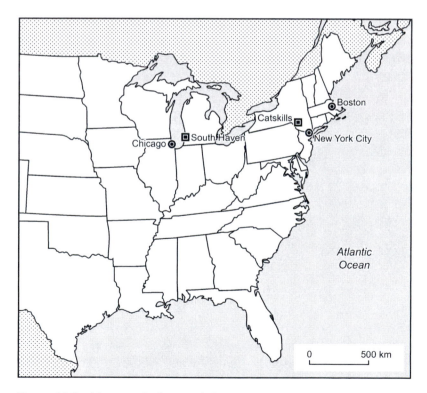

*Figure 6.5* Jewish resorts in the twentieth century.

separation from other groups and, thus, holiday areas where Yiddish – the main language of Jewish immigrants – was widely spoken and where they could be with other Jews were most appealing. Additionally and perhaps more importantly, Jews were drawn to Jewish-only resorts because these followed strict dietary regulations (kashruth) and were likely to observe the Sabbath and all religious holidays (Ioannides and Cohen Ioannides 2002).

Eventually these Jewish resort areas declined in popularity, a decline coinciding with the growing secularization of American Jews and the fact that non-Jewish resorts were forced to drop their discriminatory practices (Kraus 1999). A report published by Brown University (1995) paints a depressing picture of the once popular Catskill resort areas by describing abandoned, boarded up, or destroyed facilities. Today, most American Jews are far likelier to travel to the same destinations within the country that attract non-Jews. The southern states, especially Florida, have emerged as the principal destinations for vacationing American Jews. Also, with growing affluence, an increasing number of Jews travel overseas, especially to Israel and Europe (Gruber 2002).

### The influence of 'being Jewish' on travel trends and behaviour

Despite the decline in popularity of Jewish-only resorts in the USA over the last four decades, evidence suggests Judaism still influences the travel behaviour not only of religiously conservative American Jews (e.g. the Hassidim or Orthodox Jews), but also those belonging to liberal denominations (e.g. Reform Jews) and those who are considered secular. This influence stems largely from the aforementioned need of all Jews to perform mitzvahs of nostalgia in response to their constant quest to identify with the ways of their ancestors.

One related noteworthy trend has been the growing interest in recent years on the part of American Jews to travel either to the countries their parents or grandparents originally left as immigrants to the New World (especially Central and Eastern Europe), or to the 'old neighbourhoods' within the USA where they or their parents had once lived (e.g. parts of Brooklyn, New York). The European trips bring Jewish-Americans to the ghettos where their ancestors had once lived, such as the Old Jewish Quarter in Prague, the ancient ghetto in Venice, the Judenplatz in Vienna and Kazimierz, the ghetto in Krakow. Jews also visit Nazi extermination camps, such as Auschwitz in almost ritual-like 'nostalgic tours [that] allow the visitor a chance to see the "graves" of their forebears and perform, with others who have the same need, the required acts of public mourning for martyrs who are not necessarily family members' (Ioannides and Cohen Ioannides 2002: 19; see also Gruber 2002). Historically important synagogues, such as the Great Portuguese Synagogue of Amsterdam, draw thousands of Jewish visitors (many of them American) each year (Israelowitz 1999a; Nob Hill Travel Service 2001; World at One Destination 2001). Concurrently, museums anchor the Jewish travellers' itinerary when visiting Europe since 'as institutions that can easily be visited by the public, [they] play an education role that is of particular importance. They often become the public face of Judaism' (Gruber 2002: 126–7).

The demand for travel by American Jews has concurrently fuelled and been fuelled by the appearance of a burgeoning number of Jewish travel publications. It has also led to many businesses specializing in Jewish travel to sites within the USA and abroad

(Ioannides and Cohen Ioannides 2002). Certainly, in the case of Eastern Europe the fall of communist regimes in the late 1980s has substantially encouraged travel to that region, resulting in specialized publications and firms that organize Jewish tours to that part of the world (Ashworth 2003).

Gruber's (1994: vii) *Jewish Heritage Travel: A Guide to East-Central Europe* is 'a guide to [the] many remaining traces of Jewish culture and civilization', in a region that once was home to nearly five million Jews. Her book takes visitors on a journey 'into a ghostly, multidimensional shadowland of *Then, Now and What Might Have Been*' (Gruber 1994: 2) within a region where for the first time after 50 years, Jews have the freedom to practise their religion openly. By describing the region's rich Jewish heritage (e.g. synagogues, museums and cemeteries) and its few surviving Jewish communities, Gruber seeks to enhance the interest of her widely dispersed readers who, as Frank (1992: 12) maintains, constantly search for their sense of identity and use travel to discover the 'thread that binds' all Jews of the diaspora regardless of where they live.

Within the USA, if the increasing number of travel books serves as evidence, there is no shortage of sites considered part of the Jewish traveller's itinerary. These include, but are not limited to, attractions such as places where Jews originally passed through or settled when they first arrived in America (including Ellis Island), homes of famous American Jews, museums (including the Holocaust Museum in Washington, DC), but also sites with a religious content like historic synagogues and Jewish cemeteries (particularly the graves of well-known Jews) (Bloomfield and Moskowitz 1991; Postal and Koppman 1954). Israelowitz's (1999b: 106) *The Jewish Heritage Trail of New York*, which offers a series of walking tours through that city highlights, among others, the site of the first Jewish settlement dating back to 1660 and a synagogue built in 1939 featuring 'stone fragments from the Essen Synagogue and the Fassanenstrasse Temple (in Berlin) as well as part of a burned Torah parchment', all remnants of the Nazi destruction of German synagogues in 1938 (Box 6.1).

There are no hard statistics yet to demonstrate exactly how many Jews participate on such nostalgia-oriented tours, although the authors previously noted that American Jews travelling primarily for recreational or business purposes to the same secular travel spaces as other Americans commonly attempt to include at least one visit to a Jewish attraction in their itinerary (Ioannides and Cohen Ioannides 2002). This phenomenon indicates that like all Jews of the diaspora, American Jews, including those who are secular, retain many of the characteristics of a 'pilgrim-tourist' (Cohen 1992), that is one who is unable to escape the sphere of influence that is exerted by the centre of his/her spiritual world (Eliade 1969; Turner and Turner 1978). What attracts them to this 'centre' is their never-ending desire to re-identify with their culture and ultimately discover themselves as Jewish-Americans.

## Conclusions

David Lowenthal (1985: 4) has described the past as a 'foreign country', albeit one with the 'healthiest tourist trade of all'. Latching onto this now familiar metaphor, Eisen (1998) has argued how, wherever they are, Jews of the diaspora – even those considered part of secular society – have a deeply ingrained need to perform 'mitzvahs of nostalgia', deeds positioning them in touch with their past and enabling them to identify with the

---

**Box 6.1 Selected sites of interest to Jewish travellers in New York City**

| *Location* | *Points of interest* |
|---|---|
| Financial District | Ellis Island Immigration Museum (accessed by ferry from Manhattan or Jersey City) |
| | Museum of Jewish Heritage |
| | Jewish 'Plymouth Rock' (commemorating the first Jewish community in N. America) |
| | Site of North America's First Synagogue |
| Lower East Side | Shteeble Row, 225–283 East Broadway (mini synagogues) |
| | Jewish Mural (depicts the history of Jewish people in the Lower East Side) |
| | Jewish tenements, 137–139 East Broadway |
| | Lower East Side tenement museum |
| | Greek Synagogue and museum |
| | Shapiro's kosher winery |
| | Streits Matzoh factory |
| | Oldest Synagogue building |
| East Village | Yiddish Theatre Stars' Walk |
| Midtown West | Historic Jewish Cemeteries |
| Upper West Side | Holocaust Survivors' Synagogue |
| Upper East Side | Largest Reform Temple in the World |
| | Park East Synagogue |
| | The Jewish Museum |

Source: adapted from Israelowitz (1999).

---

ways of their ancestors. It is these deeds, which include travel to other places to witness the life of their forebears during previous times that have, according to Eisen, enabled Judaism, particularly the Judaism of the diaspora, to survive through the ages.

The primary purpose of this chapter has been to investigate whether American Jews, the largest group of the diaspora, express such a 'pining for the past' (Eisen 1998: 157) through their travel behaviour. It was argued that regardless of whether they are religiously strict or secular, Jews in the USA demonstrate a particularly strong desire to reaffirm their cultural identity. Ironically, this need appears to be strengthening precisely during a period when American Jewry is experiencing growing secularization and assimilation (through intermarriage, suburbanization and so on) within the mainstream population. That this desire exists and has blossomed over time can be attributed to numerous factors, the most important being the need to remember following the Holocaust (Sarna 2002; Shapiro 1992).

Many of the journeys performed by American Jews can be labelled 'modern day pilgrimages'. Not pilgrimages in the traditional religious sense to pray at sacred sites, but pilgrimages performed in the name of nostalgia to places, wherever these may be, that have some connection, past or present, to Jewish culture. As Boyarin and Boyarin (2002: 11) maintain, the relationship of Jews to their *homeland*, the lands of their ancestors – regardless of where these are geographically – 'is primarily commemorative, rather than kin-based or economic'. They go on to state that because a 'distinctive feature of Jewish diaspora is the repeated experience of *rediasporization'* as Jews through the centuries have settled and resettled numerous times from place to place, this means there is not just one *homeland* for them to return to, a characteristic that may be found in the case of other diasporas (e.g. Greek, Italian). In essence then, every place that has or has had a relationship to Judaism, whether it is the site of a *shtetl* (Jewish hamlet) in Eastern Europe, a cemetery in North Africa, or a neighbourhood in Brooklyn, New York, becomes a potential destination for Jewish pilgrimage. It is by performing nostalgic voyages to such places that American Jews become contemporary pilgrim-travellers (Cohen 1992), journeying to places bringing them closer to the centre of their world (Eliade 1969), not necessarily a purely religious-spiritual world, but one that embodies in a cultural sense what it means to be Jewish.

## References

Adler, C. (1895) *Museum Collections to Illustrate Religious History and Ceremonials*, Washington, DC: GOP.

Ashworth, G.J. (2003) 'Heritage, identity and places: for tourists and host communities', in S. Singh, D.J. Timothy and R.K. Dowling (eds) *Tourism in Destination Communities*, Wallingford: CAB International.

Benjamin, I.J. (1956) *Three Years in America 1859–1862*, Philadelphia: Jewish Publication Society of America.

Bloomfield, B.C. and Moskowitz, J.M. (1991) *Traveling Jewish in America: The Complete Guide for Business and Pleasure*, Lodi, NJ: Wandering You Press.

Boyarin, D. and Boyarin, J. (1993) 'Diaspora: generation and the ground of Jewish identity', *Critical Inquiry* 19: 693–725.

Boyarin, D. and Boyarin, J. (2002) *Powers of Diaspora*, Minneapolis: University of Minnesota Press.

Brown University (1995) *Jewish Scholars Study History, Cultural Significance of the Borscht Belt*. Online. Available: http://www.brown.edu/Administration/News_Bureau/1995-96/95-062i.html (accessed 9 November 2002)

Central Intelligence Agency (2002) *The World Factbook 2002*. Central Intelligence Agency. Online. Available: http://www.cia.gov/cia/publications/factbook (accessed 16 December 2002).

Chaliand, G. and Rageau, J.-P. (1995) *The Penguin Atlas of Diasporas*, New York: Penguin Books.

Cohen, E. (1983) 'Ethnicity and legitimation in contemporary Israel', *The Jerusalem Quarterly* 28: 111–24.

Cohen, E. (1992) 'Pilgrimage and tourism: convergence and divergence', in A. Morinis (ed.) *Sacred Journeys: The Anthropology of Pilgrimage*, Westport, CT: Greenwood Press.

Council of Jewish Federations (CJF) (n.d.) 'I. Demography: A. The People: 8. Origins of the Population', *CJF 1990 National Jewish Population Survey*. North American Jewish Data Bank: The Mandell L. Berman Institute. Online. Available: http://web.gc.cuny.edu/dept/cjstu/highint.htm (accessed 5 January 2003).

Eisen, A.M. (1998) *Rethinking Modern Judaism: Ritual, Commandment, Community*, Chicago: University of Chicago Press.

Eliade, M. (1969) *Images and Symbols*, New York: Sheed and Ward.

Epstein, A.D. and Kheimets, N.G. (2001) 'Looking for Pontius Pilate's footprints near the Western Wall: Russian Jewish tourists in Jerusalem', *Tourism, Culture and Communication* 3(1): 37–56.

Frank, B.G. (1992) *A Travel Guide to Jewish Europe*, Gretna: Pelican Publishing.

Gartner, L.P. (1998) 'The great Jewish migration – its east European background', *Tel Aviver Jahrbuch für Deutsche Geschichte* 27: 107–33.

Gershom, Y. (2001) *Journey to Uman: Rosh Hashanah*. Online. Available: www.pinenet. com/~rooster/uman.html (accessed 12 December 2002).

Glustrom, S. (1988) *The Language of Judaism*, Northvale, NJ: Jason Aronson.

Golden, D. (1996) 'The Museum of the Jewish Diaspora tells a story', in T. Selwyn (ed.) *The Tourist Image: Myths and Myth-Making in Tourism*, Chichester: Wiley.

Goldstein, S. (1987) 'Population trends in American Jewry', *Judaism* 36(2): 135–46.

Green, P.S. (2003) 'Jewish museum in Poland: more than a memorial', *The New York Times* Online. Available: www.nytimes.com/2003/01/09/international/europe/09POLA.html? ex=1043137490andei=1anden=2879d63f85d8fa51 (accessed 10 January 2003).

Gruber, R.E. (1994) *Jewish Heritage Travel: A Guide to East-Central Europe*, New York: Wiley.

Gruber, R.E. (2002) *Virtually Jewish: Reinventing Jewish Culture in Europe*, Berkeley: University of California Press.

Heschel, A.J. (1955) *God in Search of Man*, New York: The World Publishing Company.

Israelowitz, O. (1999a) *Guide to Jewish Europe*, 10th edn, Brooklyn, NY: Israelowitz.

Israelowitz, O. (1999b) *The Jewish Heritage Trail of New York*, Brooklyn, NY: Israelowitz.

Ioannides, D. and Cohen Ioannides, M.W. (2002) 'Pilgrimages of nostalgia: patterns of Jewish travel in the United States', *Tourism Recreation Research* 27(2): 17–26.

Kosmin, B.A., Ritterband, P. and Scheckner, J. (1986) 'Jewish population in the United States, 1986', *American Jewish Year Book* 87: 164–91.

Kraus, B. (1999) *A Time to Remember: A History of the Jewish Community in South Haven*, Allegan Forest, MI: The Priscilla Press.

Levy, A. (1997) 'To Morocco and back: tourism and pilgrimage among Moroccan-born Israelis', in E. Ben-Ari and Y. Bilu (eds) *Grasping Land: Space and Place in Contemporary Israeli Discourse and Experience*, Albany, NY: State Univeristy of New York Press.

Libo, K. (1989) 'Three centuries of migration to America', *Studia Rosenthaliana* 23: 107–18.

Lowenthal, D. (1985) *The Past is a Foreign Country*, New York: Cambridge University Press.

Nob Hill Travel Service (2001) *2001 Jewish Heritage Tour: 15-Day Tour of Eastern Europe Prague, Warsaw, Krakow and Budapest*. Online. Available: http://www.nobhilltravel. com/jht.htm (accessed 10 January 2003).

Postal, B. and Koppman, L. (1954) *A Jewish Tourist's Guide to the US*, Philadelphia, PA: Jewish Publication Society of America.

Ritterband, P. (1997) 'Geography as an element in the historical sociology of the Jews: New York, 1900–1981 and the United States, 1880–1980', *Jewish Population Studies* 27: 191–208.

Rosenwaike, I. (1989) 'Ethnic identity and the census: the case of the American Jewish population', *Jewish Population Studies* 19: 11–22.

Sarna, J.D. (2002) Marshall Sklare Memorial Lecture. Association for Jewish Studies, December.

Shapiro, E.S. (1992) *The Jewish People in America: A Time for Healing: American Jewry since World War II*, Baltimore: Johns Hopkins University Press.

Shenhav-Keller, S. (1993) 'The Israeli souvenir: its text and context', *Annals of Tourism Research* 20(1): 182–96.

Strassfeld, M. (1985) *The Jewish Holidays: A Guide and Commentary*, New York: Harper and Row.

Turner, V. and Turner, E. (1978) *Image and Pilgrimage in Christian Culture: Anthropological Perspectives*, New York, Columbia University Press.

Vukonić, B. (1996) *Tourism and Religion*, New York: Elsevier.

Weiss, A. (1991) *Death and Bereavement: A Halakhic Guide*, Hoboken, NJ: Ktav Publishing House.

Weissbach, L.S. (1997) 'Decline in an age of expansion: disappearing Jewish communities in the era of mass migration', *The American Jewish Archives Journal* 42(1/2): 39–61.

Witham, L. (2002) 'Orthodox Jews have the most synagogues', *The Washington Times*. Online. August 9, Available: http://www.washingtontimes.com/national/20020809-85213354.htm (accessed 15 January 2003).

World at One Destination (2001) *Jewish Routes*. Online. Available: http://www.jewishroutes.com/features (accessed 10 October 2002).

# 7 American children of the African diaspora

## Journeys to the motherland

*Dallen J. Timothy and Victor B. Teye*

## Introduction

A common feature of contemporary social and economic life is the enormous volume of spatial movements of people throughout the world, including a significant number of tourists who visit countries or regions they consider their ancestral homelands. This has resulted in the emergence of what can be referred to as 'roots tourism', 'diaspora tourism', 'genealogy tourism', or simply 'personal heritage tourism' (Timothy 1997). This form of travel often reflects many of the characteristics of pilgrimages, such as a personal connection to one's spiritual self. Related to this is the research in the social sciences commonly referred to as diaspora studies (Cohen 1997). Although diaspora research inherently includes understanding the historical and modern movement of people, it has been observed, 'the literature on diaspora (and hybridity) has on the whole neglected tourism, perhaps because tourists are thought to be temporary and superficial' (Bruner 1996: 290). In common with other social scientists, tourism scholars have so far paid scant attention to the notion of diaspora and tourism, although some authors have hinted at it in examining ethnicity and migration in this context (Duval 2002; Hall and Williams 2002).

One of the most momentous diasporas has been the spread of black Africans to North and South America, the Caribbean, the Middle East and Western Europe. In common with many other children of diasporas, Americans, Caribbean islanders and Europeans of African decent in recent years have begun to travel extensively throughout the world, but particularly to Africa, the land of their forebears. This has been highlighted during the past 30 years through the literary works of authors such as Alex Haley and Maya Angelou. With an increase in arrivals from the diaspora, many countries of Africa, particularly in West Africa, have begun to cater specifically to the needs of this unique group of travellers and produce tours, itineraries and infrastructures geared specifically to attracting them. White and other non-African races also have an interest in visiting West Africa for a variety of reasons and most of their itineraries and attractions are the same as those for tourists of African descent. In an effort to contribute to a broader knowledge of diaspora and tourism and African-American tourists in particular, this chapter examines their experiences at a slave route historic site on the coast of Ghana. First, however, the chapter examines the African diaspora and its manifestation today in the form of heritage attractions in Africa and the USA.

### Personal heritage and travel to the homeland

Through recent decades there has been a surge of interest among Americans in discovering their personal heritages through family history/genealogy research. This interest has led to increasing numbers of people travelling to ancestral lands in search of their identities and as a way of connecting with their deceased ancestors (Bradish and Bradish 2000; Matthiessen 1989; Parker 1989; Timothy 2001). Filial piety, or ancestor worship, has long been a part of many cultures, but other reasons have begun to influence people's decisions to travel to the lands of their ancestors. One reason is nostalgia. According to Lowenthal (1975: 5), people need the past to cope with the present, because today's complex world makes better sense to them if they share a past with it. The disappearance of traditions, cultures and historic relics has, in Lowenthal's (1979) thoughts, deepened people's need for nostalgia for the past. Nostalgia, which implies a yearning for some past condition(s), can be a motivator for individuals to travel to heritage sites and in a broader context as a reason heritage is valued, conserved and visited by collective groups and societies (Timothy and Boyd 2003). Group nostalgia, just as it does on an individual level, can evoke bittersweet yearnings within entire societies – emotions shared by people of a similar background (Baker and Kennedy 1994; Davis 1979), such as generations, cultures and nations (Belk 1990; Timothy and Boyd 2003).

Similarly, for some people it might be difficult to know where they are going until they understand where they are from (Lowenthal 1975). In this sense, travelling to ancestral lands also helps people explain and evaluate themselves (Crompton 1979). From a search for their past, they find themselves in the present. A search for one's roots and historical identity and the subsequent appreciation for one's community culture and family legacy is evidence of this pattern. Much of what is sought is something that people have learned since childhood, through stories told by their parents, grandparents, or others, about their ethnic and familial heritage. In the words of Lowenthal (1975: 6), 'the past gains further weight because we conceive of places not only as we ourselves see them but also as we have heard and read about them'. This was certainly the case for one commentator, who experienced Shanghai, the childhood home of his father, after many years of hearing about it:

> For all the times my father told me about Shanghai in my 30-plus years, my feel for it was much like . . . an image that didn't quite seem real. I'd heard about the city so often and the huge role it had played in saving my father's life and shaping his character, that it became almost mythical to me, until Sept. 25, when I stepped back in time and saw it for myself.
>
> (Compart 1999: 1)

Another clear reason is the desire to visit relatives. A few studies have examined motivations for travelling to ancestral homelands, including the desire to visit distant and close relatives (Crompton 1979; Duval 2002) and some have come to perceive such travel activities as a form of pilgrimage (Thanopoulos and Walle 1988).

Esman (1984) also viewed some ethnicities as disenfranchised groups using tourism to assert their heritage and concluded that ethnic groups that may have to travel extensively to return to the 'old country' may be engaging in a similar pilgrimage aimed at reasserting, reaffirming, or perpetuating their heritage. Taking this a step further, Thanopoulos and Walle (1988) viewed Greek-Americans as an ethnic community living far away from its cultural and geographical origins, the members of which travel to their homeland periodically as a way of solidifying their ethnicity and keeping strong ties to the motherland.

Encompassing several of these, Cohen commented:

> Old-country visitors: a peculiar type of partial tourism are trips of emigrants or of their progeny to their old country; for example, Italians or Irish from the USA visiting Italy and Ireland, American Blacks visiting Africa and ex-Corsicans vacationing in their mother island. For the generation of emigrants such trips serve primarily as reunions with kinfolk and friends, though they might also possess a touristic component, in that the returnee explores the changes which have taken place in his old home since his departure or enjoys anew the forgotten pleasures of his youth. When members of second and later generations of emigrants visit the old country of their parents, the touristic component of the trip will be considerably more pronounced.
>
> (Cohen 1974: 542–3)

Such experiences are also shared by African-Americans as they travel to Africa for reasons similar to those outlined above.

## Slavery, diaspora and African-American tourists

There were two primary regional slave trading movements from Africa several centuries before the first significant European contacts with Africa in the fourteenth century. The trans-Sahara route, for instance, was from Western and Central Africa to North Africa and the Red Sea slave trade exported slaves from Eastern Africa to Arabia and South Asia. Africa's later contact with Europe established the third area of slave traffic – the transatlantic slave trade – from West Africa to the Americas and Europe. This was the largest of the slave trading operations in terms of the number of slaves captured and sold, the dominant region of origin, the distances involved and lives lost. While figures are imprecise, some scholars suggest that the trans-Saharan and Red Sea trade involved about 6 million slaves. The transatlantic slave trade from the 1520s to the 1860s, however, involved the capture and trafficking of some 11–12 million African men, women and children, who were forced onto European vessels and sold into lives of slavery in the Americas and Europe (Appiah and Gates 1999). Over a period of some 350 years, an estimated 10 million African slaves survived the transatlantic voyage as human cargo to be purchased by slave traders and white plantation owners in the New World (Harris *et al.* 1996; Okpewho *et al.* 1999; Segal 1995).

While Africans were sold in slave markets as far north as New England and as far south as present-day Argentina, today's descendants are spread all over the world. This global dispersal of people of African heritage has come to be known as the Black, or African, diaspora. With the exception of a small number of freed slaves who, after the abolition of slavery, were returned to Sierra Leone and Liberia by England and the USA respectively, the vast majority of today's descendants of African slaves are concentrated in North America, South America, the Caribbean and a few European countries, primarily the UK and France.

In the USA, which was one of the principal perpetrators of the slave trade, most African descendants today live in rural areas of the Southeast, where slavery was the most concentrated and in urban areas throughout the country. While they were mistreated as slaves and racial discrimination occurred well into the twentieth century (and continues in some forms even today), black Americans have overcome many obstacles and have climbed to positions of power and leadership on international, national and local levels. Their standard of living has grown considerably in recent decades and more are gaining university educations than ever before.

This increased standard of living and increased awareness of their past has led to a rapid growth of domestic and international travel among African-Americans. In domestic terms, there are many heritage attractions currently being developed in the south-eastern USA to commemorate the region's slave heritage. Museums, plantations with slave cabins, historic houses, monuments and statues, slave trails and war sites are examples of attractions being developed to appeal to the African-American traveller population as well as to educate white Americans about the role of slavery and African-Americans in the nation's history (Bartlett 2001; Butler 2001; Dann and Seaton 2001; Eskew 2001; Goings 2001; Hayes 1997; Seaton 2001).

Likewise, in common with other races and ethnic groups in North America, African-Americans are becoming more conscious of their ancestral heritage (Hayes 1997; Woodtor 1993), resulting in increasing numbers of trips to 'Mother Africa' in search of their roots (Austin 1999; Goodrich 1985; Mays 2001). While a few commentaries have appeared on ethnic tourism in Africa, including that by Jamison (1999), few studies link Africa with African-Americans as a distinct travel market segment. In an exploratory study, Goodrich (1985) found that 71 per cent of African-Americans who travel would most like to visit Africa. Other studies have also confirmed a strong desire among African-Americans to visit their ancestral lands (World Tourism Organization 1997a,b). There are many travel agencies in the USA that specialize in ethnic/diaspora travel (e.g. Polish in New Britain, Connecticut; Lebanese and Syrian in Toledo, Ohio; and Scandinavian and Finnish in the Great Lakes region). Similarly, 'black' travel agencies have proliferated in the USA in cities with heavy African-American populations to assist in arranging 'black' domestic and international travel (Butler *et al.* 2002). Likewise, tours geared specifically at African-Americans to take them back to Mother Africa are being heavily promoted and developed and are labelled with attractive titles like 'coming home' tours (Mays 2001).

This movement is not surprising given that African-Americans constitute a tremendous ethnic market for Africa's tourism industry for several reasons. First, with a population of 34 million, they are the single largest group of African descent in any country. Second, their average income level has experienced an annual growth rate of approximately 16 per cent between 1990 and 2000 and they represent a market worth some US$400 billion in disposable spending. This means that the African-American market could be considered the fourteenth largest market in the world with more disposable and discretionary income than residents of Australia, Mexico or Russia (Malveaux 1998). Likewise, Philipp (1994: 479) pointed to a number of significant findings in the 1990 US Census of Population, which counted 11.5 million blacks working in occupations classified as managerial, professional and technical. Third, most African-Americans constitute a single linguistic market. Unlike peoples of African descent in many countries in the Caribbean and Latin America, who speak various languages (e.g. English, French, Spanish and Portuguese) and various derivations of local and European languages (e.g. Patois, Creole and Papiamento), the primary language of African-Americans is English. This factor tends to facilitate and enhance their travel experiences in anglophone West Africa (e.g. Ghana and Nigeria). Fourth, the growth and significance of the African-American market for both domestic and international travel has been recognized by sectors of the travel industry including hotels, airlines, cruise lines and theme parks. For example, in 1993 the Travel Industry Association of America (TIA) commissioned what it called its first-ever benchmark study on the demographic and economic characteristics and travel patterns and attitudes of African-Americans (Travel Industry Association of America 1993). Finally, the economically prosperous African-American population tends to be concentrated in large metropolitan areas, which are also gateways for international travel. For instance, some 27 per cent of the total US population of African-Americans lives in the five cities of New York, Chicago, Los Angeles, Philadelphia and Washington, DC. Other international gateway cities with significant African-American populations include Atlanta, Miami and San Francisco.

In many ways for African-Americans, going to Africa is more than simply a vacation trip. It is a journey of personal discovery that is in many respects unique only to the black race. Reflecting the earlier discussion, in the words of Austin (1999: 211),

> people of black African descent . . . share a collective identity and destiny that is centered on Africa. For those in the diaspora, an identification with the origins of the transatlantic slave trade and with Africa as the ancestral 'home' is increasingly being perceived as the missing link in their quest to find their roots and to understand their collective sociohistorical experience, an act that is considered necessary for their self-realization.

## The context: the slave route and diaspora attractions in Ghana

Probably the most popular destination country in West Africa for all Americans, but particularly African-Americans, is Ghana. The single most important category of cultural resources with significant potential for heritage tourism development in Ghana are the numerous forts, or castles as they are sometimes called, built along the coast between the fifteenth and eighteenth centuries by various European groups, including the Portuguese, Dutch, Danes, Germans, Swedes and English. The forts served several different functions, including trade, commerce, administrative rule and military activities that also have significant relevance for contemporary development. However, the forts are particularly known for the role they played in the transatlantic slave trade. Africans captured along the coast or in the interior and taken to the coast were imprisoned in some of these fortifications for weeks and months at a time and transported in shackles overseas, mostly to the Americas. The concentration of some 50 European forts, several of which are still well preserved and designated UNESCO World Heritage Sites, along Ghana's 560 km coastline (van Dantzig 1980; Essah 2001) is significant in slave heritage for three main reasons. First, it makes Ghana unique in Africa with respect to the number of such historical and cultural structures built by different European countries. Second, it is an indication of the intensity of European activities here in the part of Africa known until the late 1950s as the Gold

Coast. Third, the number of these forts and their role in the slave trade probably point to a disproportionate number of today's African-Americans having their origins from this part of West Africa.

These forts, together with villages, routes, pathways and monuments, are now being promoted and conserved by the United Nations Educational Scientific and Cultural Organization (UNESCO) and the World Tourism Organization (WTO) in their joint Slave Route Project. Their purpose is,

> to rehabilitate, restore and promote the tangible and intangible heritage handed down by the slave trade for the purpose of cultural tourism, thereby throwing into relief the common nature of the slave trade in terms of Africa, Europe, the Americas and the Caribbean.
>
> (WTO/UNESCO 1995: 1)

For the reasons noted above, Ghana is the primary destination of choice for many children of the African Diaspora, particularly those from North America, and the country has begun to plan much of its tourism development and promotional efforts around this theme. Despite a relatively recent history of political unrest, lack of adequate infrastructure and access and underdevelopment (Brown 2000; Gartner 2001; Teye 1988), Ghana's current 15-Year National Tourism Development Plan (1996–2010) recognizes the increased number of heritage tourists and the potential of the Slave Route Project for the promotion of ethnic/heritage tourism. The plan acknowledges that 'the number of African-Americans returning to West Africa and Ghana in particular to explore their cultural roots is increasing. Their demand for facilities and assistance in exploring their heritage must be a major factor in the product development such as the slave route project' (Ministry of Tourism/WTO/UNDP 1996: 78). For African-Americans, Ghana is appealing because of the large number of historic sites available that deal specifically with the slave trade, slavery and the proposed UNESCO Slave Route Project. In this regard, the following statement is an example of the sentiment of many African-Americans:

> But statistically speaking, since historians believe that more than half of the approximately 60 most active African slave transport sites were in the Gold Coast area of what is now Ghana, there is a good chance that my ancestors are from here, or at least passed through en route to the Americas. That explains why I feel so connected to Ghana.
>
> (Kemp 2000: 12)

One of the most popular heritage tourist attractions in Ghana is Elmina, or St George's, Castle. The site is more a fort than a castle and its role in history was significant as a major slave holding location, trading site and point of departure for the Americas (Bruner 1996; Essah 2001; Gartner 2001). In the early 1990s, it was estimated that between 5,000 and 10,000 tourists visited the fort each year, but the number is said to have increased to between 15,000 and 20,000 people a year in the mid- and late 1990s (Austin 1999; Bruner 1996).

Today, Elmina Castle functions as a museum dedicated to educating visitors about the events that transpired there during the transatlantic slave era. Artefacts, including shackles, photographs, balls and chains and whips comprise many of the tangible elements of history on display to visitors. While there are interpretive signs, the most common medium is guided tours. Independent travellers can visit the fort and wander at their own pace, while groups of African-Americans are given special tours and performances, such as the 'Through the Door of No Return – The Return' performance without the presence of white tourists. At other times, when mixed groups include whites and blacks, all tour participants are allowed to be involved in the demonstration (Bruner 1996).

At the conclusion of their visit to Elmina Castle, visitors are invited to record information about themselves and their experiences in a guest book. The book provides space for their names, addresses, nationalities/races and comments about their experiences. The contents of these books were made available to the authors from throughout the 1990s and form the basis of the discussion in this section. The comments revealed a great deal about visitors' feelings, emotions and impressions of the heritage site itself and how it was interpreted, but the most significant pieces of information were those that described how people felt about slavery and the emotive sentiments the visit elicited. After sorting through a total of some 14,120 entries, US residents were targeted and a qualitative content analysis was conducted on the expressions that they (US residents) logged into the book, which were categorized by race to see if there were any notable differences or similarities in their experiences at the site. The races were broken down into African-Americans and white Americans. The bulk of the findings for white Americans are reported in Teye and Timothy (2004).

## African-American experiences at Elmina Castle

Many visitors were impressed by the architecture, the conservation projects and the tours. Comments about these were among the most common. However, the core of the experience for many people was reflected in their written statements as they expressed their emotions, feelings and thoughts during their visit to the castle. No distinction was made between independent visitors and tour participants.

It became clear that the experiences were personal and individual, spiritual and emotional, educational and full of discovery, cause for apprehension and anxiety, a homecoming and a finding of cultural identity and an opportunity for closure, revenge and new beginnings.

Seven themes were identified in this analysis: grief and pain, good versus evil, revenge, forgiveness and healing, coming home, in memory of our ancestors and God and holy places. The range of emotions among African-Americans was considerable, given the fact that it was their ancestors who were forcibly sold, held captive and shipped from these sites. This is also where they can trace to their ancestors, their own long history of being mistreated by whites. Thanapoulos and Walle (1988: 12) recognized that people who travel to their homelands sometimes experience various forms of emotional trauma. This is clearly so in the case of African-Americans when they visit slave heritage sites in Africa.

## Grief and pain

In common with their white counterparts (Teye and Timothy 2004), African-Americans' experiences at Elmina were sad, painful and sometimes emotionally overwhelming. A similar observation led Austin (1999) to suggest that such heritage trips might not be leisure-oriented, for they are of a more serious nature. Aside from many mentions of sadness, the element of grief was commonly reflected in the shedding of tears. One woman simply wrote 'Tears, tears, tears'. One African-American man was so moved that he wrote Elmina Castle was 'A place which can bring an unemotional man to tears of a child. I was deeply moved'. Many black Americans expressed pain: 'This is painful', 'I'm hurting'. These experiences reflect the very deep and personal nature of the encounter with the place where their ancestors were tortured, killed and forcefully moved from their homes. No doubt this is magnified by the slave history in the USA, the conditions of which were rarely better that those at Elmina.

### Good versus evil (black versus white)

Understandably, it became clear that African-Americans viewed whites as evil in terms of the slave trade. Many African-Americans mentioned the evil nature of the events there between the 1500s and the late 1800s. The Africans were rightfully seen as the innocents in this occasion and the white foreign slave traders were the evil, devils themselves, or under the influence of wickedness. Sentiments like 'This is the work of the devil' demonstrate this emotion. For some people the experience highlighted the plight of blacks as the victim and the role of whites as the perpetrators. One woman commented that she was 'An African, stolen and victimized by the USA'.

### Revenge

For the tremendous malevolence and evil committed by the whites, which reflects the European traders in Africa and the slave owners in the USA, several African-Americans understandably demonstrated a sense of hatred and malice toward Caucasians.

'Damn you white man!' and 'Never trust the white man any more', are statements indicative of this mood. Several visitors also indicated the desire to pay back the 'white man' for his transgressions. 'Let's return the favour' and 'Kill the white people' are perfect illustrations of this attitude. Clearly there is some degree of danger with the notion of revenge and such feelings may be stirred by destination environments, tour guides and other interpretive media:

> The utilization of the historic events of the [slave] trade in tourism has to contend with the ethical dilemma relating to the sensitivity of those events to the descendants of black African slaves and its potential encouragement of racism amongst black people, towards whites. . . . Depending on how it is presented and interpreted, it could create an even wider rift between blacks and whites.
>
> (Austin 1999: 211–12)

### Forgiveness and healing

Despite strong feelings among several people for revenge, even more expressed notions of forgiveness and coming to terms with slavery. These people had a tendency to recognize that what happened has happened and there is not much that can now be done to change history. As part of the healing process, several people emphasized the need to forgive, but not forget. Quotes like 'Forgiveness is the key' and 'We must forgive, but not forget', were common among African-American visitors. Some visitors emphasized the need to look forward not backward (e.g. 'Our people need healing'), while others recognized that African-Americans have already come a long way in this regard (e.g. 'I feel good about the strength of my people').

## Coming home

According to Austin (1999: 213), 'Africans in the diaspora on visits to the African continent see themselves as "coming home". This feeling of "coming home" and the reconnection with the land of their fathers represents the essence of the visit'. The evidence left by African-American visitors supports Austin's claim. Many people wrote 'Africa is my home' and 'I'm home at last!' For many, the trip to Ghana was a very personal journey wherein they discovered themselves and filled a part of what they felt was missing in their lives (e.g. 'A new discovery for a missing part of my life'). One woman wrote that she was 'One of Africa's daughters who survived', in a sense reflecting the struggle of living outside of Africa in a world that has mistreated her people for centuries.

## In memory of our ancestors

Related to coming home was the notion of remembering ancestors. Some people felt as though their progenitors were speaking or crying to them (e.g. 'My ancestors cried to me'). Others commented on the pain they felt for their ancestors. One person wrote the following emotional appeal: 'I've learned a lot about the history of my ancestors, their suffering, etc. I almost broke down and cried for their pain'. Several people pronounced blessings upon the heads of their forebears. 'My ancestors' spirits – may they remain beautiful' and 'Peace to our ancestors', are two prime examples. Many of these feelings are summarized by Kemp (2000: 12), who experienced Ghana, the probable origin of her ancestors:

> I don't find it disturbing. At least not anymore. Because I understand that I have been singled out by the spirits of my ancestors to tell their stories. It is their presence I feel beside me and it is why they whisper incessantly from the moment I arrived in the country (Ghana), 'where have you been daughter? There is work to do'.

## God and holy places

According to Bruner (1996), the forts and other slave route sites are sacred ground among African-Americans. The data from Elmina Castle support this claim (e.g. 'This is holy ground'). Many of the diaspora visitors found comfort in their faith in God. Many turned to their faith while at the fort as a way of dealing with their pain. One woman commented, 'God's love would have changed this history' and another prayed, 'God grant us strength. Truth shall set us free'.

**Conclusions**

This study has highlighted many of the experiences recorded by African-American tourists at Elmina Castle, a slave route heritage site in Ghana. Four key reasons for homeland/diaspora travel were identified at the outset of the chapter: nostalgia, explanation or understanding of self, visiting relatives and the assertion of heritage by a disenfranchised group. For African-Americans, the trip to Ghana was laden with nostalgia. The trip assisted a major segment of diaspora blacks in coming to terms with the tragedies of the past. Visiting the sites of the slave trade, as demonstrated in the analysis here, is key in helping African-Americans make sense of the troubled past and understand themselves from an historical perspective. For African-Americans the visit is clearly one of closure, a homecoming of sorts, which helps them understand themselves better in the context of the USA and slavery. Also, travelling to Africa enables African-Americans to assert their heritage (which has traditionally been overlooked in the USA) in both their current homeland and the land of their ancestors. Visiting relatives is not a major consideration for African-Americans, as the forced migration of slavery took place so long ago, few children of the early African Diaspora have close family members in Africa.

The experiences of African-American visitors to Elmina Castle were profound. All felt a great sense of sadness and pain. They displayed a terrible sense of pain on behalf of their ancestors. The experience led many African-Americans to exalt the slaves and damn the white traders with an accompanying sentiment of revenge. Nonetheless, the evidence also suggests that some people gained closure and were able to begin the process of healing and forgiveness and many suggested that God was nearby, consecrating the ground upon which they walked.

These findings have significant management implications. In common with the Holocaust heritage of Jews, such a heritage of atrocity is political and extremely sensitive (Tunbridge and Ashworth 1996), as the notion of revenge against and condemnation of the white race would indicate. Such conflicts, within the context of interpretation in combination with location, history and attitudes may accentuate situations that must be dealt with delicately and thoughtfully. Careful planning, preparations and training must be involved. As Austin (1999: 214) states, '. . . as a result of the sensitive nature of the events of the [slave] trade to various groups, intergroup conflicts are inevitable at sites and other presentations associated with it. These conflicts over time may shape the future market and the viability of the tourism development. Management . . . must seek to identify, understand and manage these conflicts'. For example, local tour guides must be trained to handle the array of very difficult questions posed by African-Americans, such as those regarding the role of African tribes themselves in rounding up and selling Africans, which constituted an important function in the transatlantic slave trade. Management at slave heritage sites must also anticipate the motivations and the emotions of African-American visitors. This sub-group sometimes arrives in a mixed-race group tour and may undergo intense emotional experiences that transform their attitudes toward white tour group members. Indeed, they may potentially become downright hostile at the slave heritage site and/or on the journey from the site.

Clearly, visiting heritage sites that have personal and emotional connections is a complex issue that needs to be addressed in further research. This chapter has attempted to provide an initial examination of the personal connections in diaspora, or personal heritage, tourism with one of the largest groups and, what many agree is the most important group involved in the transatlantic slave trade: African-Americans. It is hoped that this discussion and these findings can be fruitfully brought to bear on additional research into diaspora group travel, ethnic tourism and the experiences and emotions tied to personal heritage travel.

## References

Appiah, K.A. and Gates, H.L. (1999) *Encarta Africana: Comprehensive Encyclopedia of Black History and Culture*, Redmond, WA: Microsoft Corporation.

Austin, N.K. (1999) 'Tourism and the transatlantic slave trade: some issues and reflections', in P.U.C. Dieke (ed.) *The Political Economy of Tourism Development in Africa*, New York: Cognizant Communications.

Baker, S.M. and Kennedy, P.F. (1994) 'Death by nostalgia: a diagnosis of context-specific cases', *Advances in Consumer Research* 21: 169–74.

Bartlett, T. (2001) 'Virginia develops African-American tourism sites', *Travel Weekly*, 4 June 2001: 16.

Belk, R.W. (1990) 'The role of possessions in constructing and maintaining a sense of past', in M.E. Goldberg, G. Gorn and R.W. Pollay (eds) *Advances in Consumer Research*, vol. 17, Provo, UT: Association for Consumer Research.

Bradish, C. and Bradish, P. (2000) 'Doing genealogy on the road', *Everton's Genealogical Helper* 54(2): 44.

Brown, D.O. (2000) 'Political risk and other barriers to tourism promotion in Africa: perceptions of US-based travel intermediaries', *Journal of Vacation Marketing* 6(3): 197–210.

Bruner, E.M. (1996) 'Tourism in Ghana: the representation of slavery and the return of the Black Diaspora', *American Anthropologist* 98(2): 290–304.

Butler, D.L. (2001) 'Whitewashing plantations: the commodification of a slave-free antebellum south', *International Journal of Hospitality and Tourism Administration* 2(3/4): 163–75.

Butler, D.L., Carter, P.L. and Brunn, S.D. (2002) 'African-American travel agents: travails and survival', *Annals of Tourism Research* 29(4): 1022–35.

Cohen, E. (1974) 'Who is a tourist? A conceptual clarification', *The Sociological Review* 22(4): 527–55.

Cohen, R. (1997) *Global Diasporas: An Introduction*, London: Routledge.

Compart, A. (1999) 'Genealogy travel: TW writer's road to Shanghai', *Travel Weekly*, 13 December 1999: 1, 18.

Crompton, J.L. (1979) 'Motivation for pleasure vacations', *Annals of Tourism Research* 6: 408–22.

Dann, G.M.S. and Seaton, A.V. (2001) 'Slavery, contested heritage and thanatourism', *International Journal of Hospitality and Tourism Administration* 2(3/4): 1–29.

Davis, F. (1979) *Yearning for Yesterday: A Sociology of Nostalgia*, New York: The Free Press.

Duval, D.T. (2002) 'The return visit–return migration connection', in C.M. Hall and A.M. Williams (eds) *Tourism and Migration: New Relationships Between Production and Consumption*, Dordrecht: Kluwer.

Eskew, G.T. (2001) 'From Civil War to civil rights: selling Alabama as heritage tourism', *International Journal of Hospitality and Tourism Administration* 2(3/4): 201–14.

Esman, M.R. (1984) 'Tourism as ethnic preservation: the Cajuns of Louisiana', *Annals of Tourism Research* 11: 451–67.

Essah, P. (2001) 'Slavery, heritage and tourism in Ghana', *International Journal of Hospitality and Tourism Administration* 2(3/4): 31–49.

Gartner, W.C. (2001) 'Issues of sustainable development in a developing country context', in S. Wahab and C. Cooper (eds) *Tourism in the Age of Globalization*, London: Routledge.

Goings, K. (2001) 'Aunt Jemima and Uncle Mose travel the USA: the marketing of memory through tourist souvenirs', *International Journal of Hospitality and Tourism Administration* 2(3/4): 131–61.

Goodrich, J.N. (1985) 'Black American tourists: some research findings', *Journal of Travel Research* 24(2): 27–8.

Hall, C.M. and Williams, A.M. (eds) (2002) *Tourism and Migration: New Relationships Between Production and Consumption*, Dordrecht: Kluwer.

Harris, J.E., Jalloh, A. and Maizlish, S.E. (eds) (1996) *The African Diaspora*, College Station, TX: Texas A and M University Press.

Hayes, B.J. (1997) 'Claiming our heritage is a booming industry', *American Visions* 12(5): 43–8.

Jamison, D. (1999) 'Tourism and ethnicity: the brotherhood of coconuts', *Annals of Tourism Research* 26: 944–67.

Kemp, R. (2000) 'Appointment in Ghana: an African American woman unravels the mystery of her ancestors', *Modern Maturity*, July–August: 1–17.

Lowenthal, D. (1975) 'Past time, present place: landscape and memory', *Geographical Review* 65(1): 1–36.

Lowenthal, D. (1979) 'Environmental perception: preserving the past', *Progress in Human Geography* 3: 549–59.

Malveaux, J. (1998) '"Black power" means economic clout today', *USA Today*, 6 February 1998: 11A.

Matthiessen, P. (1989) 'The captain's trail', *Condé Nast Traveller* 24(1): 106–34.

Mays, B.J. (2001) 'Black heritage tour to focus on Ghana', *Travel Weekly*, 26 July 2001: 25.

Ministry of Tourism/WTO/UNDP (1996) *Tourism Development Plan for the Central Region*, Accra: Ministry of Tourism, Integrated Tourism Development Program.

Okpewho, I., Davies, C.B. and Mazrui, A.A. (eds) (1999) *The African Diaspora: African Origins and New World Identities*, Bloomington, IN: Indiana University Press.

Parker, J.C. (1989) *Going to Salt Lake City to Do Family History Research*, Turlock, CA: Marietta Publishing.

Philipp, S.F. (1994) 'Race and tourism choice: a legacy of discrimination?' *Annals of Tourism Research* 21: 479–88.

Seaton, A.V. (2001) 'Sources of slavery – destinations of slavery: the silences and disclosures of slavery heritage in the UK and US', *International Journal of Hospitality and Tourism Administration* 2(3/4): 107–29.

Segal, R. (1995) *The Black Diaspora*, New York: Farrar, Straus and Giroux.

Teye, V.B. (1988) 'Coups d'etat and African tourism: a study of Ghana', *Annals of Tourism Research* 15: 329–56.

Teye, V.B. and Timothy, D.J. (2004) 'The varied colours of slave heritage in West Africa: white American stakeholders', *Space and Culture* 7(2).

Thanopoulos, J. and Walle, A.H. (1988) 'Ethnicity and its relevance to marketing: the case of tourism', *Journal of Travel Research* 26: 11–14.

Timothy, D.J. (1997) 'Tourism and the personal heritage experience', *Annals of Tourism Research* 34: 751–4.

Timothy, D.J. (2001) 'Genealogy, religion and tourism', paper presented at the Association of American Geographers Annual Conference, New York City, 1–3 March.

Timothy, D.J. and Boyd, S.W. (2003) *Heritage Tourism*, Harlow: Prentice Hall.

Travel Industry Association of America (1993) *African-American Travellers: Dimensions of the African-American Travel Market*, Washington, DC: TIA.

Tunbridge, J. and Ashworth, G.J. (1996) *Dissonant Heritage: The Management of the Past as a Resource in Conflict*, Chichester: Wiley.

van Dantzig, A. (1980) *Forts and Castles of Ghana*, Accra: Sedco.

Woodtor, D.P. (1993) 'African-American genealogy: a personal search for the past', *American Visions* 8(6): 20–3.

World Tourism Organization (1997a) *Tourism Market Trends: Africa 1988–1999*, Madrid: World Tourism Organization.

World Tourism Organization (1997b) *Tourism Market Trends: Africa 2000*, Madrid: World Tourism Organization.

World Tourism Organization/UNESCO (1995) *Accra Declaration on the WTO-UNESCO Cultural Tourism Programme 'The Slave Route'*, Madrid: World Tourism Organization.

# 8 Preparation, simulation and the creation of community

## Exodus and the case of diaspora education tourism

*Erik H. Cohen*

### Introduction: the Israel experience and the Exodus Program

Every year, thousands of young Jews from youth groups around the world visit Israel on educational tours. These tours are a kind of spiritual pilgrimage with two primary, explicitly stated goals: namely, to instil a sense of connection with Israel; and to help participants develop and strengthen their 'Jewish identity'. With its first appearance in 1994, a new dimension was added to some of the trips to Israel. Rather than flying into Ben Gurion airport, the groups first visit sites of Jewish history in Europe, only then to sail by boat, docking at port in Haifa four days later. The 'Exodus Program', as it has become known, is designed as a quasi-simulation of the famous Exodus voyage half a century ago, during which boatloads of Holocaust survivors ran the British blockade to reach British Palestine. Both take their symbolic name from the biblical story of the Children of Israel coming out of Egypt to the Promised Land.

The Exodus boat tour and the Israel Experience to which it is allied, are among the most consciously organized, well-reported and carefully evaluated examples of diaspora tourism. Since the inception of the Exodus Program, over 12,000 young Jews from various diaspora countries have taken part in the programme (More 2002). This represents a tremendous achievement in nurturing educational heritage tourism among the Jewish diaspora. Based on the evaluations of the participants themselves as well as the comments of the staff, reporters and other observers on board, the Exodus Program has become one of the most powerful parts of an already highly rated and strongly recommended tour programme. Young tourists who took the Exodus boat trip have been found to have had a significantly different experience in Israel than those who flew directly to Tel Aviv. When the evaluations of all the Israel Experience participants were compared, those who arrived by boat were more satisfied with their time in Israel than those who arrived by the more common air route. This provoked a series of research questions which this chapter seeks to address. For instance, what has made the experience onboard so powerful for participants? How and why did the Exodus Program have such a significant impact on a visitor's subsequent time in Israel? And how has the original historical event been presented in simulation to groups of teenagers two generations later?

## The setting for Exodus: the Israel Experience

Before delving into the paradoxes and intricacies associated with these questions, a brief description of the Exodus Program and the subsequent educational tours to Israel, which it prefaces, is warranted (see also Cohen 1999). The wider programme of educational tours to Israel, for which the Exodus boat trip is an optional introduction, is known as the 'Israel Experience' and is sponsored by the Education Department of the Jewish Agency. The Israel Experience has been in existence as long as the State of Israel itself. The curriculum of each tour is largely designed by youth groups, which range from orthodox-nationalist groups to secular groups focused on Jewish continuity in the diaspora. The tours include social activities, visits to religious and historical sites, hikes, seminars, volunteer work and meetings with Israelis. They last between two and eight weeks. Participants usually range from 14 to 18 years old. A 10-year, on-going survey has shown that the demographic profile of the Israel Experience has changed very little from year to year (see also Chazan 1997). A team of specialized counsellors, called *madrichs*, who comprise individuals both from Israel and the group's home community, accompanies each tour group (see also Cohen *et al*. 2002).

### *Exodus in profile*

Most of the groups on the Exodus boats are also more or less homogenous, with all participants coming from the same home country and youth movements affiliated with the same denomination of Judaism (orthodox, conservative, reform, secular). The itinerary onboard consists of seminars, discussion groups, ceremonies, religious services and social activities. Upon reaching Israel, the students take part in a ceremony that includes speeches by survivors of the original Exodus voyage (Joint Authority for Jewish Zionist Education 1994, 1995).

All educational acts must have aims, objectives and learning outcomes and a key feature distinguishing educational tourism from recreational tourism is the presence of articulated messages, which the tour leaders are meant to convey to the visitors. According to Boyd (1996), the Exodus Program has five main goals: namely, to

- create an awareness of the brief existence of the State of Israel
- instil a feeling that the existence of the State of Israel should not to be taken for granted
- begin to address the question of an individual participant's relationship to Israel
- familiarize participants with modern Israeli culture
- create a dynamic for arrival in Israel.

Ultimately, the journey is meant to prepare participants for their time in Israel. Although learning history through detailed texts and 'facts' is *not* a primary goal of the trip, the Exodus Program does attempt to bring to life a largely overlooked period in the struggles to obtain sovereign Jewish statehood. Although many participants have studied the Holocaust and the creation of the State of Israel in their Jewish educational programmes, the period between the liberation of the concentration camps and the declaration of statehood is largely overlooked in the curriculum. The quasi-simulation offered by the Exodus Program uses this relatively powerful, yet poorly known episode to make more abstract notions of past and self, in particular the making of the Jewish state and its impact on diaspora identity, more concrete and part of the participants' personal experience and identity-building in post-event reflection. More widely, this has the potential to impact on the social politics of memory. As recent research has demonstrated, the Holocaust and the founding of the State of Israel do not have the same immediacy for young diaspora Jews as they did a generation ago (Tye 2001).

For the purposes of this narrative, it is not important to enter into a detailed historical account of the original Exodus voyages and the subsequent events surrounding it. Instead, the salient features are summarized in Box 8.1. Contemporary Exodus trips take many of their cues from the original voyage. In the context of trip as preparation, a thematic approach to the subject matter was deemed a more appropriate mechanism than a literal simulation or a faithful chronological reproduction. As the sample itinerary in Box 8.2 reveals, the metaphor of the sea voyage from Europe to Israel has been used to encourage participants to explore themes of identity, power and powerlessness and independence (physical, cultural and spiritual). For Boyd (1996: 5), 'such an approach [has] enabled each [participant] to examine his/her own personal feelings and responses to the various issues raised during the trip, which meant that each one clarified for him/herself why they were travelling to Israel'.

---

**Box 8.1  Key moments in the voyage of *Exodus 1947***

| Date | Event/episode |
|---|---|
| 29 March 1947 | A ship covertly purchased and manned by the Hagana (Jewish guerilla army) from a US scrap yard, set sail, purportedly to China. |
| 10–11 July 1947 | At the French port Séte, the crew smuggled aboard approximately 4,500 Jewish Holocaust survivors (men, women and children) and headed for British Palestine. |
| 17 July 1947 | In a ceremony at sea, the ship was renamed *Exodus 1947*. |
| 18 July 1947 | After ignoring warnings to turn back, the *Exodus 1947* was attacked by British destroyers in the Mediterranean. Several passengers and a crewman were killed. The refugees were forced onto transport ships to France. Members of the UN Special Committee on Palestine observed the incident. |
| August 1947 | Most refugees refused to disembark in France. |
| 8 September 1947 | The refugees were forced off the ships and into a 'Displaced Persons' camp in Germany. |
| 29 November 1947 | The UN Assembly voted to divide the British Mandate of Palestine into an Arab and a Jewish state. |
| 14 May 1948 | The State of Israel declared Independence. Most *Exodus 1947* refugees came to Israel when immigration limits were lifted. |

Sources: adapted from Gruber (1948), Henrik (1999), Holly (1969), Ohef Shalom Temple Archives (2002).

---

**Box 8.2 Sample itinerary of Exodus Program experience**

Day one     Embarking on ship; taking of aliases; search by 'British soldiers'; lessons in conversational Hebrew; Jewish folk dancing; lecture by *Exodus 1947* survivor; groups begin work for a performance about the illegal immigration.

Day two     Small groups create broadcasts for a radio show; group discussions of issues faced by Holocaust refugees; Hebrew lessons; Friday night (beginning of *Shabbat*) festive meal, singing and dancing.

Day three    *Shabbat* on board – folk dancing and singing, festive meal, group prayer services; discussions of Jewish identity and Israel; evening presentation of groups' performances; during the performance the ship is leafleted by a 'British plane' demanding they turn back.

Day four    Approaching the shores of Israel; film of UN vote to partition British Palestine; reading of Israel's Proclamation of Independence; discussions of issues in Israel (absorption of immigrants, Israel–Arab conflict, etc.); ship is ordered to turn back, temporarily turns back then docks in Haifa port; singing and dancing; welcome ceremony with speeches by surviving refugees.

---

## Simulation, education and community

### *Diaspora tourism and longing for the homeland*

Committment to Israel among diaspora Jews is not a simple matter of political or financial support. Young people are meant to relate to Israel as their ancestral homeland (Cohen 1997). Throughout the two millennia of the diasporic condition, Jewish people have maintained a strong sense of Israel as their spiritual homeland. This is a theme that runs throughout the traditional liturgy, but until the recent establishment of Israel, few had the opportunity to make pilgrimages to the Holy Land. In this sense, diaspora tourism of the Jews to Israel differs from tourism among other migrants or exiles whose families left their homelands within the past few generations. Nevertheless, there are some characteristics of diaspora tourism and the idea of longing for the homeland that are similar across many populations. One of these is an underlying contradiction within diaspora tourism for Jews. Diaspora tourism, by definition, relies on the continued existence of the diaspora. Few of the tourists searching for their ancestral homeland want to actually live there. This includes Jewish tourists to Israel and represents a fundamental change in the nature of the Jewish diaspora. As Levine (1986: vii) opines, 'for 2000 years exile has been the quintessential, normative Jewish condition. However, since 1948 diaspora is no longer an inevitability but an option'.

This inherent paradox presents a problem for the organizers of the Exodus Program. In classic Zionist ideology, which has guided many of the agencies and organizations sponsoring the Israel Experience tours, the successful creation of the modern State of Israel should mean, in theory, the eventual end of the Jewish diaspora. Consistent with this message, in the early Israel Experience tours immigration into Israel was strongly encouraged. However, over time it has become apparent that the majority of Jews living in Western countries are not going to emigrate to Israel. Therefore, the emphasis has shifted to strengthening emotional commitments and ties to Israel and to Judaism among Jews who visit Israel and then return to their homes in the diaspora.

### *Propædeutics: the need for preparation in diaspora tourism*

How can a sense of longing for and attachment to homeland be conveyed to such short-term visitors? If the tour is to be meaningful, the process of building such feelings of attachment must begin before the visitors arrive at the destination. Preparation may be short-term and specific to the trip; an actual orientation programme is organized for the travellers. It also may be long term and structural. Involvement in a community, or with a culture that instils in members a sense of longing for and duty to, the homeland serves as structural preparation for an eventual pilgrimage. The Exodus boat trip represents short-term preparation, but its participants, almost without exception, have been involved in structural preparation for the trip for most of their lives. They and their families are active in their local Jewish communities which have provided them with a context for the trip to Israel. The on-going survey of the Israel Experience shows that the impact of the programme is greatly enhanced if it takes place in a context of involvement with the Jewish community and Jewish education both before and following the time in Israel (Cohen 1999). Without such structural preparation, the short-term preparation given during the four days of the Exodus voyage is unlikely to be successful, as the symbolic keys to understanding may be missing. In conjunction with such structural preparation, the Exodus Program enhances the experience in Israel even further.

## Informal education and diaspora tourism

The Exodus Program is clearly a form of tourism consumed by diaspora members and distinctive by virtue of its delivery of informal educational experiences. It is impossible to understand the boat tour solely as tourism without examining its nature and aspirations as a vehicle for informal diaspora education.

### *The use of simulation as a pedagogic device*

The Exodus Program utilizes a particular educational tool in preparing the participants for their time in Israel, that of simulation. In classrooms, simulation games and role-playing, especially those using computers have gained popularity. 'Instead of history being a body of received information and ideas, it can be viewed more as an activity by the pupil, who learns his [sic] history as a voyage of discovery, experiencing historical evidence in a visual, aural and tactile manner' (Birt and Nichol 1975: 4). In their explanation of historic simulation, it is noteworthy that the authors use the term 'voyage'. The Exodus is a voyage both on the physical and on the symbolic levels. In general, simulation games can increase students' motivation; allow them to empathize with historical characters; show history as a process which had many possible outcomes; improve learning and memory; develop social skills; and introduce variety into the educational experience (Birt and Nichol 1975: 6–7). These advantages can be seen among the Exodus participants. The voluntary nature of the programme and its elements of fun increase motivation. Participants identify with the Holocaust refugees in a very personal manner. The idea that history could have had other outcomes – for instance that perhaps the State of Israel might not have come into existence – is a key idea which the participants are meant to understand. Almost inevitably, therefore, this learning experience will be long-remembered by participants and social skills and group dynamics are an important part of the journey.

The uniqueness of the Exodus Program as an educational programme and as a diaspora tourism experience begins to become apparent. Nevertheless, there are concerns with simulation in which everything is constructed as an educational tool. Simulations always involve simplification of complex relationships and processes (Grabe and Grabe 1996; Morgan 2002). Also, simulations of specific situations may not impart knowledge which students are able to apply to other contexts (Alessi and Trollip 1991; Der-Thang and David 2002). For this reason, the term 'quasi-simulation' is used here best to describe the Exodus Program. Although the trip contains elements of historical re-enactment and simulation, they do not comprise the entire programme. There is some simulation-drama involved, particularly upon approaching the Israeli coastline when the boat is 'ordered' to turn back (which happened to the original ship), but the historical event is not literally re-enacted. The organizers recognize the impossibility of faithfully and precisely recreating a historical event which took place over half a century ago. The participants come from an entirely different set of circumstances and settings than the refugees on the voyage of the *Exodus* in 1947. Moreover, they are arriving in an Israel that has also changed drastically since that time. Elements of simulation are used where they are felt to be effective, but contemporary realities are allowed to enter as necessary.

Since its inception, the Exodus Program has been the most publicized aspect of youth trips to Israel. Reporters are invariably on board and the trips are usually recorded on video. Activities on the ship tend to be large-scale and public, consisting as they do of numerous group celebrations, performances and ceremonies, each of which involves hundreds of participants. The time in Israel, on the other hand, is relatively intimate, with smaller groups and little press coverage. The curriculum includes few large ceremonies or performances. The public versus private (or intimate) dimension is one of the primary differences between the Exodus Program and the Israel Experience tour itself.

### *Issues affecting the authenticity of the Exodus simulation*

As understood by the organizers and the participants in any re-enactment, many factors differentiate the contemporary simulation from the actual event. 'Authenticity' in education is generally described as a similarity between the created learning environment and the real-world situation it is meant to simulate (Petraglia 1988). In any planned educational activity with the outcome given, authenticity is diminished (Herman and Mandell 2000). Authenticity can be approached on three levels: the perceptual, or sights, sound, etc.; the manipulative, that is actions and consequences of the learner; and the functional, or the internal structure of the lesson's content (Levin and Waugh 1988). In the case of the Exodus Program, there are several organizational and logistical issues which affect the authenticity of the re-enactment.

The first of these is the eventual return to the diaspora. The temporary and voluntary nature of the participants' stay in Israel presents an ideological challenge. The original

Exodus passengers were coming to Israel to stay, whereas the programme participants will visit for a few weeks. This aspect is compounded by the relatively uniform nature of the groups. Exodus boat trip participants are more consistent in terms of age and background than the passengers on the refugees' ships. In the first Exodus Programs, a variety of youth groups travelled together, but in recent years, however, the boat trips have become more homogeneous. Religious services conducted on board have been adapted to the affiliation of the majority of participants, or more than one service has been available where required. The refugees themselves did not encounter such harmonization, nor did they enjoy such a plurality of services when they required them. Indeed, there were no Israelis on board, unlike the current voyages. During some of the Exodus re-enactments structured meetings with Israeli youths have taken place. These are not a new features, existing as they have as part of the Israel Experience tours for many years. However, they contrast with the historical chain of events. At the time of the voyage, Israel did not exist as a sovereign state, merely an aspiration. Unlike the original passengers, contemporary travellers are exposed to the ideologies of those (currently) living in a (contemporary) territory that has endured a complex and contested history since Independence.

A fourth issue relates to the potential presence of Isrealis and the values they imbue into the Exodus Program tours. The Exodus tours are set in a contemporary educational context with the ability to impact markedly on the authenticity of the re-enactment. In recent decades, Jewish identity has come to the forefront of the Jewish education system. In the past, Jewish education meant almost exclusively learning religious texts and laws. Jewish identity was learned and formed through a reading of religious and rabbinical texts. In the current system, Jewish identity-building is more pro-active and given a privileged position at the top of the agenda in the education system. Such an immediate presence of the Jewish identity in the curriculum that guides both the Exodus Program and the Israel Experience would not have been recognizable to the refugees themselves. To make an event like the escape from Europe to Israel comprehensible to Jews living in the USA or Britain today, educators attempt to shape an understandable narrative with a contemporary edge. The inflections used by the contemporary narrators speak in a very different way to their audience, their perspectives on Judaism and their relationship with the Jewish nation.

Two further issues surround the management of the visitor experience from a supply-side perspective. Logistics and security are key considerations. Since the trip is an educational experience, not a flight of war refugees, the organizers cannot take unnecessary risks to make the simulation more realistic. In recent years, security concerns over global events and world terrorism have become increasingly serious, not least with such acts as the *Achille Larou* tragedy and more recently the September 11 attacks. During the last two years, the number of participants in Israel Experience programmes has fallen drastically and at times this has led to the suspension of the Exodus Program. This represents another paradox inherent in this case but reflective of wider diaspora tourism concerns. Diaspora tours often attempt to recreate emotions and experiences from politically and socially turbulent times, but if war actually breaks out, tourists, even those allegedly purporting to visit a place for an 'authentic experience', tend to take their vacations elsewhere (cf. Sönmez 1998). This inherent need to address the sensibilities of contemporary consumers also manifests itself in the need to adapt the programme to different groups or audiences. Just as any teacher adjusts the subject matter to a particular classroom and its pupils, this subject is translated into specific curriculum based on the demographic character of the participants. In trips organized for youth groups from the UK, the role of the British in obstructing the refugees' attempt to reach Israel is de-emphasized to minimize conflict with British Jews' loyalty to their home country. Similarly, the programme for groups of modern orthodox participants differs slightly from that given to groups of secular or reform participants. This raises the question of which story is the most accurate? Again, it is important to recognize that the purpose of the programme is not to present an unbiased, historically faithful account, but to impress upon the young travellers certain attitudes and feelings towards Israel and their upcoming time there.

Finally, the passage of time has delivered one fundamental change from the original voyage, but one which will eventually diminish: namely, on contemporary voyages there are often veterans of the original event present. This is perceived by tour managers and tourists alike as a positive influence. People, who were part of the crew of the refugee ships, as well as surviving refugees, have been brought on board to help make the quasi-simulation feel more 'real' to the participants. Talks, discussion groups and presentations with people who were on the refugee boats have become important parts of the programme itinerary. Veterans add to the authenticity of the experience by bringing to their trips a wealth of personal experience, knowledge and emotion that is inevitably otherwise lacking in younger tutors.

## Components of informal education

The Israeli sociologist Reuven Kahane (1997) advanced the theory that informal education is increasingly important in a post-modern world defined by rapid change and choice. He outlined a code of informality and defined its major components (see also Cohen 2001). Many of these components of informality can be recognized within the curriculum of the Exodus Program. *Moratorium* is the suspension of the roles of everyday life, allowing participants to explore different roles and identities, secure in the knowledge that such roles can be abandoned at the end of the educational experiment. *Pragmatic symbolism* is the conversion of symbols into deeds and deeds into symbols (Kahane 1997: 26). The quasi-simulation enacted on the boat permits participants to try on a different kind of Jewish identity and a different relationship to Israel than the ones they experience in their daily lives. Rather than being financially comfortable and living in a relatively safe and free environment (though levels of anti-Semitism vary depending on the host country), they can try to imagine what it meant to be a Jew who survived the Holocaust and fled Europe with nothing. Rather than seeing Israel as a strong country which has existed for their entire lives, or as place to visit and to send charity, participants can try to see Israel through the eyes of the refugees, who saw it as hope for a new life, a place from which they were being barred and as the dreamed-of Jewish state, which did not yet exist. The link between symbol and deed is strong in this simulation game. The boat itself is a symbol and every act performed on it is overlaid with symbols of freedom, of fear and of hope.

*Dualism*, the parallel existence of conflicting ideas, further allows participants to explore the ideas such as life in the diaspora versus life in Israel, or power versus powerlessness in modern Jewish history. *Symmetry*, the replacement of hierarchical student–teacher relations with less authoritarian leadership of peers, describes the relationship between participants and the *madrichs*. These counsellors are close to the participants in terms of age (most being less than a decade older than the participants), background and culture.

Admittedly, the boat represents an unusual setting for an informal educational programme. However, of particular importance is its all-encompassing nature. In this way, in Goffman's words (1961: 6–7), it can be described as a *total institution*; that is,

> A basic social arrangement in modern society is that the individual tends to sleep, play and work in different places, with different co-participants, under different authorities and without an over-all rational plan. The central feature of total institutions can be described as a breakdown of the barriers ordinarily separating these three spheres of life. First, all aspects of life are conducted in the same place and under the same single authority. Second, each phase of the member's daily activity is carried out on in the immediate company of a large batch of others, all of whom are treated alike and required to do the same thing together. Third, all phases of the day's activities are tightly scheduled, with one activity leading at a prearranged time into the next, the whole sequence of activities being imposed from above by a system of explicit formal rulings and a body of officials. Finally, the various enforced activities are brought together into a single rational plan purportedly designed to fulfil the official aim of the institution.

Within the Exodus Program, we can see all of these aspects of the breakdown of barriers that occurs within total institutions. More accurate would be to call the boat a *temporary total institution*, since the programme lasts only a few days. Some of the other details may be modified in addition. A boat is an extreme example of a place in which all aspects of life take place in one setting that the participants literally cannot leave for the duration of the programme. The authority on the boat is somewhat diffuse since there are many groups, each of which has its own team of *madrichs*. Similarly some activities are not mandatory and there is sometimes a selection of options offered. Based on the feedback from years of Israel Experience participants, organizers have attempted to allow for free time. More important, however, is that activities are being carried out among a large group of people, all of whom are involved in the same basic set of activities and who are equal within the social structure of the programme. And certainly, as described earlier, the activities (both scheduled and free social time) are deliberately intended to fulfil the goals of the programme.

## Tour guides as role models: the *madrichs*

The *madrichs* are a special kind of counsellor/guide. They function as informal educators, group leaders, logistics organizers and tour guides. However, their most important function, which distinguishes them both from commercial tour guides and from leaders of traditional religious pilgrimages, is that of role model for the participants (Cohen *et al.* 2002). Their success in being positive and well-received role models for the participants has been found to be one of the key differences between a mediocre and an excellent programme. During the days at sea, participants have the opportunity to develop a relationship with the *madrichs* who will be guiding their onward trip through Israel. It also enables the Israeli *madrichs* and the *madrichs* from the participants' home countries to begin their work relationship on neutral territory.

## Transmission of social memory

Though memorizing dates, places, names and events from this period of history (1945–1948) are not crucial to the Exodus Program, the transmission of a social memory of that time period is. The attempt to create a collective or social memory is reflected in the choice of themes for the programme. The French sociologist and anthropologist Maurice Halbwachs described the concept of history as a collective or social memory. For him, memories may be constructed by social groups as well as by individuals. Social memories may be transmitted through oral traditions, written records, images, ritual actions and the creation of space (Halbwachs 1938, 1939). The Exodus Program utilizes all of these media to give the participants a feeling of empathy for the refugees who immigrated to Israel (Box 8.3).

This type of multi-media transmission of history through a simulated event is not without precedent in Jewish culture. As pointed out by Danny Levine in a personal communication, the ritual telling during the Passover holiday of the biblical Exodus from Egypt and the *seder*, the ritual meal which accompanies it, are also symbolic recreations of a collectively remembered event. It is a classic example of Halbwachs' creation of social memory because every Jew is supposed to feel that s/he personally came out of Egypt and to the Land of Israel.

Telling out loud, with embellishment and discussion, the story of the Exodus from Egypt, based on a written text preserved through the centuries, in books often decorated by artwork, is intended to help participants 'remember' the bondage of slavery and the subsequent liberation from it as is the attendant ritual eating of foods to represent events in the story, such as dipping foods in salt water to represent tears and the particular spatial arrangement of the meal, such as reclining while eating to symbolize emancipation. Similarly, participants in the Exodus Program are meant to feel that they personally were delivered from the Holocaust and were underway to the emergent State of Israel. In fact, in 1997 the Exodus ship of NFTY members held a 'Freedom Holiday' which intentionally mimicked the Passover *seder*, including symbolic objects and foods (Steinberg 2001).

As in any collective memory, certain events are emphasized and others downplayed or ignored, thereby reflecting the perspectives of the organizers and the audience. The period of history with which the Exodus Program deals was one of great conflict, but the conflict has not yet, in fact, ended. There are, undoubtedly, radically different readings of this period of time. Even within the Jewish world, the presentation of the Exodus voyage has mutated over the past half century. This can be seen in the various novels and movies (such as Otto Preminger's famous film about the Exodus, 1960) that have been released on the subject. Today, young Jews have to come to terms with their feelings about the events surrounding the creation of Israel in light of the Palestinian nationalist movement and widespread international sympathy with it. In the Israel Experience–Exodus Program, the story is unequivocally told from the point of view of the Jewish immigrants and the voyage as an act of heroic nationalism and survival. Other perspectives, such as the Arab population of British Palestine or Jews who immigrated to other countries, are often dealt with in the privacy of the small groups

---

*Box 8.3* **The use of different media to transmit social memories on the Exodus boat trip**

| Medium deployed | Manner of usage |
| --- | --- |
| Oral traditions | Veterans from the original refugee ships speak at a ceremony welcoming the visitors upon their arrival in Israel. This powerful experience of hearing these stories will become part of the participants' own memories. In this way, the memories of the refugees become a part of the students' own memories on a collective level. |
| Written records | Material given to participants includes excerpts from diaries and headlines and clippings from newspaper articles from the 1940s. Such documentation is important in grounding the experience in historical record. |
| Images | Photojournalists recorded the immigrants in 1947 leaving Europe, at sea, approaching Israel, battling with the British and at the detention camps. These images (still and video) are a powerful tool in shaping how young Jews today 'remember' this period of history. |
| Ritual actions | The drama acted out on the ship, the choosing of new names upon boarding, the evasion of 'British officers', the temporary turning back before finally reaching the shore of Israel, are ritual actions performed to recreate an historical event so that participants may 'remember' it themselves. Of course, since the event is staged, the organizers control the memory and choose what is remembered and what is forgotten. |
| Creation of space | The ship itself is a space in which the lesson of history can be learned. The arrival in Israel by sea rather than by air, over the course of days rather than hours, is far more memorable. An airplane is an anonymous, generic space. There is limited ability to move around and talk to other passengers. The ship is a completely different kind of space. The students have discussions, sing and dance together, have lessons about the refugee-immigrants and the creation of the State of Israel. They scan the horizon for land and watch the cost of Israel slowly come into view. The boat is a controlled environment and an ideal space for the transmission of ideas and emotions. |

---

travelling around Israel together. The public event of the Exodus boat trip is the creation of a social memory for young Jews from the diaspora of the refugee-pioneers' journey to the future State of Israel.

## Between diaspora and homeland

### *Creation of community*

As noted in Steinberg's (2001: 1) study of the Exodus Program, the ship provides a space which is 'already not America, but not yet Israel'. In this sense, it may be seen as a metaphor of the biblical story of the Exodus from Egypt. The nation of the Israelites was forged in the desert between Egypt and Israel. Their journey was prolonged to give them time to prepare themselves for coming into the 'promised land'. The Exodus boat ride prolongs the journey to Israel, four days in an unlived-in space, which is both public and communal. Here, the participants and the *madrichs* are given time to bond together as a group.

This opportunity to create a community on the ship has been found to be particularly important to the groups from France. The political culture of France does not allow for the creation of minority ethnic communities. Individuals are granted the rights of citizens, but ethnic groups as such are not recognized. Observers of the French groups' voyages have all noted that other group divisions between youth movements or denominations have vanished in the enthusiasm to create a French-Jewish community that cannot technically (de jure) exist in France. For French groups, the preparation and Jewish identity aspects of the trip are emphasized over the historical simulation aspect.

## Heightening expectations

It is not uncommon for Israel Experience alumni to say that the tour was one of the most significant events in their life (Cohen 1994). Without diminishing the effectiveness of the programme itself, these responses must be understood in the context of the excitement and anticipation for visiting the 'promised land', which is part of these young people's cultural upbringing. The time spent on the boat, in the space between leaving home and arriving in the 'homeland' increases the anticipation for the tour in Israel and adds extra value to the experience. Even without the educational programme itself, a simple lengthening of the trip would have had this effect. In using the time to give the participants a well-planned and executed curriculum specifically designed for the purpose, the excitement and expectations of the participants is further built up. The young participants dock in Israel literally singing, dancing and crying.

## The Exodus Program and Jewish identity

Ethnic identity is group identity and the impact on group dynamics is one of the key reasons that the Exodus Program has been successful in enhancing the subsequent weeks spent in Israel. The Exodus Program fosters a sense of collective Jewish identification not only with the large group of peers on the boat, but also with their ancestors on the original boats of refugees and with Israeli Jews. Visiting Jewish cemeteries and old synagogues in Europe, participating in a quasi-simulation of Holocaust refugees' voyage and the emotional arrival in Israel intensify the participants' sense of identification with the Jewish people and with Israel. The Exodus tour helps to prepare the participants for the questions they will encounter in Israel. The name of the programme itself is inextricably linked with the concept of Jewish identity. The

Exodus from Egypt is the defining moment in the story of the Jewish people, the birth of the nation and the origins of its culture (Atlan 1979; Fredman 1981). Thus, though the simulation game focuses on the more recent and smaller-scale Exodus from post-Second World War Europe, the participants are simultaneously involved in a symbolic simulation of the original, national Exodus from Egypt.

## Conclusion

A study of the Exodus boat tour to Israel links three fields: tourism, diaspora studies and informal education. Without giving such names to their expectations, the students expect all three of these complementary facets to be present. They want travel; the same curriculum would fall flat if taught in a classroom setting. They want to learn during their travels; this is a self-selected population and they choose *not* to spend their vacation at the beach or in nightclubs and bars. And they expect the educational tour of Israel to help them in their personal searches, as diaspora Jews, for identification with Judaism and with Israel. They would not be satisfied with a straight historical tour, however educational, such as they might enjoy if travelling to another destination with no personal connection for them. The leg of the journey that takes place on the boat enhances all three aspects. The travel itself, by ship rather than by air, is more romantic, more exotic and more fun. It gives participants time to contextualize the place they are about to visit by learning some more of its history. The educational features of the trip, which are delivered primarily through quasi-simulations, makes the lessons seem very real. They clearly address Jewish identity, diaspora–Israel relations and a host of other questions with which many of these young people have already been wrestling and with which they will continue to grapple during and after their time in Israel. For these reasons, the specific tool of simulation is highly effective for educational diaspora tourism. Travel away from home and out of one's daily routine is the ideal context for trying out new roles. The dramatic elements bring together the elements of fun, learning and emotion that the participants expect.

The Exodus boat trip, as an introduction to a later tour of Israel, has been found to increase the participants' satisfaction with the latter. Continuing surveys of youth tours to Israel have already shown that long-term, structural preparation is essential (Chazan 1992, 1997; Cohen 1986, 1992; Cohen and Wall 1993; Mittleberg 1994, 1999). Involvement in a local Jewish community, family commitment to Israel and to Judaism and Jewish education all create a cultural context that gives the tour personal meaning. The Exodus Program has shown that short-term orientation is also important. The young travellers were given time to bond as a group and even to form a temporary community. The *madrichs* had a chance to get to know the participants and to work on group dynamics in a neutral setting. Anticipation was built up and ultimately expectations were heightened.

These findings may be useful to other tours designed to bring diaspora populations to a cultural or spiritual homeland. Although structural preparation is generally not in the hands of tour organizers, market analyses may take the presence or absence of it into consideration. Short-term orientation is more feasible for tour companies to organize. By offering several days prior to the final destination to tour in another location, or to travel by boat or overland, especially if combined with learning and emotionally powerful programmes, the subsequent tour experience may improve and with it visitor satisfaction. Experimenting with historical simulation games as part of such an orientation would be worthwhile, where appropriate and possible.

# References

Alessi, S.M. and Trollip, S.R. (1991) *Computer Based Instruction: Methods and Development*, New Jersey: Prentice Hall.

Atlan, H. (1979) *Entre le cristal et la fumée: Essai sur l'organization du vivant*, Paris: Editions du Seuil.

Birt, D. and Nichol, J. (1975) *Games and Simulations in History*, New York: Longman.

Boyd, J. (1996) *Exodus '96: Evaluative Comments*, Jerusalem: Joint Authority for Jewish Zionist Education, Youth and Hechalutz Department.

Chazan, B. (1992) 'The Israel trip as Jewish education', *Agenda-Jewish Education* 1(1): 30–3.

Chazan, B. (1997) *What We Know About The Israel Experience*, New York: Israel Experience Inc.

Cohen, E.H. (1994) *A Compilation of Direct Quotes of Participants' Personal Comments and Evaluations* (13 Volumes), Jerusalem: Israel Experience (ongoing survey and evaluation).

Cohen, E.H. (1999) *Israel Experience: A Sociological and Comparative Analysis*, Jerusalem: The Birthright Foundation and the Department for Jewish Zionist Education, JAFI.

Cohen, E.H. (2001) 'A structural analysis of the R. Kahane code of informality: elements toward a theory of informal education', *Sociological Inquiry* 71(3): 357–80.

Cohen, E. H., Ifergan, M. and Cohen, E. (2002) 'A new paradigm in guiding: the *madrich* as a role model', *Annals of Tourism Research* 29(4): 919–32.

Cohen, R. (1997) *Global Diasporas*, London: Routledge.

Cohen, S.M. (1986) *Participation in Educational Programs in Israel: Their Decision to Join the Programs and Short-term Impact of their Trips*, Jerusalem: Native Policy and Planning Consultants.

Cohen, S.M. (1992) 'The good trip to Israel', in *The Israel Experience*, Jerusalem: CRB Foundation.

Cohen, S. M. and Wall, S. (1993) *Excellence in Youth Trips to Israel*, a report, Montreal and Jerusalem: CRB Foundation.

Der-Thanq, C. and David, H. (2002) 'Two kinds of scaffolding: the dialectical process within the authenticity–generalizability (A–G) continuum', *Educational Technology and Society* 5(4): 148–53.

Fredman, R. G. (1981) *The Passover Seder: Afrikoman in Exile*, Philadelphia, PA: University of Pennsylvania Press.

Goffman, E. (1961) *Encounters: The Studies in the Sociology of Interactions*, New York: Bobbs-Merrill.

Grabe, M. and Grabe, C. (1996) *Integrating Technology for Meaningful Learning*, Boston: Houghton Mifflin.

Gruber, R. (1948) *Destination Palestine: The Story of the Haganah Ship, Exodus*, New York: Current Books.

Halbwachs, M. (1938) 'The collective psychology of the reasoning', *Review Zeitschrift fur Sozialforschung*: 357–74 (originally published in French).

Halbwachs, M. (1939) 'Individual conscience and collective spirit', *American Newspaper of Sociology* 44: 812–22 (originally published in French).

Henrik, J. *et al.* (1999) *Poeppendorf instead of Palestine, Documentation of an Exhibition*, Hamburg: Doelling and Galitz Publishing House (translated from German).

Herman, L. and Mandell, A. (2000) 'The given and the made: authenticity and nature in virtual education', *First Monday* 5(10). Online. Available: http://www.firstmonday.dk/issues/issue5_10/herman/

Holly, D. (1969*) Exodus 1947,* Boston: Little, Brown.

Joint Authority for Jewish Zionist Education, Youth and Hechalutz Department (1994) *The Exodus Bulletin: Exodus 1994 Voyage Exclusive Story,* Jerusalem: The Jewish Agency.

Joint Authority for Jewish Zionist Education, Youth and Hechalutz Department (1995) *Exodus 1995: The Road to Eretz Israel,* Jerusalem: The Jewish Agency.

Kahane, R. (1997) *The Origins of Postmodern Youth: Informal Youth Movements in a Comparative Perspective,* New York and Berlin: Walter de Gruyter.

Levine, D (2002) Senior educational consultant of the Exodus project, personal communication.

Levine, E. (1986) 'Confronting the *aliyah* option', in E. Levine (ed.) *Diaspora: Exile and the Contemporary Jewish Condition,* Tel Aviv: Steimatzky Shapolsky.

Levine, J. and Waugh, M (1988) 'Educational simulations, tools, games and microworlds: computer based environments for learning', *Journal of Educational Research* 12(1): 72–9.

Mittleberg, D. (1994) *The Israel Visit and Jewish Identification,* New York: Institute on American Jewish–Israeli Relations, American Jewish Committee.

Mittleberg, D. (1999) *The Israel Connection and American Jews,* Westport, CT: Praeger.

More, D. (2002) Director of the Israel Experience programs, personal communication.

Morgan, S. (2002) 'The postmodernist possibilities of computer-based instructional simulations', English department, Vista University.

Ohef Sholom Temple Archives (2002) *Faces of the Past – Voices of the Future.* Norfolk, VA. Online. Available: archives@ohefsholom.org.

Petraglia, J. (1988) *Reality by Design: The Rhetoric and Technology of Authenticity in Education,* Mahwah, NJ: Erlbaum.

Sönmez, S.F. (1998) 'Tourism, terrorism and political instability', *Annals of Tourism Research* 25(2): 416–56.

Steinberg, P. (2001) *Contact-Zone en-route: Nationalism and Ethnicity on a Voyage to Israel,* paper presented at the research seminar on 'Education and Belonging? The Experience of Israeli and Diaspora Jews', Jerusalem, Department of Jewish Zionist Education, The Jewish Agency.

Tye, L. (2001) *Homelands: Portraits of the New Jewish Diaspora,* New York: Henry Holt and Company.

# 9 'To stand in the shoes of my ancestors'

## Tourism and genealogy

*Kevin Meethan*

### Getting connected: mobility, home and self

This chapter examines a form of tourism, roots or genealogy tourism, which has until now been a neglected topic with only a few exceptions (Nash 2002; Stephenson 2002; Timothy 1997; see Ch. 7). The quote that forms the title of this chapter was provided by one informant in this study, explaining what had motivated her to travel over 3,000 miles from her home in the USA, to visit a small village on the west coast of Ireland, from where her ancestors had emigrated some 150 years earlier. Her response was both direct and, in its own way, self-evident and challenging – how could it be otherwise? What this also signalled was a neglect of a related topic, that of accounting for individual motivations and experiences of travel and how these relate to notions of self and identity (Desforges 2000; Edkins 2001; Elsrud 2001; Galani-Moutafi 2000; Li 2000; Nash 2002; Stephenson 2002; Suvantola 2002). This does not include defined genres of travel writing, travel documentaries or other media forms, or for that matter issues concerning consumer choice or satisfaction of the tourism product. Rather of interest here is the role of travel in the biographical and narrative construction of self-identity. However, to begin to address these issues, they need to be seen in the context of a globalized and increasingly mobile world.

There are several ways in which the problem of globalization can be approached, many of which are arguments concerning the definition of the term and the actual or possible consequences (Eade 1997; Friedman 1994; Held *et al.* 1999). In terms of the materials discussed here, there are two factors that are of particular interest: first, increased spatial mobility, which in turn contributes to the creation of diasporas and other forms of translocational identity; and second, the rapid spread of information technology as both a source of information and as a means of communication.

To begin, the following proposition is a good starting point: namely, that the contemporary global condition is characterized by flows and mobility rather than stasis (Beck 2000; Clifford 1997; Friedman 1994; Kraidy 2002; Tomlinson 1999; Urry 2000; 2003; Welsch 1999). There are of course a number of consequences that follow from this, not least of which, as Cohen (1997: 74–5) points out, is the notion of exclusive citizenship of defined territories as the *sine qua non* of identity, is being replaced by 'an increasing proliferation of subnational and transnational identities'. A similar point is made by Friedman (1994: 79), who argues that forms of identity that assumed an unproblematic linkage between place, people and nation, have tended to be replaced from the 1970s onwards by a 'search for roots' and the apparent emergence of 'hybrid' forms of cultural identity in which individuals or social groups are perceived to be 'between' cultures (Clifford 1997; Hannerz 1996; Kraidy 2002; Nederveen-Pieterse 1995; Werbner and Modood 1997). As Beck (2000: 74) remarks, humankind is now faced with a situation where 'one's own life is no longer tied to a particular place' and perhaps, as Giddens (1990) argues, social relations have become 'disembedded' across space and time.

The material presented in this chapter supports this claim in many respects, but it is also crucial that the central importance of place, either real or imagined, should not be discarded from the ways in which people construct a sense of being in the world. Place, as Archibald (2002: 65) writes, 'is the crucible of memory' and as Sarup (1994: 97) reminds us, 'we are born into relationships that are always based in a *place*. This form of primary and 'placeable' bonding is of quite fundamental human and natural importance' (emphasis in original). Indeed, it is this form of bonding, with its associated emotive appeal, that is commonly associated with the idea of 'home' or 'homeland'. While these forms of connection between family, friends and place show considerable variance across cultures, the idea of 'home' as an inalienable focus for self and group identity invokes a set of attitudes and beliefs that link a people to a place, or even a place at a particular time.

Those who are uprooted, willing or unwilling migrants and refugees, may continue to see their 'home' not as the place in which they live, but as one that is elsewhere. For these people, the point of origin – the homeland – is a place to return to after displacement or even death. Such diasporic identities are forged around the idea of a homeland that is elsewhere in both space and time and it may also be that the idea of return itself, even if deferred by several generations, acts as a focus for collective identity (Lovell 1998; Stephenson 2002). This may be so even if, as Van Hear (1998: 48) claims, many of those will not know their homeland. It appears then, that the issue at hand is a form of diasporic identity.

To claim, however, that one is a member of a diaspora is in part a conscious choice, perhaps also requiring adherence to an associated ideology and belief system (Climo and Catell 2002). Having demonstrable proof that one's ancestry can be traced to a particular place may in some cases authenticate and legitimize such a claim, yet it must also be borne in mind that to identify with a culture, sub-culture, nation-state, region or locality in this way is but one of the forms of transnational identity that the processes of globalization reveal. As both Van Hear (1998) and Cohen (1997) point out, one feature of the contemporary world is the way in which new possibilities of diasporic identity have emerged. There is, as Cohen (1997: 175) writes, 'no longer any stability in the points of origin, no finality in the points of destination' to the extent that diasporic identities are not linked to single, but multiple, points of origin and consequently multiple allegiances to place (Van Hear 1998: 4). Although the forms of mobility that characterize globalization then appear to challenge the notion of stasis and replace it with that of change and flux, it is also apparent that this is not leading to a 'placeless' world. Rather, what is evident is that the relationship between transience, mobility and belonging are being recast and new opportunities are emerging from which a sense of place in the world can be forged.

As well as the physical mobility engendered by globalization, another significant factor is the growth and spread of computer technology and the ways in which this has transformed spatial relationships (Castells 1996). This has created the possibility of new forms of social interaction and connectivity (Foster 1997; Mitra 1997). For example, Miller and Slater's (2000: 58–60) ethnography of Internet use in Trinidad found that the Internet enabled long-term relationships between distant kin, both genealogically as much as spatially, to be viable in new ways. Foster (1997) and Mitra (1997) noted similar trends. This is what Beck (2000: 72–7) calls 'place polygamy' where technologies enable both time and place to be bridged in new ways so that people's lives are no longer circumscribed by a locality, but rather are linked to multiple localities. These multiple localities are as much imagined as they are physical. People's sense of who they are in the world, the biographical narratives around which they weave their sense of identity, is itself globalized, so that 'the world's oppositions occur not only out there but also in the centre of people's lives' (Beck 2000: 73). Self-identity, then, rather then being simply prescribed by roles and institutions and bounded by place (or nation), are now organized in a more reflexive and negotiated fashion that may span many different places and indeed nations (Giddens 1991; Hall 2001).

In turn, this necessitates a consideration of what is meant by identity. The intention here is not to enter into an assessment of this literature, but rather to outline an approach that focuses on the ways in which people create and maintain a sense of who they are through forms of biographical narrative. As Giddens (1990) argues, one way of approaching the idea of self-identity is to consider it as the capacity to maintain a coherent narrative of self in light of contingent circumstances. Such narratives are a continuous process of making and remaking the self, which is always 'work in progress' (Hall 2001: 23). Hearn (2002: 748) writes that narrative is 'seen as crucial for orienting and guiding behaviour, making both practical and moral sense of reality'. A narrative then is the organizing framework around which the stories of lives are woven. It is the ordering of memory and experiences into a coherent and explicable pattern (Cavarero 2000; Miller 2000; Roberts 2002; Skultans 1998)

Today's situation is more fluid and mobile than in the past, both in a literal and metaphorical sense. In turn this mobility recasts the terms, the opportunities and constraints, the context in which we all strive to make sense of our lives. Tourism, involving as it does the mass and temporary movement of people, exemplifies the changes and challenges of such mobility.

Although travel is often seen as an escape from the daily grind of work into a world of hedonistic excess, which for many it clearly is, another dimension is to see travel and tourism as a voyage of self-discovery (Robertson *et al.* 1994). As an example, Suvantola (2002) describes the motivations of those who undertook backpacking trips as a search for new perspectives and personal development, a *rite de passage*, that can also apply to more ordered and controlled forms of tourist experiences. To travel away, he argues, is to break with routine and this abandonment of the daily routine allows space for retrospection. Desforges (2000) in his study of long-haul travellers noted that the decision to travel is often taken when issues of self-identity are questioned and may result in a challenge or rejection of ascribed social roles. Such travel, he adds, 'is imagined as providing for the accumulation of experience, which is used to re-narrate and represent self-identity' (Desforges 2000: 943). On a similar theme, Galani-Moutafi (2000: 205) draws attention to a rather neglected topic in tourism research, that of the importance of self-reflexivity: 'Through their descriptions which structure and give meaning to their experiences in the process of narration, travellers can reflect upon their journeys in ways that produce images of self and identity'.

## Methods

The information that informs this chapter is part of a wider research project that focuses on genealogy and has two aims and three distinct phases. The first aim was to define some general and common characteristics of those engaged in genealogical research using the Internet. The second aim was to select a sub-sample of those who had travelled and could participate in a qualitative enquiry. The first phase was an exploratory study with fourteen volunteers recruited via an e-mail discussion list. The purpose in this stage was to explore the characteristics of these people, test the feasibility of using e-mails as a method of qualitative fieldwork (see Hine 2000) and exploit the connectivity of the Internet by recruiting the sample on a global basis. A question/response approach was adopted by initially asking informants to write a biographical account of how they first came to research their family history. When the replies arrived, additional questions were asked based on the first set of responses. The intention here was to make the whole process as interactive, and therefore as close to an interview, as possible. Initially this yielded a large amount of rich data, but keeping the dialogue going after four or five exchanges proved to be difficult and the responses simply stopped, indicating that keeping a remote dialogue going over an extended period of time is problematic.

Despite this setback, the material yielded enough data to carry out phase two, which was the construction of an online questionnaire. E-mail list owners were contacted and asked to circulate a message from the author that contained a hyperlink to the survey, held on a secure server. This resulted in over 1,000 responses. Additionally, volunteers were sought who had travelled for the purpose of genealogy research and who were willing to answer more questions about their travel experiences. As noted, the problem of keeping a sustained dialogue via e-mails even with a small number of informants ruled that out as well. Instead, following a preliminary analysis of the data, it was decided that this phase would involve sending informants a topic list and asking them to write a narrative account, describing how they began their research, why it was important, why they decided to travel and what difference their travel experiences had made to them. The analysis presented draws primarily from sixty of those accounts.

## Starting out

Genealogists have to rely in great part on the construction of a logical and coherent pattern of events from an often incomplete and fragmentary record. In this sense, tracing one's roots is a form of historical research that involves authentication through documentary and other forms of evidence (Bevan 1998; Colwell 1996; Nash 2002), a process that was often compared to detective work. As one informant put it:

> The hobby has the appeal of detective work, but not just any detective work as the subjects are people of my flesh and blood (or vice versa). The puzzle pieces don't turn up in any particular order. The whole process is long term, never ending and fascinating.

Although it can involve different kinds of data, family history overwhelmingly relies on the bureaucratic mechanisms that underpin modern nation-states, the records of births, marriages and deaths and other recorded data that identify and locate an individual in time and space and also defines his/her rights and entitlements. In this sense the existence of an archive memory that provides the means by which individuals can be tracked through the complex structures of modern state institutions, the means through which the officially sanctioned material trace of an individual's life course, can be authenticated and legitimized for legal purposes.

Whatever the motivations for undertaking genealogical research and related travel, the aim is to create an accurate timeline that links people to their ancestry through verifiable sources such as the institutionally defined significant moments of their births, marriages and deaths and the details available in various censuses. However, the existence of these publicly available sources of official information is often matched to the existence of private and family sources. One of the more significant findings to emerge from the data was the importance that was attached to family memorabilia (Hallam and Hockey 2001) and in particular old photographs (Edwards 1999). The latter were seen as important in two ways: not only did they often provide names and dates and as such are a source of information in themselves, but also because to many informants, it was the 'discovery' of these items that led them to begin their search in the first place. As two informants noted,

> I did not set out to do genealogy. I 'came by' almost all of our family photos, documents, etc. as family members died off. My family has kept much and I am swamped in it. What else was I to do with it all?

> My mother inherited a box of photos that her grandfather had left to her mother. My grandmother had the instincts of an archivist and had everything sorted and labelled. These photos not only had the names, dates and locations (where known) written on the back, but also the person's relationship to my grandmother.

As most genealogists know, no source of information on its own can be taken at face value and needs be checked against other evidence, as these two quotes illustrate:

> I found as I got into the research, however, that (1) it's a lot more complicated than it appeared and I got into trying to find out the REAL facts and (2) there were errors in [name removed]'s research, as she had pretty much contented herself with family stories, published sources (which have errors in them) and other non-verified sources. I'm trying to correct those errors.

> . . . so I had all these names and dates and spent so much time chasing after them until I realized a lot weren't accurate! Some just didn't exist and some were two people who had gotten scrambled together.

Trying to unscramble such connections and false leads can of course be costly and time consuming, but it is greatly aided by the fact that records can now be traced with relative ease via the Internet. In turn, this has resulted in the emergence of a large number of both commercial and non-commercial Internet sites relating to genealogy. Online search engines now provide access to material that was previously available only to the most dedicated of researchers with both time and money. The extent of the popularity of online searching can be gauged by the fact that when the pay-to-view 1901 census of England and Wales went online in January 2001, the system crashed in a matter of days owing to the high level of demand and it remained unavailable for 5 months while it was redesigned. Whereas this area of activity has created its own niche of specialist subscription suppliers of data, as well as PC packages (for example see http://www.ancestry.com), many remain free and are often created by unpaid volunteers (see http://freebmd.rootsweb.com). The most valuable archive in terms of its general extent and availability, however, is that provided by the Church of Jesus Christ of Latter-day Saints (also known as the Mormon Church) which, for a variety of reasons connected to its beliefs, has amassed the largest single archive of vital and historical records related to genealogy. The church's website also allows users to access its library catalogue free of charge to locate specific records and sources. It also provides free online searches that include the entire 1880 Canadian and US censuses and the 1881 England and Wales census (http://www.familysearch.org). Also of importance are the large number of e-mail lists devoted to particular surnames and places of interest (see http://www.rootsweb.com for a comprehensive index of over 18,000 lists) where names, dates and other information can be freely exchanged and advice sought and given.

## Getting the facts: travelling for data

Despite the growth of digital archives and their associated networks, which appear to be growing at an exponential rate, the amount of material available is still only a fraction of the archival material that can be found, which is uneven in its coverage. To many people then, there is no option but to travel to gain access to archival material. For all the connectivity that the Internet and e-mail lists offer and despite the fact that information can be and often is freely exchanged, access to overseas archives and records is still very patchy. Knowing a document exists and is in a particular location is one thing. Getting hold of it, if it is not in the form of a certificate, is quite another, as the following quotes makes clear:

> I first travelled to Salt Lake City and spent four days in the archives [of the Mormon Church] . . . armed with that information, I then sought out other archives up and down the state.

> If given a choice of two business trips, I would always choose the one most likely to offer a local library with local unpublished or small run histories, or local data bases, so that I could do research in the evenings.

Travel for evidence does not just involve the verification of official records, as many of the respondents also travelled to gather oral testimony from kin. For example,

> I began to travel and talk to cousins, aunts and uncles, getting them to tell me all they could remember, names, places and other stuff that helps fill in the gaps.

There was also a strong sense of loss and regret among some informants, that they had not begun this process earlier.

My grandparents used to tell me stories of their parents and grandparents when I was a child. Then I lost interest. So when I got interested again many years later they were no longer around, so all those memories are gone forever.

What needs to be considered here then are the temporal dimensions of biographical memory. In general, this does not often go beyond two generations of parents and grandparents, at least in the Western World. In this sense there is always a finite but constantly shifting horizon of family life events, within which recollections and family stories act as the bridge between living generational memory and the archived past (Miller 2000: 38). If this is no longer available, then to go beyond, contextualization was carried out through forms of more general and public history. For instance,

My reading of Irish history, for me, provides a context for understanding the lives of the people whose names and dates turn up on my tree. In some cases, I have been lucky enough to obtain some old letters, possessions, or records to provide some documentation for their lives. But where no such records exist, history fills in a lot of blanks.

I had nothing to go on so then I started reading history and found out all kinds of stuff about the social conditions in England at that time [nineteenth century] which was a bit of a shock. Then I really knew why you can find whole families in the death indexes that had died more or less at the same time.

This contextualization of biography through the reading of general history leads into a consideration of the relationship between the archive, or 'public' memory, as scholarly history and heritage and the realm of 'private' memory in the form of family experiences and memories. As Roper (2001: 319) notes, the construction of personal narratives occurs 'via the available public languages and their associated store of "public imaginaries"'. It may be tempting to see these family tales as the narratives of history writ small as it were, as family histories, as the reflection of a discourse within which people position themselves. It is here perhaps that claims of ancestry are found being linked to wider forms of collective identity and culture that are 'rediscovered' as a result of genealogical research such as those described by Nash (2002). Among informants in this study there was very little sense of people identifying with 'another' culture in the present, even if the links to the past were a motivation for travel:

I think the bottom line reason is that I wanted to see the ancestral homes of my family for myself. I was raised in a household [in Canada] which looked firmly to Scotland and England as the source of its roots and the need to return was ever there.

Yet as Nash (2002: 47) points out, the tracing of roots can have complex and even contradictory relationships to both collective and individual identities. This is certainly born out in the narratives received where often people expressed surprise at the 'mix' of their ancestry, as the following quote demonstrates:

There are times, though, when researching one's family one finds the invariable family skeleton. I found a few on both sides. When it was discovered that my

American Grandfather was a Cherokee Indian, I was almost thrown out of the family reunion. A similar discovery for the India connection. My English grandfather married an Indian women as did his son and his son's son. There was a time this horrified my aunts. I don't think anyone has told one of the aunts and the other was 'prepared' for it. I love it. It explains my penchant for curry.

In part the complexity of the problem that Nash (2002) outlines can be explained if the nature of biographies is considered. As noted earlier, rather than see these as a form of reflection of dominant discourses, the construction of narrative biography is not a passive process and that while some narratives are 'undoubtedly a vehicle for shared cultural representations', they also involve '. . . imaginative truth and creativity' (Skultans 1998: 27). To be intelligible, narratives like other forms of experience, must be rendered into forms that are both culturally specific and common, a process in which people interpret, negotiate and create their own particular meaning from what materials are available (Skultans 1998: 63; see also Miller 2000: 140–2). Similarly, Gable and Handler (2000) in their analysis of the heritage museum at Colonial Williamsburg (USA), note that the production of memory among visitors is not the simple acceptance of one dominant 'reading', but instead is produced through a complex process of interaction, often involving family and memories of past visits. Doing family history means working with not only the material traces that can be located within the archive memory, but also with the discourses of public history, as well as the more personalized and idiosyncratic memories of family and kin.

## Witnessing: travel as authentication

Motivations for travel may then be driven by the need to access or acquire specific information, which corresponds more to an academic model of fieldwork and research, rather than what may be regarded as conventional tourist behaviour. Although some tour operators offer organized 'roots tours' (e.g. http://www.mircorp.com) most people undertaking such trips are better thought of as 'independent travellers' who organize their own itineraries and book their own tickets. In part, this is due to the specificity of individual family lineages which are often more mixed and transcultural and indeed multi-locational, than may be first imagined:

> . . . then I started looking and found that going back not that far – 3 generations – there were Irish, German and Polish ancestors, then the Canadian side of the family that were [sic] from Scotland.

However, what also came across strongly in the narratives was the importance of actually standing and witnessing, as some put it, the places where their ancestors had come from:

> I'm luckier than most researchers I've encountered, since I have so much original and very personal material to work with. I look forward to making more trips in the future and want to follow the westward migrations of two sets of great-great-grandparents. I find that simply standing in a spot they did and looking at what they saw really helps me understand their experience.

Finding land records gave me some idea about the locations of the homesteads. Reading books that mentioned contributions some of my ancestors made to the areas where they settled caused me to feel as if I had known them. Court proceedings records gave me some insight into the character of some of them. I have surmised why some things occurred in their lives, because of what I found about their character.

In these cases, we can see an emotive, imaginative and creative reconstruction of past events where history becomes re-made as a demonstrable and personalized link to the people and places of other times and the point at which the narrative of the self is reflexively organized, re-written and re-thought. This is most apparent in the ways in which people reflected on how their ideas of who they were had changed as a result of their research and travel. 'My sense of who I am has changed' and 'I am more of a mixture than I ever thought I was' are good examples. This can also be seen in the emotive way in which people described the linkages between themselves and the wider historical record:

> I see all those names [of ancestors] written in the records and think they were all people once! It's a bit frightening at times, they have all disappeared – who were they? What were they really like? What did they think? Were they like me?

> Why did we decide to travel? Because I wanted to sit at the graveside of my ancestors and I hoped to find parish and town records that might shed light on their origins (I did all of these things!).

In many cases another more existential means of verification by personal experience was seen not so much as a necessity but rather as an added extra. The subjective element – I was there – while not a guarantee of objectivity nonetheless ascribes a form of authority akin to that granted in a court of law to an eyewitness. Either singly or in combination, it is through such processes of authentication that social or collective values and institutions are linked experientially to the individual (Eakin 1999: 39), as a combination or 'entwinement' of history and memory (Gable and Handler 2000: 246), the personalization of what may appear to be the more abstract forces of history as one of the means by which individuals can negotiate and position themselves in a rapidly changing world.

## Conclusions

The kinds of travel described here are rather different to those described by Galani-Moutafi (2000) and Desforges (2000). In those cases the place or destination seems to have no significance in its specificity, being generalized by the travellers into a rather more abstract 'other', set apart from the mundane and the ascription of social roles. The act of travel itself became the means by which a sense of self could be reassessed. To roots tourists, however, the very specificity of the places visited, is a very important element (Stephenson 2002). As noted above, family history and its insistence on fidelity to and accuracy of, the material trace of the archive memory that legitimizes kinship, is located at a particular point, or points, of origin. Set alongside this essentially bureaucratic model of legitimacy, the act of travel legitimizes, but in a different way through the immediacy of emotions and embodied experiences of place, creating a form of existential authenticity that is both unique and inalienable (Ning 1999). The findings in this chapter reinforce a number of points made in the introduction. First, the current condition of global mobility and connectivity is leading to the emergence of new forms of identity that cut across both time and space. Second, these forms of identity tend to encompass multiple, rather than single, points of origin and in that respect correspond to the kinds of diasporas best described as multi-locational. Third, the development of digital technology plays a crucial role, as to many family history researchers and all of those included in this study the rediscovery of roots and their legitimation through the documentary sources of the archive memory is achieved through the Internet and other media. Fourth, while such technology is often cast as an agent of change that will break down traditional forms of connectivity, there appears to be a movement in the opposite direction: the rediscovery of history, not in the form of grand narrative, but in a more individual and personalized form. Rather than leading to a placeless world, the opportunities for interconnectivity offered in these cases are creating forms of identity that combine notions of locality, belonging and home into a reflexively constituted biographical narrative that extends beyond the immediate constraints of space and time.

## References

Archibald, R.R. (2002) 'A personal history of memory', in J.J. Climo and M.G. Catell (eds) *Social Memory and History: Anthropological Perspectives*, Walnut Creek, CA: AltaMira Press.

Beck, U (2000) *What is Globalization?* Cambridge: Polity Press.

Bevan, A. (ed.) (1998) *Tracing your Ancestors in the Public Record Office*, 5th edn, London: Public Records Office.

Castells, M. (1996) *The Rise of the Network Society*, Oxford: Blackwell.

Cavarero, A. (2000) *Relating Narratives: Storytelling and Selfhood*, London: Routledge.

Clifford, J. (1997) *Routes: Travel and Translation in the Late Twentieth Century*, Cambridge, MA: Harvard University Press.

Climo, J.J. and Catell, M.G. (eds) (2002) *Social Memory and History: Anthropological Perspectives*, Walnut Creek, CA: AltaMira Press

Cohen, R. (1997) *Global Diasporas: An Introduction*, London: Routledge

Colwell, S. (1996) *Teach Yourself Tracing Your Family History,* London: Teach Yourself Books.

Desforges, L. (2000) 'Travelling the world: identity and travel biography', *Annals of Tourism Research* 27: 926–45.

Eade, J. (ed.) (1997) *Living the Global City*, London: Routledge.

Eakin, P.J. (1999) 'Autobiography and the value structures of everyday experiences: Marianne Gullestad's everyday life philosophers', in R. Josselon and A. Lieblich (eds) *Making Meaning of Narratives*, Thousand Oaks, CA: Sage.

Edkins, J. (2001) 'Authenticity and memory at Dachau', *Cultural Values* 5(4): 405–20.

Edwards, E. (1999) 'Photographs as objects of memory', in M. Kwint, M. Breward and J. Aynsley (eds) *Material Memories: Design and Evocation,* Oxford: Berg.

Elsrud, T. (2001) 'Risk creation in travelling: backpacker adventure narration', *Annals of Tourism Research* 28: 597–617.

Foster, D. (1997) 'Community and identity in the electronic village', in D. Porter (ed.) *Internet Culture*, London: Routledge.

Friedman, J. (1994) *Cultural Identity and Global Process*, London: Sage.

Gable, E. and Handler, R. (2000) 'Public history, private memory: notes from the ethnography of Colonial Williamsburg', *Ethnos* 65(2): 237–52.

Galani-Moutafi, V. (2000) 'The self and the other: traveller, ethnographer, tourist', *Annals of Tourism Research* 27: 203–24.

Giddens, A. (1990) *The Consequences of Modernity*, Cambridge: Polity Press.

Giddens, A. (1991) *Modernity and Self-Identity: Self and Society in the Late Modern Age*, Cambridge: Polity Press.

Hall, S. (2001) 'The Multicultural Question', *Pavis Papers in Social and Cultural Research* 4, Milton Keynes: Open University.

Hallam, E. and Hockey, J. (2001) *Death, Memory and Material Culture*, Oxford: Berg.

Hannerz, U. (1996) *Transnational Connections: Culture, People, Places*, London: Routledge.

Hearn, J. (2002) 'Narrative, agency and mood: on the social construction of national history in Scotland', *Comparative Studies in Society and History* 44(4): 745–69.

Held, D., McGrew, A., Goldblatt, D. and Perraton, J. (1999) *Global Transformations: Politics, Economics and Culture*, Cambridge: Polity Press.

Hine, C. (2000) *Virtual Ethnography*, London: Sage.

Kraidy, M. (2002) 'The global, the local and the hybrid: a native ethnography of glocalization', in S. Taylor (ed.) *Ethnographic Research: A Reader*, London: Sage.

Li, Y. (2000) 'Geographical consciousness and tourism experience', *Annals of Tourism Research* 27: 863–83.

Lovell, N. (ed.) (1998) *Locality and Belonging*, London: Routledge.

Miller, D. And Slater, D. (2000) *The Internet: An Ethnographic Approach*, Oxford: Berg.

Miller, R.L. (2000) *Researching Life Stories and Family Histories*, London: Sage.

Mitra, A. (1997) 'Virtual commonality: looking for India on the Internet', in S.G. Jones (ed.) *Virtual Culture: Identity and Communication in Society*, London: Sage.

Nash, C. (2002) 'Genealogical identities', *Environment and Planning D: Society and Space*, 20(1): 27–52

Nederveen-Pieterse, J. (1995) 'Globalization as hybridization', in M. Featherstone, S. Lash and R. Robertson (eds) *Global Modernities*, London: Sage.

Ning, W. (1999) 'Rethinking authenticity in tourism experiences', *Annals of Tourism Research* 26: 349–70.

Roberts, B. (2002) *Biographical Research,* Buckingham: Open University Press.

Robertson, G., Mash, M. Tickner, L., Bird, J., Curtis, B. and Putnam, T. (eds) (1994) *Traveller's Tales: Narratives of Home and Displacement,* London: Routledge.

Roper, M. (2001) 'Splitting in unsent letters: writing as a social practice and a psychological activity', *Social History* 26(3): 318–39.

Sarup, M. (1994) 'Home and identity', in G. Robertson, M. Mash, L. Tickner, J. Bird, B. Curtis and T. Putnam (eds) *Traveller's Tales: Narratives of Home and Displacement*, London: Routledge.

Skultans, V. (1998) *The Testimony of Lives: Narrative and Memory in Post-Soviet Latvia*, London: Routledge.

Stephenson, M. (2002) 'Travelling to the ancestral homelands: the aspirations and experiences of a UK Caribbean community', *Current Issues in Tourism* 5(5): 378–425.

Suvantola, J. (2002) *Tourist's Experience of Place*, Aldershot: Ashgate.

Timothy, D.J. (1997) 'Tourism and the personal heritage experience', *Annals of Tourism Research* 24: 751–54.

Tomlinson, J. (1999) *Globalization and Culture*, Cambridge: Polity Press.

Urry, J. (2000) *Sociology Beyond Societies*, London: Routledge.

Urry, J. (2003) *Global Complexity*, Cambridge: Polity Press.

Van Hear, N. (1998) *New Diasporas: The Mass Exodus, Dispersal and Regrouping of Migrant Communities*, London: UCL Press.

Welsch, W. (1999) 'Transculturality: the puzzling form of cultures today', in M. Featherstone and S. Lash (eds) *Spaces of Culture: City, Nation, World*, London: Sage.

Werbner, P. and Modood, T. (eds) (1997) *Debating Cultural Hybridity: Multi Cultural Identities and the Politics of Anti-Racism*, London: Zed Books.

# Part II

# Settings and spaces for diaspora tourism

# 10 The 'isle of home' is always on your mind

## Subjectivity and space at Ellis Island Immigration Museum

*Joanne Maddern*

[W]hat I, Georges Perec, have come here to examine is dispersion, wandering, diaspora. To me Ellis Island is the ultimate place of exile, that is, the place where place is absent, the non-place, the nowhere.

(Perec and Bober 1995: 58)

### Introduction

Ellis Island Immigration Museum is located on a small island in New York harbour between the iconic Statue of Liberty and the omnipotent skyscrapers of Manhattan's financial district. Formerly the buildings on Ellis Island were used as a federal immigration station as well as a detention and deportation facility. Between the years 1892 and 1924 over 12 million people passed through the buildings on Ellis Island (Moreno 2001) on their way to a new life in a new land.

Ellis Island is now a well-known tourist attraction and commemorative landscape that is visited by several million people every year as a result of its 'symbolic importance' (Foner 2000: 2; see also Smith 1992). Indeed, Ellis Island's popularity means it has become a metonym for immigration in the USA, due to its complex history as an immigration, detention and deportation facility (see Table 10.1). Ellis Island is a unique space where multiple readings of diasporic identities and histories are possible; it is a site where the many types of collisions between the ambitions of cultural tourism producers, national histories and the individual subjectivities of a range of diasporic identities may be observed.

This chapter explores how the entangled tensions of different local, national and extra-national histories at Ellis Island have been managed by the heritage professionals there. In other words, considering the complexity of possible histories that could be evoked at a place like Ellis Island, by whom and for whom has Ellis Island Immigration Museum been restored? These represent crucial research questions when exploring the complex relationships between diasporas and tourism. All too often in the existing literature there has been a willingness to put one diasporic group under the microscope exclusively. For instance, travel and tourism consumption patterns in the Jewish diaspora have been extensively treated, while groups from Asia and the Pacific have increasingly been the focus of discourse (Feng and Page 2000; Kang and Page 2000; Ioannides and Cohen Ioannides 2002; Lew and Wong 2002; Nguyen and King 2002). Such studies, however, largely overlook that multiple diasporas may occupy and contest a single space at any given time. Indeed, it is the dynamics of contestation that often lead to the explicit and surreptitious definition of (tourism) place. Moreover, multiple occupation has implications on two further levels: in a more abstract sense, interaction and contestation has clear resonances for the mediation of diaspora identity; on a more functional, managerial level, it forces museum operators to consider their target markets more effectively and to understand their demands more thoroughly.

The answers are unpacked in five substantive sections and are based on the results of a largely ethnographic programme of fieldwork with rich observation, participation and structured interviews at the core. In the first section, a brief overview inclusivity and 'othering' of diasporas in museums and tourism are explored. A brief history of Ellis Island followed by a virtual tour of the exhibit are crucial preludes to discussions of representation of immigrant identity, both through meta-narratives and individual, destabilizing discourses. Mainstream 'European' diasporic readings are compared with African-American and Japanese-American texts.

## Migration, diasporas and tourism

There has so far been a relative lack of academic discourse on the ways in which the subjectivities of diasporic identities and histories are appropriated during the production of national monuments and heritage sites. Historically, the production of museums has been seen as closely intertwined with the fortunes of the nation-state. The heritage industry is often viewed as a mechanism for re-inscribing nationalist narratives in the popular imagination (see Wright 1985; Sherman and Rogoff 1994; Johnson 1995; Lowenthal 1996).

As Bender (2001: 5) observes, the powerful histories told at heritage sites are often ones that 'stress stability, roots, boundaries and belonging' through the supposed mists of time (Bhabha 1990). Museums and national monuments, drawing on a kind of sedentarist metaphysics (cf. Malkki 1992) as a means of legitimization have traditionally pathologized the extra-national loyalties of migrants and other mobile groups. Simultaneously such sites have advanced national aspirations and expressed a national sense of identity or character precisely by defining the nation *in opposition* to those whom it excludes through its hermetically sealed ideological borders. For Deleuze and Guattari (1987: 23), 'history is always written from a sedentary point of view and in the name of a unitary State apparatus . . . even when the topic is nomads. What is lacking is a Nomadology, the opposite of a History.'

Even where the transnational histories and geographical connectivities of mobile groups have been successfully incorporated into national creation myths at heritage sites, this narrative 'assimilation' has not been total and absolute. As Nash (2002: 32) contends, some migrant groups find their memories are more successfully incorporated into public meta-narratives than others:

> National histories in . . . the 'new world' valorise specific genealogies. In settlers contexts the ancestry of particular groups – the first arrivals and founding families and their descendants in New England Puritan, Daughters of the American Revolution, or First Fleet genealogies – define the nation-state. More recent migrants can be assimilated into the grand narrative but their contributions are always subordinated to the story of the founding people.

As Creswell (2001: 20) recognizes, 'few modern nations are so thoroughly infused with stories of wandering, of heroic migrancy and pilgrimage . . . than the Americans'. Thus, in a similar vein, Chavez (2001: 4) points to how immigrants are reminders of how Americans as a people came to be. Their cultural and linguistic difference and *otherness* often raise questions and concerns about population growth, economic competition and various perceived threats to the prevailing order. In the discursive construction of national identity, immigrants are categorized simultaneously as liminal, marginal, or even pathological subjects, but also as central objects of crisis (Chavez 2001; Dahlman 2002).

Diasporic groups, (both the traditionally recognized diasporas and other transnational dispersions), frequently have as defining aspects of their collective identities, legacies of marginalization or exclusion both from *space* and from *official* versions of national history. Mirzoeff (2000: 3) notes how 'diaspora peoples have been marginalized by . . . [the] visualization of national cultures in museums, whilst consistently using visual means to represent their notions of loss, belonging, dispersal and identity'.

Geographies of diaspora are characterized by networks, flows and connections which link multiple locations in complex local and global connections (Said 1979, 1986; Appadurai 1987, 1990; Clifford 1997, 1998; Dwyer 1999). Not surprisingly, the inclusion of these extra-national histories at national heritage sites often tends to be a highly contested process that involves 'dwelling in language, in histories and in identities that are constantly subject to mutation' (Chambers 2001: 5).

In informing the production of 'official histories' at heritage sites, diasporic knowledges are able to reach outside of the normative territory and temporality of the nation-state, exceeding and criticizing its structures by mobilizing critical versions of historical recovery and return which challenge the idea of a bounded nation-state as a fixed site of belonging and a site for the production of a national culture (Clifford 1997: 250–1; Dwyer 1999: 228; Ifekwunigwe 1999; Nash 2002: 33). Essentially therefore, migrants and exiles are subjects that cross borders and in doing so, 'break barriers of thought and experience' (Said 2000: xi).

## Ellis Island immigration station: a brief history

Until around 1875, immigrants landed on the soil of the USA without restriction. Progressively stricter measures were introduced to control immigration. At first, these were by municipal and harbour authorities; later by the Secretariat for Immigration, a federal body (Perec 1999: 134). To deal with increasing numbers of nineteenth-century European migrants, the first federal immigration station was opened at Ellis Island in 1892 (Table 10.1). This marked the beginning of an 'official, institutionalized and industrialized emigration' (Perec 1999: 35). Between 1892 and 1924, approximately twelve million people passed through at rates of five to ten thousand per day (Moreno 2001).

Ellis Island delivered contrasting experiences. For the successful it was the gateway to a new life in the land of hope. In contrast, those with suspected diseases, abnormalities or undesirable political afflictions were either deported or detained at the island. Episodes of alleged mismanagement, cruelty, poor conditions and financial exploitation caused repeated scandals in European newspapers (see Kraut 1982).

*Table 10.1* Timeline of salient episodes and events in Ellis Island's history

| Date | Description of episode/event |
| --- | --- |
| 1892 | Ellis Island opens as a federal immigration station. |
| 1907 | Peak year of immigration to the USA – 1.2 million 'aliens' examined at Ellis Island. |
| 1917–1918 | Ellis Island's hospitals are used for wounded servicemen. Enemy aliens detained during the war. |
| 1921 | Quota systems introduced to limit immigration. |
| 1924 | National Origins Act passed. Dramatic reduction in immigration. |
| 1939 | Coast guard training station opens on Ellis Island. |
| 1942 | Approximately 1,000 German, Italian and Japanese 'enemy aliens' held at Ellis Island from May onwards. |
| 1943 | Ellis Island now used only for detention purposes. |
| 1954 | Last detained 'aliens' are removed and Ellis Island is abandoned. |
| 1965 | Ellis Island is proclaimed part of the Statue of Liberty National Monument, administered by the US National Park Service. |
| 1973 | Ellis Island clean up campaign inaugurated by Dr Sammartino of Fairlie Dickenson University. |
| 1982 | In a press conference, President Reagan announces plans to restore Ellis Island, with the help of celebrity businessman Lee Iacocca. |
| 1990 | Ellis Island Immigration Museum opens to the public. Restoration project cost: US$156 million. |
| 1992 | 1 January 1992 is officially declared 'National Ellis Island Day'. |

Source: abridged from Moreno (2001).

After a short spell at a hospital and political detention centre at the end of the First World War, in 1920 Ellis Island re-opened as an immigration centre. Stringent measures were introduced to stem immigration, with the result that the 'great wave' of (European) immigration was reduced to a trickle (Kirshenblatt-Gimblett 1998). More especially, such legislation effectively precluded entrants from southern and eastern Europe as well as Asia (Kraut 1982; Zinn 1996; Daniels 1997b; Perec 1999).

As formalities were handed to consulates, the role of Ellis Island diminished to a prison for 'enemy aliens'. It finally closed in 1954. For over 20 years Ellis Island became a source of contention for federal, state and local governments, commercial developers and historic preservationists (Johnson 1984: 157). In the 1970s interest in uncovering ethnic heritages and tracing genealogies stimulated interest in transforming the ruins into a museum (Seitz and Miller 2001). Complex debates surrounded its funding and potential legal status. Johnson (1984) notes how, as a pioneering public-private enterprise, many commentators were concerned about the influence of the private sector in packaging the past. For instance, they contended that:

> In a worst case scenario, Ellis Island could become a Disney-like 'Immigrant land' – with smiling, native garbed workers selling Coca-Cola to strains of 'It's a Small World After All'.
>
> (Johnson 1984: 161)

Instead of Ellis Island becoming a dynamic space to commemorate diaspora history and culture, there were concerns that immigrant cultures would be essentialized and commodified under the instrumentality of the tourist industry (Bodnar 1995).

Finally, in 1990 at a time when 70 per cent of the American public was said to be in favour of reduced immigration and alarmist images of the 'immigrant problem' proliferated the popular US media (Chavez 2001: 20), part of Ellis Island opened. This was at a cost of US$156 million (Holland 1993). Two years later, legislation designating 1 January 1992 'National Ellis Island Day' proclaimed it to stand as '. . . a reminder of the hope for freedom and prosperity that the United States offered to the poor, tired, hungry and downtrodden of the world'.

## Re-enacting the immigrant experience

After queuing patiently in line and negotiating the rows of National Park Service police manning security gates in Battery Park, I am transported past the Statue of Liberty to the twenty-seven acre island by a commercially operated boat named 'Miss New York'. I debark along with hoards of fellow recreational immigrants (Kirshenblatt-Gimblett 1998) from all over the world.

On arrival at Ellis Island, I am confronted with the former immigration centre, a striking and ornate redbrick building with turrets and scalloped edges faced in white stone (Plate 10.1). The museum contains photographs, texts, models, oral histories, graphic representations and artefacts such as immigrant clothing and possessions on three floors of 'self-guided' exhibits.

Upon entering the vast arrival hall otherwise known as the 'baggage room', I am met head-on by a large display of replica immigrant baggage and luggage (Plate 10.2). The display is nearly thirty feet long, cordoned off and accompanied by plaques and black and white vintage photographs. Rogoff (2000: 41) notes how it 'virtually blocks the visitor's progress, along the vast arrival hall . . . [and] condenses the experience of immigration to a single visual metaphor . . . produc[ing] a concrete borderline for a national culture, to embody the moment of crossing over to America'.

Here, in the corner of the hall, I pick up a special *Acoustiguide*. By putting on the headset and pressing a button, I receive my own personal tour of the museum narrated by the NBC newsreader Tom Brokaw, who dramatically informs me that I will be 'inhabiting history, walking among the shadows of my parents and grandparents and great grandparents' during my visit. This narrative strategy creates a vertical bond between today's living and yesterday's dead (Namer 1991). The choice of a famous newsreader as storyteller serves to lend a calculated air of familiarity and gritty authenticity to the historical plot.

Upstairs, the huge Great Hall where immigrants were processed in large 'cattle pen'-type structures has purposely been left empty (except for two large American flags), in an attempt to invoke the visitors' imagination, a deliberate absence with presence. The famous 'American Immigrant Wall of Honor' can be found in the museum grounds. The wall consists of a series of stainless steel plates attached to a large stone circle (Plate 10.3). By paying a minimum of US$100 dollars families can 'honour their immigrant ancestors' by having their names inscribed alphabetically.

Throughout my visit, I am repeatedly invited to 'step back in time' and self-consciously re-enact the experience of immigration. Through a sense of the 'embodied, the dramaturgical and the performative' (Crang 2000: 158), the museum uses various mechanisms to paradoxically 'bring the past to life' and resurrect it before the eyes of contemporary tourists.

*Plate 10.1* The imposing façade of the Ellis Island Immigration Museum in New York harbour.
Source: Timothy Coles.

As it is peak season, I am able to sit in on a play called 'Embracing Freedom', which attempts to dramatically recreate the 'immigrant experience'. 'Decide an Immigrant's Fate' is another organized re-enactment whereby the audience takes on the role of the 'Board of Special Inquiry' and decides the fate or fortune of an 'alien' who had for one

*Plate 10.2* The poignant display of replica immigrant baggage and luggage at the Ellis Island Immigration Museum.

*Plate 10.3* The 'American Immigrant Wall of Honor' in the grounds of the Ellis Island Immigration Museum.

reason or another aroused the suspicions of officials (based on actual cases heard during the 1908–1918 period).

At the end of the visit, I am invited to pick up genealogical mementoes from the on-site souvenir shop selling 'products that reflect the heritage of people from all over the world who passed through Ellis Island, Gateway to America'. The gift shop is located adjacent to the 'Global Fare' café where hoards of tourists can be seen ordering fries, burgers and cola from the quintessentially American fast food menu against the backdrop of large black and white photomurals of Ellis Island immigrants at long tables with their modest bowls of stew and crusts of bread (Kirshenblatt-Gimblett 1998: 185).

## Representing immigrant identity

At least two conflicting storylines can be disentangled from the landscapes of Ellis Island Immigration Museum. The first of these is a popular meta-narrative of immigration history, which concentrates on the professional and industrial achievements of successfully 'assimilated' immigrants who are seen as nation-builders and defenders of the land and its 'democratic ideals' from immoral or dangerous outsiders (Bodnar 1995). Such accounts inevitably rely on the simplification, exclusion, substitution, assimilation and disciplining of memories and aestheticization of migrant landscapes to create 'official histories' (Duncan and Lambert 2002: 264).

The second narrative is polysemic and uncertain, comprised of the unruly subjectivities of diasporic knowledges, which are defined not against the other, but through continuities, complexities, linkages and hybridities. These rhizomatic networks, entanglements, divergences and differences produce transcultural histories which focus on 'routes' rather than 'roots' thereby contesting the simplicity of patriotic 'nation-building' narratives (Hall 1990; Gilroy 1993; Massey and Jess 1995; Clifford 1997).

## Meta-narratives: immigration and the nation

Many critics have argued that the histories presented at Ellis Island are 'bogus' histories, designed to foster national unity through mythologizing immigrants and their 'national' achievements (Ball 1990: 59). Kirshenblatt-Gimblett (1998: 9), for example, has called the museum 'a repository of patriotic sentiment . . . [and] an exemplar of institutional memory under the aegis of corporate sponsorship'. For Bodnar too 'the popular understanding of Ellis Island tends to be directed more towards notions of patriotism than justice'. In both 'the popular imagination and official mind' he protests that 'immigrants emerge in the timeless and sacred abstraction of patriots' (Bodnar 1995).

Patriotic versions of history at Ellis Island have thus been dependent on a careful use of imagery in the museum and strategic aesthetic manipulation of the historical landscape in order to police the meanings extracted from the site. For instance, during an interview an Immigration and Naturalization Service (INS) historian suggested to me that the strategy of returning the building to its appearance during the narrow 1892–1924 period of mass immigration had cleverly avoided dealing with questions of immigration *restriction*. A series of deliberately encoded aesthetic erasures in the historic landscape (including the removal of 'prison-like' features such as high chain link fences, outdoor detention pens and 'cattle pen'- like structures in the 'Great Hall'), serves to deter uneasy visitor questions about 'alternative' uses of the building and the treatment, detention and deportation of immigrants. For Frisch and Pitcaithley (1990: 224) then, 'the political deportation of radicals, such a substantial and disgraceful feature of Ellis Island's history during and after World War I [is] . . . barely mentioned'.

A careful use of fabricated exhibitory also plays a crucial role in reiterating a standardized conceptualization of migration to America. So for instance, a clever 'optical illusion' called the *Flag of Immigrant Faces* is seen as a series of individual monochrome faces of many ethnic backgrounds when viewed from one angle but looks like an American flag when looked at from another (Plate 10.4; Smith 1992). Kirshenblatt-Gimblett (1998: 184) sees this as a representation that effectively subordinates individuality and difference by appropriating immigrants into a larger project of nation-building:

> Unhuddled, each . . . face distinct, the tired and the poor of 'The Great Colossus' are disciplined by the orderly . . . grid, [an] instrument . . . that appear[s] to treat all citizens equally [whilst also] . . . neutralizing significant differences by virtue of [its] arbitrary and repetitive structure . . .

The American Immigrant Wall of Honour found in the grounds of the museum has also been a source of criticism. Conceptualized by Lee Iacocca as a fund-raising tool, the wall has been seen as a major intrusion into the historic landscape, an extraordinary example of visitors being 'duped' into 'purchasing history on credit':

> American Express invites its cardholders to honour an ancestor by inscribing the persons name on the American Immigrant Wall of Honour . . . Not death on the battlefield of war or disease, but only cardholder status and a minimum of one hundred dollars charged to Amex is required.
>
> (Kirshenblatt-Gimblett 1998: 181)

Similarly, *Village Voice* journalist Edward Ball (1990: 87) compares the wall to a 'voluntary version of the Vietnam War Memorial' for those who 'can't be inscribed anywhere else in history'. 'Like battlefield casualties on a war memorial', Bodnar (1995) too, complains that the 'complex lives (of immigrants) are simplified, abstracted and connected to ideals of patriotic sacrifice' so that 'the citizen is linked to the nation (and) private history becomes public memory'.

Patriotic discourses at Ellis Island are assimilatory and abstract, failing to distinguish between individual migrants, instead creating a generic immigrant stereotype. Patriotic narratives draw on an imaginative ordering and disciplining of time (Crang 1994: 29). To focus on Ellis Island as a place of beginnings is a strategy that effectively erases the complex memories of the Old World by cutting off histories from their mythological and primeval origins abroad.

Ellis Island can be viewed psychoanalytically as a holding space, a substitute for family memories and sets of social relations, interconnections and affiliations that stretch across the globe: a place of 'roots' for the manifestly 'uprooted' (see Handlin 1951). Patriotic immigrant histories at Ellis Island have also been found guilty by critics of valorizing 'white' European-American discourse, whilst subordinating the stories of people who came to America under different circumstances in ways that obscure important differences (Lippard 1997; Kirshenblatt-Gimblett 1998).

Just as the Flag of Immigrant Faces and the Wall of Honour essentializes immigrants, historian Mike Wallace has argued that former US president Ronald Reagan (with the help of celebrity philanthropist Lee Iacocca) tried during the restoration to 'fix the meaning of the 'immigrant experience' . . . in a campaign of narrow ideological self-justification . . . (which included) defin[ing] the historical immigrant experience in a way that facilitated contemporary anti-immigrant politics' (Wallace 1996: 57).

Political appropriation of the 'immigrant experience' continues long after the museum opened to the public. Ellis Island has routinely been used as a 'global public stage' (Urry 2000: 151) on which public figures including Hilary Clinton, Mayor Giuliani and President George W. Bush have, with the help of the media, mobilized the nation through spectacle, delivering speeches linking the past to the present through invocations of progress and manifest destiny.

John Urry (2000) has recently argued that a feature of the contemporary mobile world (characterized by global scapes and flows) has been the need for universal perfor-mativity on specially manufactured global public stages upon which nations must compete by soliciting public support through continually effecting performances. This was nowhere more evident than when President George W. Bush addressed the world's citizens from the Great Hall at Ellis Island on the first anniversary of 9/11. In a rousing televised speech he equated the 'war on terror' with the patriotic ideals of 'freedom and hope [that] drew millions [of Immigrants] to this harbo[u]r' (*The Guardian* 2002). Such accounts nostalgically confine the practices and experiences of migration to the past where they can be rendered legible. In reality immigration continues as Asians, Latin Americans and West Indians seek refuge from instabilities caused by an uneven globalizing market economy and America's political and economic penetration worldwide (Foner 2000: 18).

A careful reworking of the structures of feeling associated with Ellis Island is particularly apparent here. In the 1970s, Ellis Island was described as 'The New York immigration station notorious for the insolence and inhumanity of its staff' (Payton 1970: 218, in Funk 1984: 1). Frederic C. Howe, US commissioner of Immigration at the port of New York, wrote that Ellis Island was 'a storehouse of sob stories for the press; deportations, dismembered families [and] unnecessary cruelties, [making] it one of the tragic places of the world' (Howe 1925: 253, in Funk 1984: 1). However, Ellis Island is now routinely described by interpretive staff, park rangers and promoters as 'a symbol of the promise of a new beginning in a new land . . . [that] represents the hopes and dreams of all people who came and are still coming in search of the American Dream' (Statue of Liberty – Ellis Island Foundation 1990). Critical perceptions of the workings of Ellis Island have been seamlessly replaced, historically reinterpreted in the service of patriotism so that Ellis Island now officially represents a 'golden gateway to a life of freedom and opportunity' (Statue of Liberty – Ellis Island Foundation 1990).

## Diaspora, subjectivity and space at Ellis Island

In contrast to this rationalizing set of narratives, this final section turns to explore how the 'alternative' diasporic knowledges of key social groups and individuals, including

African-Americans and Japanese-Americans, have been central in critiquing and contesting the patriotic versions of immigration history presented at Ellis Island. By invoking links to multiple places and cultures, these groups have been able to assist in the critical recovery and return of formerly marginalized immigrant histories, presenting versions of ethnic identity which cannot easily be confined to singular forms of national identification and belonging.

The recovery of diasporic histories by certain influential African-American actors has been central in challenging facets of the 'elaborate mythology' that has grown around immigration at the turn of the century at Ellis Island (Foner 2000: 2).

## Forced migrations and the African-American diaspora

America's immigrant history has often wrongly been collectively imagined as having white or European roots (Dyer 1997; Chavez 2001: 25; Forest 2002). Official histories of migration have often ignored the mostly forced immigration of Africans, which was underway well before the 'great wave' of European migration touched the shores of the USA, peaking in the early decades of the twentieth century (Foner 2000). The majority of forced immigrants – numbering nearly 300,000 – arrived in the USA between 1741 and 1808, the year in which the importation of slaves was banned. As a legacy of an enforced passage to America, many African-Americans have endured long histories of discrimination and exclusion (Agnew and Smith 2002: 231). This has included the use of silence and various forms of *forgetting* rather than confrontation of the issues surrounding the Atlantic slave trade (Chivallon 2001: 350).

During museum production, the National Park Service attempted to reflect the complexity and ambiguity of this multi-faceted history of migration and its impact on American national identity. This was largely a result of the revised National Park Service thematic framework introduced in 1987, which in turn was influenced by the academic insights of the 'New American History' (Foner 1991). Because of this, the Historians' Advisory Committee, a voluntary group of prominent new social historians from across America was consulted over the intellectual content of the exhibitory from the very beginning (Blumberg 1985; Wallace 1996).

The historians petitioned that the museum should focus on the larger history of mobility, rather than focussing exclusively on the memories of immigrants who passed through Ellis Island. Not to include the histories of groups such as 'forced migrants' or contemporary migrants who arrive from Africa and Asia by plane rather than steamship would be 'educationally devastating' the historians argued:

> I remember one morning we were discussing all of . . . [the museum themes] . . . and someone in the group . . . said 'now you know, let's envision a class of New York City public school students coming to Ellis Island, including many little African-American kids, what's here for them? What are they to make of all of this?' And it was a wonderful question well put, because it got us to think about who would be coming to Ellis Island and how could we present this in a way that would be inclusive and accurate and at the same time engaging and engaging a very broad public.
>
> (Alan Kraut, Historians' Advisory Committee, October 2001)

*Plate 10.4* The 'Millions on the Move. Worldwide Migration' exhibit at the Ellis Island Immigration Museum.

Historians and Park Service interpreters decided to include graphics and accounts of the transportation of slaves to America in two displays called *Forced Migration: The Atlantic Slave Trade* and *Settlers, Servants, Slaves: Immigration before 1780*. For some African-American members of staff, such as Daniel Brown, Chief of Interpretation, however, much of the museum narrative remains largely incongruous for the large numbers of school children from non-European-American backgrounds that visit Ellis Island as part of their schooling (cf. Plate 10.4):

> It is a challenge for me to know that when (school groups from predominantly Native-American, Hispanic or African-American areas) are asked to respond to certain questions that we give out, there is no connection to them and their family heritage at Ellis Island . . . [T]hat bothers me a little because I know that it is talking about a narrow perspective of immigration and not . . . the whole big picture. But I'm the type of person that I know that that is a challenge and if I can make any inroads in presenting that here at Ellis Island or at the Statue of Liberty, that's a responsibility for me as an African-American. But I also have a responsibility as a National Park Service employee to tell the 'untold story' and then tell the 'real story,' so it's a dichotomy you bring.
> (Danni Brown, Chief of Interpretation, April 2002)

The *double consciousness* (see DuBois 1989; Gilroy 1993) Daniel feels personally as an African-American and professionally as a National Park Service employee is also experienced by Diane Dayson, Superintendent of the Statue of Liberty – Ellis Island site:

> One of the things that I said when I got here was that I am really not excited about Ellis Island because I'm African-American. I know how my relatives got here – my decedents and it was a struggle and they came under duress. It's not that they wanted to come here. They had no choice . . . [T]here were many Africans who came through Ellis Island and many Caribbean [immigrants] and that is a story that is not told and needs to be told. One of the goals for next year for African-American month . . . will be planning an exhibit around the Caribbean immigrants and African immigrants who came through Ellis Island to make that connection because the majority of the African-American community are not aware of that . . . I think it is important for people to be educated on a much higher level and a much broader spectrum no matter how uncomfortable it is for them.

She continues:

> When we have exhibits that . . . don't focus in on the European experience . . . I know that that creates lots of controversy . . . History is not a happy story. There is the good side of our history and then there is the bad side of our history and everyone likes to look at the good side because it's warm, it's easy there's no confrontation. But however I like to push for the bad history and the not so comfortable [history] because I strongly feel it helps people to grow and it helps people to get out of their comfort zones and to appreciate others who also helped to build this country and make it what it is. . . .
>
> (Diane Dayson, Superintendent of the Statue of Liberty –
> Ellis Island National Monument, February 2002)

Similarly, Charles, an African-American guide, employs his own personal subjectivities in his tours, weaving diasporic knowledges cleverly and subtly into the official park 'themes' outlined in the site's disciplining comprehensive interpretive plan. For instance, he brings in discussion about 'the American tragedy called the Civil War' as well as a discussion of the civil rights movement and the changing attitudes of 'Uncle Sam' (NPS guided tour 2002). Essentially, Charles' tours are unique in that they offer a much more critical and nuanced interpretation of American history than those given by many of the other guides. Thus collectively, the African-American employees have been able to begin to deploy particular strategies and tactics with which to fragment the museum's dominant European bias.

## American concentration camps: the Japanese-American experience

Curators at the Japanese-American museum in Los Angeles have also been influential in challenging the kinds of histories presented at Ellis Island. In 1997, an exhibit called *America's Concentration Camps: Remembering the Japanese-American Experience* was temporarily installed at Ellis Island. National Park Service officials and some anonymous Jewish groups were said to object to the use of the term 'concentration camp', suggesting that most Americans would immediately associate this term with Nazi death camps during the Holocaust. It was argued that using it to describe the internment of Japanese-Americans could diminish the horror of the Nazi slaughter (Sengupta 1998).

In a letter to Ms Ishizuka (a senior curator at the Japanese-American National Museum), Ellis Island superintendent Diane Dayson expressed concern that because 'concentration camp' connotes death camps, the 'very large Jewish community' in New York City 'could be offended by or misunderstand' the title (see Dubin 1999). The curators, however, objected to any modification of the exhibition title. This prompted US Senator Daniel K. Inouye, a Democrat from Hawaii and a member of the Japanese-American museum board, to appeal to the Secretary of the Interior, Bruce Babbity, who then oversaw Ellis Island as a national historic site. Eventually the curators were told that the exhibit could go on, with the original title.

> 'We need to call them what they were', said Karen Ishizuka, a third generation Japanese-American whose parents and grandparents were imprisoned during the war. 'The exhibit depicts an episode in American history that too few people understand . . . This happens to be our experience and it is our responsibility to tell it the way we experienced it'.
>
> (Sengupta 1998: 35)

The exhibit described the experiences of 120,000 Americans of Japanese ancestry incarcerated due to an executive order signed by President Roosevelt two months after the attack on Pearl Harbour in December 1941. The order required Japanese-Americans, mostly on the west coast to leave their homes. Most spent the rest of the war years in one of ten camps, from Manzanar in California to Jerome, Arkansas. Many lost the properties and businesses that they left behind (Zinn 1996; Lippard 1997: 67). It was not until the passage of the Civil Liberties Act of 1988 that former inmates were granted reparations and offered an official government apology. Given the experiences of Japanese-Americans, curators at the Japanese-American National Museum in Los Angeles defended the term 'concentration camp'. They argued this was a 'historically accurate' description of groups, who were treated unconstitutionally and who were forced to live in desolate camps in seven states west of the Mississippi solely on the basis of their ancestry (Sengupta 1998: 35).

Crucially, in an *Ellis Island News* fact sheet prepared by the Statue of Liberty – Ellis Island Foundation, Irene Hirano, President of the Japanese American National Museum, observed how the histories of mass incarceration remained relatively *unknown* to most Americans, especially those living on the east coast, the South and Midwest: 'Presenting the exhibition at Ellis Island provid[ed] a unique opportunity to make . . . information available to many new audiences – national, international and school groups – that may not be familiar with the mass incarceration of Japanese-Americans during WWII' (Statue of Liberty – Ellis Island Foundation, 1998).

While this temporary display was exhibited without contestation on the west coast, at a hegemonic monument such as Ellis Island, it was rendered controversial (Dubin 1999). Significantly, the exhibit also explored Ellis Island's role as a detention facility for Japanese-Americans, as well as German-Americans and Italian-Americans in the

Second World War, thereby uncovering important *alternative* histories of Ellis Island not included in the official interpretive plan.

## Conclusions

Many diasporic groups have sought to claim Ellis Island as a *sacred space* or 'terrain of belonging' (Fortier 2000: 175) where their wide-ranging autobiographical experiences and memories of migration may be narrated to an (inter)national audience. Whilst much recent literature has explored the multifarious ways in which patterns of movement, exile and diaspora inform constructions of place and locality (Dawson and Johnson 2001: 319), such studies have usually been confined to the study of one diasporic group and its inhabitation of one particular space.

This chapter has explored the difficulties of attempting to represent *more than one* hyphenated community at a single site. As we have seen, Ellis Island is qualitatively different from many other diasporic spaces in that it is one of the few tourist sites which attempts an ontological reconciliation of complex multiple readings of diasporic movements. Along the way difficult value judgements over inclusion and exclusion have been grappled with, decisions which in many ways *reproduce* the literal exclusions of groups such as Asian and African migrants (Kirshenblatt-Gimblett 1998: 181).

The Ellis Island Immigration Museum attempts to correct the exclusions of the past by finding a place for those who were never allowed to pass through its doors when the island was still in the immigration business; the stories of most Americans bypass the island.

Thus, while advocates of patriotic versions of history within the museum have attempted to discipline the subjectivities of diasporic knowledges and incorporate them into singular and official immigration meta-narratives, they have not always succeeded.

For example, key African-American actors have been central in contesting the 'nation of immigrants' approach and have been instrumental in overseeing the critical recovery and return of formerly suppressed memories of less glorious types of migrations. Japanese-American exhibit producers have also raised important political questions about the (mis)treatment of immigrants on the basis of race. For these groups, to have their ancestral memories incorporated into the museum is to claim space for their identities within the imaginary boundaries and terrains of belonging (see also Sengupta 1997a,b on the Armenian Diaspora exhibit at Ellis Island). In a post 9/11 perspective, where there has been a tendency towards a very real 'hardening of geographies and identities' (Smith 2002: 100), this is particularly important.

For Hardt and Negri (2000: 213), '[e]ven the most significant population movements of modernity (including the black and white Atlantic migrations) constitute Lilliputian events with respect to the enormous population transfers of our times'. If commemorative sites are to engage with and represent adequately the raft of new diasporic subjectivities created by the epic 'population transfers' predicted by Hardt and Negri, they will need to excavate not just histories, but *nomadologies* in which mobile, multicultural identities are seen through the perspective of geographical connectivity as central to the production of national identities. It is these complex, fragmented and extra-national narratives, constructed from networks, flows and mobile genealogies that heritage professionals must reveal in the future if they are to avoid marginalizing the new hyphenated communities of the twenty-first century.

## References

Agnew, J. and Smith, J. (eds) (2002) *American Space/American Place: Geographies of the Contemporary United States*, Edinburgh: Edinburgh University Press.

Appadurai, A. (1987) 'Putting hierarchy in its place', *Cultural Anthropology* 3(1): 36–49.

Appadurai, A. (1990) 'Disjuncture and difference in the global cultural economy', *Public Culture* 2(2): 1–24.

Ball, E. (1990) 'Museum of tears', *Village Voice*, 11 September, 35(37): 59, 87.

Bender, B. (2001) 'Introduction', in B. Bender and M. Winer (eds) *Contested Landscapes: Movement, Exile and Place*, Oxford: Berg.

Bhabha, H.K. (1990) *Nation and Narration*, London: Routledge.

Blumberg, B. (1985) *Celebrating the Immigrant: An Administrative History of the Statue of Liberty National Monument, 1952–1982*, New York: Institute for Research in History.

Bodnar, J. (1995) 'Remembering the immigrant experience in American culture', *Journal of American Ethnic History* 15(1): 3–27.

Chambers, I. (2001) *Migrancy, Culture, Identity*, 2nd edn, London: Routledge.

Chavez, L.R. (2001) *Covering Immigration: Popular Images and the Politics of the Nation*, London: University of California Press.

Chivallon, C. (2001) 'Bristol and the eruption of memory: making the slave-trading past visible', *Social and Cultural Geography* 2(3): 347–63.

Clifford, J. (1997) *Routes: Travel and Translation in the Late Twentieth Century*, London: Harvard University Press.

Clifford, J. (1998) *The Predicament of Culture: Twentieth-Century Ethnography, Literature and Art*, Cambridge, MA: Harvard University Press.

Crang, M. (1994) 'Spacing time, telling times and narrating the past', *Time and Society* 3(1): 29–45.

Crang, M. (2000) 'Playing nymphs and swains in a pastoral idyll?', in A. Hughes, C. Morris and S. Seymour (eds) *Ethnography and Rural Research*, Cheltenham: Countryside and Community Press in association with the Rural Geography Study Group of the RGS-IBG.

Creswell, T. (2001) *The Tramp in America*, London: Reaktion Books.

Dahlman, C. (2002) 'Reviews: covering immigration: popular images and the politics of the nation by L.R. Chavez', *Environment and Planning D: Society and Space* 20: 499–504.

Daniels, R. (1997a) *Not Like Us: Immigrants and Minorities in America, 1890–1924*, Chicago: Ivan R. Dee.

Daniels, R. (1997b) 'No lamps were lit for them: Angel Island and the historiography of Asian American immigration', *Journal of American Ethnic History* 17(1): 3–18.

Dawson, A. and Johnson, M. (2001) 'Migration, exile and landscapes of the imagination', in B. Bender and M. Winer (eds) *Contested Landscapes: Movement, Exile and Place*, Oxford: Berg.

Deleuze, G. and Guattari, F. (1987) *A Thousand Plateaus: Capitalism and Schizophrenia*, Minneapolis, MN: University of Minnesota Press.

Dubin, S.C. (1999) *Display of Power: Memory and Amnesia in the American Museum*, New York: New York University Press.

DuBois, W.E.B. (1989) *The Souls of Black Folk*, London: Penguin (reprinted from 1903 original edition).

Duncan and Lambert (2002) 'Landscape, aesthetics and power', in J. Agnew and J. Smith (eds) *American Space/American Place: Geographies of the Contemporary United States*, Edinburgh: Edinburgh University Press.

Dwyer, C. (1999) 'Migrations and diasporas', in P. Cloke, P. Crang and M. Goodwin (eds) *Introducing Human Geographies*, London: Arnold.

Dyer, R. (1997) *WHITE*, London: Routledge.

Feng, K. and Page, S.J. (2000) 'An exploratory study of the tourism, migration–immigration nexus: travel experiences of Chinese residents in New Zealand', *Current Issues in Tourism* 3(3): 246–81.

Foner, E (ed.) (1991) *The New American History*, Philadelphia, PA: Temple University Press.

Foner, N. (2000) *From Ellis Island to JFK: New York's Two Great Waves of Immigration*, London: Yale University Press.

Forest, B. (2002) 'A new geography of identity? Race, ethnicity and American citizenship', in J. Agnew and J. Smith (eds) *American Space/American Place: Geographies of the Contemporary United States*, Edinburgh: Edinburgh University Press.

Fortier, A. (2000) *Migrant Belongings: Memory, Space, Identity*, Oxford: Berg.

Frisch, M. and Pitcaithley, D (1990) 'Audience expectations as resource and challenge: Ellis Island as a case study', in M. Frisch (ed.) *A Shared Authority: Essays on the Craft and Meaning of Oral and Public History*, Albany, NY: State University of New York Press.

Funk, S.M. (1984) 'Ellis Island, the island of tears: a new look', unpublished paper Washington, DC: The American University.

*The Guardian* (2002) 'We will not relent until justice is done', *The Guardian*, 12 September.

Gilroy, P. (1993) *The Black Atlantic: Modernity and Double Consciousness*, Cambridge, MA: Harvard University Press.

Hall, S. (1990) 'Cultural identity and diaspora', in J. Rutherford (ed.) *Identity: Community, Culture, Difference*, London: Lawrence and Wishart.

Handlin, O. (1951) *The Uprooted*, Boston: Little, Brown.

Hardt, M. and Negri, A. (2000) *Empire*, London: Harvard University Press.

Hayley, A. (1976) *Roots: The Saga of an American Family*, Garden City, NY: Doubleday.

Hewison, R. (1987) *The Heritage Industry: Britain in a Climate of Decline*, Andover: Methuen.

Holland, R. (1993) *Idealists, Scoundrels and the Lady: An Insider's View of the Statue of Liberty-Ellis Island Project*, Chicago: University of Illinois Press.

Howe, F.C. (1925) *Confessions of a Reformer*, New York: Charles Scribner's Sons.

Ifekwunigwe, J. (1999) *Scattered Belongings: Cultural Paradoxes of 'Race,' Nation and Gender*, London: Routledge.

Ioannides, D. and Cohen Ioannides, M.W. (2002) 'Pilgrimages of nostalgia: patterns of Jewish travel in the United States', *Tourism Recreation Research* 27(2): 17–25.

Johnson, L. (1984) 'Ellis Island: historic preservation from the supply side', *Radical History Review* 28/29/30: 157–68.

Johnson, N. (1995) 'Cast in stone: monuments, geography and nationalism', *Environment and Planning D: Society and Space* 13: 51–65.

Kang, S.K.-M. and Page, S.J. (2000) 'Tourism, migration and emigration: travel patterns of Korean-New Zealanders in the 1990s', *Tourism Geographies* 2(1): 50–65.

Kirshenblatt-Gimblett, B. (1998) *Destination Culture: Tourism, Museums and Heritage*, London: University of California Press.

Kraut, A. (1982) *The Huddled Masses: The Immigrant in American Society, 1880–1921*, Arlington Heights, IL: Harlan Davidson.

Lew, A. and Wong, A. (2002) 'Tourism and the Chinese diaspora', in C.M. Hall and A. Williams (eds) *Tourism and Migration: New Relationships between Production and Consumption*, Dordrecht: Kluwer.

Lippard, L. (1997) *The Lure of the Local: Senses of Place in a Multi-centered Society*, New York: The New Press.

Lowenthal, D. (1996) *The Heritage Crusade and the Spoils of History*, London: Viking.

Malkki, L. (1992) 'National geographic: the rooting of peoples and the territorialization of national identity among scholars and refugees', *Cultural Anthropology* 7(1): 24–44.

Massey, D. and Jess, P. (eds) (1995) *A Place in the World*, Oxford: Oxford University Press/ Open University Press.

Mirzoeff, N. (2000) *Diaspora and Visual Culture: Representing Africans and Jews*, London: Routledge.

Moreno, B. (2001) *Ellis Island Chronology Timeline (1674–2001)*, New York: Statue of Liberty – Ellis Island National Monument.

Namer, G. (1991) 'Mémoire sociale, temps social, lien social', Actes du XIIIe Colloque de l', *AISLF* (l'Association internationale des sociologues de langue française) 1: 214–22.

Nash, C. (2002) 'Genealogical identities', *Environment and Planning D: Society and Space* 20: 27–52.

Nguyen, T.H. and King, B. (1998) 'Migrant homecomings: Viet kieu attitudes towards travelling back to Vietnam', *Pacific Tourism Review* 1: 349–61.

Nguyen, T.H. and King, B. (2002) 'Migrant communities and tourism consumption: the case of the Vietnamese in Australia', in C.M. Hall and A. Williams (eds) *Tourism and Migration: New Relationships between Production and Consumption*, Dordrecht: Kluwer.

Payton, G. (1970) *Webster's Dictionary of Proper Names*, Springfield, MA: G and C Merriman Co.

Perec, G. (1999) *Species of Spaces and Other Pieces*, London: Penguin Books.

Perec, G. and Bober, R. (1995) *Ellis Island*, New York: The New Press.

Rogoff, I. (2000) *Terra Infirma: Geography's Visual Culture*, London: Routledge.

Said, E.W. (1979) 'Zionism from the standpoint of its victims', *Social Text* 1: 7–58.

Said, E.W. (1986) *After the Last Sky: Palestinian Lives*, New York: Pantheon.

Said, E.W. (2000) *Reflections on Exile and Other Literary and Cultural Essays*, London: Grant Books.

Seitz, S. and Miller, S. (2001) *The OTHER Islands of New York City*, Woodstock, VT: The Countryman Press.

Sengupta, S. (1997a) 'At Ellis Island museum, dispute on Armenia show: massacre photographs deemed 'too gory'', *New York Times*, 11 September: B3, 5.

Sengupta, S. (1997b) 'Ellis Island, yielding, permits photos of Armenian massacre', *New York Times*, 14 October: B2, 5.

Sengupta, S. (1998) 'What is a concentration camp? Ellis Island exhibit prompts a debate', *New York Times*, 8 March: 35.

Sherman, D.J. and Rogoff, I. (1994) *Museum Culture: Histories, Discourses, Spectacles*, London: Routledge.

Smith, J. (1992) 'Exhibition review: celebrating immigration history at Ellis Island', *American Quarterly* 44(1): 82–100.

Smith, N. (2002) 'Scales of terror: the manufacturing of nationalism and the war for US globalism', in M. Sorkin and S. Zukin (eds) *After the World Trade Center: Rethinking New York City*, New York: Routledge.

Statue of Liberty – Ellis Island Foundation (1990) Archived promotional leaflet.

Statue of Liberty – Ellis Island Foundation (1998) *World War II Incarceration of Japanese-Americans Featured in Ellis Island Exhibit*, New York: Statue of Liberty – Ellis Island Foundation.

US Government Printing Office (1992) 'Joint resolution designating 1 January, 1992, "National Ellis Island Day"', Washington, DC: US Government Printing Office.

Urry, J. (2000) *Sociology Beyond Societies: Mobilities for the Twenty-First Century*, London: Routledge.

Vecoli, R. (1983) Private communication with Mr James Watt, Department of the Interior, 19 May.

Wallace, M. (1986) 'Ronald Reagan, Ellis Island and history of immigration', *Organization of American Historians Newsletter*, November: 1–3.

Wallace, M. (1987) 'Hijacking history: Ronald Reagan and the Statue of Liberty', *Radical History Review* 37: 119–30.

Wallace, M. (1991) 'Exhibition Review: Ellis Island', *Journal of American History* 78: 1023–32.

Wallace, M. (1996) *Mickey Mouse History and Other Essays on American Memory*, Philadelphia: Temple University Press.

White, E.J. (2002) 'Forging African diaspora places in Dublin's retro-global spaces: minority making in a new global city', *City* 6(2): 251–70.

Wright, P. (1985) *On Living in an Old Country*, London: Verso.

Zinn, H. (1996) *A People's History of the United States: From 1492 to the Present*, London: Longman.

# 11 The culture of tourism in the diaspora

## The case of the Vietnamese community in Australia

*Thu-Huong Nguyen and Brian King*

### Introduction

Over the past 50 years, the number of migrants and migrant communities has increased significantly. Diaspora has become a potent force in shaping the linkages between the First and Third Worlds. Notwithstanding, relatively little research has examined the travel characteristics of non-European diasporic communities and the wider implications of such travel. In particular, the connection between Southeast Asian diasporic culture and travel behaviour has remained largely unexplored (cf. Nguyen 1996, 2002; Nguyen and King 1998, 2002; Nguyen *et al*. 2002; Lew and Wong 2002, Ch. 13). This knowledge gap drives the current chapter.

For members of diasporic communities, issues of identity and meaning are ever present as they negotiate their existences. Given the fundamental and self-reinforcing interface between culture, identity and meaning and travel, we argue that discourse on diasporas and tourism should be informed by more nuanced readings of 'Eastern' worldviews. More generally, King (1994: 174) argues that migrant travellers display 'a sense of belonging to or identifying with a way of life that has been left behind'. The role of family connections and shared cultural values has been identified as one aspect to develop clearer understanding of the relationship between ethnic tourism and migration. Family connections and shared culture do not have even influence between and among diasporic communities. For instance, as Nguyen (1996) notes, trips may be prompted by a sense of obligation or compulsion, not least in Eastern cultures. However, little subsequent research has examined the cultural settings and underlying cultural factors which propel the process of diaspora travel, especially in Asian cultures that emphasize duty and piety.

The Vietnamese diaspora experience is under the microscope here. The *Viet kieu* are those people who carry with them Vietnamese cultural heritage, who have taken citizenship of other countries and who live (permanently) outside the territory of Vietnam. Such migrants share a strong sense of common origin, history and culture through the experience of the physical and emotional traumas of migration. This chapter examines the influence of cultural factors on the travel behaviour of the *Viet kieu* in Australia with particular reference to return visits to Vietnam. It explores the factors that motivate trips and that contribute to the shape a trip takes. A conceptual framework is proposed which provides a basis for examining the relationship between *Viet kieu* 'adapted culture' and travel behaviour.

## Vietnamese diaspora and major migrations

The *Viet kieu* and their ancestors were separated from Vietnam in two major waves. As a precursor, after the French occupation of Vietnam in 1858, a stream of Vietnamese, many of whom completed their education in France, travelled to Europe in search of business and commercial contacts with Europeans (Karnow 1983).

Civil war in 1954 and the ensuing period of truce prompted the first main wave of out-migration. Many Vietnamese decided to flee the country after the defeat of the French. Those who had assisted the French had an opportunity to move to South Vietnam when the country was partitioned along the Seventeenth Parallel. Though precise numbers are unknown, many Vietnamese moved to other countries altogether (Nguyen 1996). Some left because of their dislike of a Marxist-oriented government, which was antagonistic to private wealth and external cultural influences. Many others chose to leave because they assisted the French colonial government, they held strong allegiances to France and they feared reprisals by the new regime. In fact, through the process of dislocation, most refugees sought safety in France and its territories. The outcome of the process was, however, a stark one; many of these migrants have lived overseas for nearly 50 years and there have been at least two subsequent generations of *Viet kieu*, whose migratory roots are associated with this period and who have been born outside Vietnam (Nguyen 1996).

The second major wave of out-migration resulted from the civil war which ended in 1975. About 2 million Vietnamese left Vietnam and are now living in different countries scattered around the world (Hiebert 1993). Many are in the USA, Canada, the UK and Australia. Small communities also exist in Norway, Japan, Hong Kong and Germany. State-sponsored suspicion of Vietnamese expatriates and discouragement of large-scale visitation virtually precluded them from Vietnam in the 15 years after the war ended. Many feared return because entering Vietnam may have led to prosecution for having left the country 'illegally', i.e. without permission (Nguyen 1996). Notwithstanding, those who fled up to three decades ago have also subsequently raised families overseas. They and their children are now of an age where they are able to travel 'freely' to Vietnam and hence they constitute a potentially large and lucrative travel market for Vietnam.

## The *Viet kieu* community in Australia: a snapshot

Australia's Vietnamese population forms an important part of the country's evolving multi-cultural history as well as the wider Vietnamese diaspora. Most Vietnamese migrants arrived after the civil war ended. This coincided with the growing realization of Australia's active geopolitical participation in Asia and the value of providing assistance with the maintenance of ethnic identification after settlement. Australia's *Viet kieu* proved to be a test case for both the policy of accepting large numbers of Asian refugees and for the dream of multi-cultural harmony (Viviani 1984). In fact, they constitute one of Australia's newest ethnic groups and one of six migrant communities exceeding 100,000 people (BIMPR 1996). They are the largest Asian-born group to have arrived under the Australian refugee programme and now form one of the largest non-English-speaking background (NESB) communities in Australia. The total number of *Viet kieu* is estimated at around 197,800, or approximately 1 per cent of Australia's population. Of these 151,100 are first-generation and 46,800 second-generation. When compared with the much larger numbers of Italians, Greeks and Chinese, they are a small group. Over 90 per cent of the group has arrived in a single concerted wave since 1975 in contrast to other migrant groups who arrived over extended periods. The *Viet kieu* are highly urbanized in comparative terms, with most members concentrated in Sydney and Melbourne.

### The basic influences of home culture on travel

Shared Vietnamese cultural values provide a first step in unravelling the linkages between those who live in the diaspora, those who have remained in Vietnam and their potential connection through the medium of travel and tourism.

The commonly held assumption that transformation in identity, values and beliefs only occurs after Vietnamese people arrive in the West is a popular misconception. Culture in the 'home country' is in constant flux. Vietnamese traditions have been exposed to a range of Asian cultural influences as well as Western ones (e.g. French, American). Nevertheless, primary importance is attached to family, kinship ties and obligations; the considerations and expectations that surround marriage; and the maintenance of religious duties to ancestors and relevant spiritual beliefs. The precepts of Confucianism are reflected in rules of behaviour that guide all relationships. They establish clear and binding responsibilities and decision-making protocols. Autonomy and individualism is virtually absent from this world. The self is negotiated through relationships, obligations and protocols. Meaning, identity and social stability are dependant upon, and derive from, the maintenance of this primary structural context (Nguyen 2002).

Religion and spirituality play an important role. The 'three religions' exert a profound influence on both individual and communal behaviour (Nguyen and King 2002). Confucian teaching sets out moral and ethical ideas on relationships. Equally, the pervasive nature of ancestral spirits allows *Viet kieu* to feel connected with each other and with the homeland while in diasporic communities. Viviani (1984) has noted that for those who are away from their homeland, having an ongoing sense of connection with ancestors may help the process of adjustment to the new environment. Thus, a range of travel behaviours and attitudes may be influenced by spiritual beliefs and religious practices, personal interests, family ties and obligations (Nguyen 1996). These include the reasons for travel; the time of travel; the choice of destination; decisions about who should travel first; where and when to travel; and who should make travel decisions.

Family and community play a crucial role in providing a sense of 'home' for the *Viet kieu*. There is a noticeable sense of a shared cultural identity and an obligated desire to maintain and foster their cultural traditions. Living in the diaspora, the *Viet kieu* have also faced cultural alienation, feelings of insecurity and the need to cope with trauma and nostalgia for family and homeland. These feelings are reinforced by self-doubt and their unease with cultural difference. Family and community fulfil cultural needs by acting out the role of surrogate homeland to reinforce traditional family values and self-identity (Viviani 1984). The crucial place of kinship ties in providing security and order, as well as the norms surrounding marriage and family obligations, often precipitate trips and influence destination choices, particularly where these incorporate visiting friends and relatives. Thus, a return trip to Vietnam is often interpreted as helping *Viet kieu* to maintain a balanced life based on the home country and to resolve identity-related issues in the diaspora.

## The search for meaning in diasporic communities

Diaspora refers to collective trauma, or a banishment into exile and longing to return home (Cohen 1997). A person's adherence to a diasporic community is demonstrated by an acceptance of an inescapable link with past migration history and a sense of co-ethnicity with others from a similar background. The 'old country', defined in terms of language, religion and customs, exerts some claim upon one's loyalties, emotions and identity. All diasporas are, therefore, in part cultural (Urry 2000) and cannot persist without much corporeal, imaginative and increasingly virtual travel both to that homeland as well as to other sites with the diaspora (Kaplan 1987). Clifford (1997) argues that dispersed people effectively find themselves in what he terms 'border relations' with the old country. The ambiguity of diasporic identity is further emphasized by Cox (1980), for whom migrants exist on the borders of two cultural worlds, but full members of neither (cf. Ch. 2). They linger at the intersection of self-identification between the former and adopted countries, wondering which side offers them greater acceptance and are sometimes confused about their own feelings of belonging. Most migrants go through a period of tentative evaluation and culture shock, when they are constantly comparing the present with the past in unfavourable terms. They experience instantaneous time connected to a lack of confidence about the future, but a remarkable appeal of the past that is related to nostalgia. Migrants may regard the country that they have left behind with a mixture of nostalgia and anxiety and such opinions may reflect or compound the attitudes to travel prevalent amongst migrant communities.

### *Nostalgic yearnings*

Nostalgia is a widespread phenomenon among migrants and can colour the images that potential travellers have towards their homeland. Commenting on migrants generally, Prevot (1993: 240) writes,

> For a long time immigrants' needs may centre on keeping in touch with the home country through nostalgic festivities, patriotic commemorations or even temporary trips home. Sometimes traditions and rituals that have disappeared in the home country are kept alive in the migrant communities. Migrants are increasingly torn between the desire to preserve their culture and the need to come to terms with the standards and customs of the host society.

Nostalgia is more than memory; it is memory with the pain removed. It provides a sense of continuity which helps migrants to overcome feelings of separation. It focuses on what is considered the best time of one's life and may serve as a coping mechanism when times are hard. It involves a bittersweet longing for an idealized past which no longer exists (Davis 1979; Chase and Shaw 1989). The past for which one feels nostalgia belongs overwhelmingly to personal experience and is a positive perspective (Davis 1979; Holbrook and Schindler 1991, 1994). It may reflect a negative appraisal of self in the present (Davis 1979; Kamptner 1989). Its rise has been ascribed to a pervading sense of alienation and fragmentation (Haraven and Langenbach 1981; Kaplan 1987; Strauth and Turner 1988; Laenen 1989; Kasinitz and Hillyard 1995). Many migrants downplay the present and focus on the past for a sense of security, control and confirmation of identity. Trust and confidence in the future is undermined; social life in the present appears profoundly disappointing; and so the past appears preferable in many ways.

Baker and Kennedy (1994) distinguish between 'real' nostalgia, nostalgia for some remembered past time and 'stimulated' nostalgia, a form of vicarious nostalgia evoked by stories, images and possessions (Belk 1988; Stern 1992). Nostalgic reactions depend on variables such as the role of family and friends and the availability of nostalgic stimuli (Holak and Havlena 1992). They differ due to 'stimulus' and one's time of life (Baker and Kennedy 1994). Davis (1979) argues that nostalgia is deeply implicated in the sense of who you are, what you are about and, to some degree, where are you going. It can be used as a lens when making, re-making or simply maintaining one's identity.

Nostalgia trips have therefore been characterized as surreptitious and ambivalent by Urry (1995); that is, reflective of a reluctance to lose hold of the present and belief in the future. Such visits may offer an escape from the realities and anxieties of a world that sometimes feels out of control. The visits may highlight things that are missed and bring them back with happy memories. A visit to the homeland may replenish the sense of self and provide empowerment, belonging and direction, even if it is only temporary. Equally, some visits may be disconcerting and destabilizing as emergent diasporic orthodoxies are challenged.

## Vietnamese in Australia and nostalgia for the homeland

In understanding the relationship between the diaspora and attitudes towards travel, a thorough understanding of how migrants view events and people in their former homeland must be developed. Like other migrant groups, the *Viet kieu* have to cope with the trauma of change as well as with a sense of nostalgia that they experience for family and homeland (Viviani 1996). The feeling of nostalgia was particularly intense during the early period of Vietnamese migration, when there was minimal prospect of return to their homeland (Nguyen 1987). Cultural alienation was compounded by insecurity and by the absence of the support networks and systems prevalent in Vietnam. The challenge of cultural adjustment was most acute for the elderly who are confronted by a loss of status and respect and by the erosion of their role and authority within the traditional parent–child relationship (Viviani 1984).

Hence, for many *Viet kieu*, diasporic life is frequently described as having two faces: one looking forward and one looking back (Viviani 1996). Many older *Viet kieu* express their desire to retire in Vietnam, but are torn between the wish to remain with their family in their adopted country and the desire to return to their homeland. Concurrently, many children no longer define themselves as being Vietnamese and think of Vietnam only as the place where their parents were born (Viviani 1996; Thomas 1999). Members of the *Viet kieu* connect with one another predominantly via the medium of ties with family in their former homeland. Both Thomas (1996) and Viviani (1996) use the term 'home' as interlinked with the concept of homeland. For *Viet kieu*, the conceptions of the 'home' is inexorably linked with the past and with their identity. The concept of 'home' is emblematic of empowerment and assists people in dealing with the sense of their profound nostalgia (Thomas 1996). There are, however, enormous contradictions in the notion of homeland for the *Viet kieu*. Despite their sense of alienation and marginality, many *Viet kieu* gain a sense of empowerment by reflecting on differences evident from their homeland and their people (Thomas 1999). In certain respects, *Viet kieu* identity involves crossing the boundaries of understanding between countries with different histories, different social values and different cultural mythologies. There is an ongoing tension and dissonance between 'home' and 'away' and between the Vietnamese in Australia and in Vietnam. According to Thomas (1996), diaspora identities are bound up in the constraints and opportunities of the present. Such identities give rise to historical memories and images of the future.

Connections with the homeland and a desire to maintain Vietnamese identity can be well understood through *Viet kieu* participation in the *Tet* (Lunar festival) celebrations. *Tet* symbolizes Vietnamese identity, a desire to 'belong' and to feel comfort with others who share similar cultural meanings and who feel a common nostalgia for the homeland (Nguyen *et al.* 2002). The *Tet* celebrations constitute a display and creation of shared understanding of what it is to be a Vietnamese. *Tet* is emblematic of an array of associations with a mythical past as well as aspirations for a prosperous future. *Tet* in the diasporic community represents family wholeness in Vietnam as well as a vehicle for replacing what is viewed as lacking in the host society (Thomas 1996).

## Diasporic culture and travel behaviour

The *Viet kieu* cling to the largely collectivist character of Vietnamese culture and subscribe to its core values of family, kinship, marriage and religious duties. Nevertheless, their culture is located somewhere between the home and host cultures in three principal ways: the retention, albeit weakening, of traditional Vietnamese values; the movement towards, and sometimes the adoption of, mainstream values; and the creation of something distinct from both cultures that is transitional and meets their unique 'in-between' context. In what remains of this chapter, the extent to which these general concerns impact on the specifics of *Viet kieu* travel and tourism patterns is discussed. Specific motives for travel are identified, as well as the nature of *Viet kieu* travellers and non-travellers. The particularities of travel help to develop a more subtle understanding of the relationship between adapted culture and tourism consumption, not least for Southeast Asian diasporic communities that may share some of the cultural traits of the *Viet kieu*.

### Motives for travel to the homeland

*Viet kieu* travel between Vietnam and other countries that have significant Vietnamese communities has significant growth potential (Nguyen 1996; Thomas 1996; Viviani 1996). The desire to return to Vietnam is strong, propelled by a desire to reinforce family and ethnic ties and affirm Vietnamese identity (Blaine *et al.* 1995; Nien-chu Kiang 1995; Nguyen 1996; Nguyen and King 2002). The common incidence of return travel during the period when visits were illegal is testimony to the strength of the impulse (Blaine *et al.* 1995).

Though socially determined factors drive much *Viet kieu* travel, the use of traditional socio-economic market segmentation techniques using culturally neutral indicators such as age and wealth is unlikely to explain the phenomenon adequately (Nguyen 1996). A less Eurocentric reading is required that avoids an assumption that consumers enjoy complete freedom of choice. In societies influenced by Confucian principles, the network of obligations penetrates to almost every aspect of daily life and involves the extended family (Te 1962; Muzny 1985; Nguyen 1994). The form and extent of social obligations vary on the basis of religion, age and place in the social hierarchy and birth order. *Viet kieu* travel involves the fulfilment of family obligations, a strong element of compulsion and may require reinforcement of one's social circumstances. Hence, Confucian-based philosophy with its emphasis on obligations and family offers some explanation for the behaviour, direction, type and purpose of *Viet kieu* travel (Nguyen 1996).

An example of such cultural influence is the desire by many *Viet kieu* men to marry an 'authentic' and 'untainted' wife. This may motivate them to travel to Vietnam. Such travel involves more than who should be selected as the bride but is also a reflection of the Confucian view of the male dominance over the female. This is associated with family obligations such as filial piety and the desire to have and maintain a 'truly Vietnamese' family life. Those who fail to perform such roles may be subjected to a range of social sanctions and be regarded as in breach of central and fundamental family and religious rules. Though the consequences differ according to norms and culture of each family, they share the common feature of obligations and compulsions. A crucial issue is the extent to which the potential *Viet kieu* traveller considers him or herself to be under an obligation to make a particular trip. Compulsion has as yet been given little consideration in the literature on diaspora and tourism (Nguyen 1996).

The *Viet kieu* generally express positive attitudes towards travel back to the origin country, an impulse driven by a desire for family togetherness and the affirmation of Vietnamese identity (Nguyen 1996; Nguyen and King 1998, 2002). Explanations include nostalgia and attachment to the homeland as well as alienation and problems in adjusting to the new culture. Such powerful motive forces mean that a trip to Vietnam is perceived as desirable and attainable by most *Viet kieu*.

### *The characteristics of* Viet kieu *travel*

The *Viet kieu* started to travel to Vietnam in earnest after 1989 when the state introduced an 'open door' policy to encourage home visits by overseas Vietnamese. Based on Nguyen's (2002) work, it is possible to unravel the characteristics of *Viet kieu* travellers. This study consisted of 435 respondents, of whom 51 per cent were males and 49 per cent were females. The majority of respondents were born in Vietnam (99 per cent). They were well educated, such that over 60 per cent had tertiary qualifications and a further 30 per cent had completed secondary schooling. Almost 40 per cent described themselves as workers, with the remainder split roughly evenly between manager/administrator and professional (13 per cent), trades and salesperson (13 per cent) and students (13 per cent).

Nearly 90 per cent of respondents reported maintaining close contact with family, friends and/or relatives in Vietnam; a third claimed to have very close contact. Primary ties that linked first generation *Viet kieu* to their family and relatives in Vietnam accounted for a substantial component of the home travel market. When respondents were asked about their identity, most (over 90 per cent) felt Vietnamese, while a third expressed the feeling 'very highly'. Many expressed a strong desire to maintain the Vietnamese language as part of their Vietnamese culture. Most were happy with their lifestyle in Australia, albeit three-quarters believed they would have a 'somewhat' better life in Vietnam were they to return. This result appears contradictory. Though the majority of respondents lived happily in Australia, they were convinced sentimentally that a better life awaited back in Vietnam.

Away from the homeland, it is understandable that the *Viet kieu* 'feel attached to the soil and to the ancestral tradition of Vietnam' and 'always think about the ways of life they or their family left behind'. They have mixed feelings of sadness and pleasure, of good and bad memories. Memories of the past are sometimes evoked and nostalgia becomes more prevalent in the face of difficulties and failures, perhaps even with success.

Gift-purchasing for family, friends and relatives is a typical phenomenon amongst *Viet kieu* travellers. They travel to places suggested by family in the homeland or in the diaspora. Like other east Asian communities, this indicates the enduring influence of the extended family and obligation in the decision-making process. Beyond such motives, *Viet kieu* travel to Vietnam is mainly for the purposes of 'cultural heritage', 'family maintenance' and 'marriage', with 'cultural heritage' being considered most important. This reflects a strong cultural need among the *Viet kieu*: the need to maintain tradition and to preserve cultural, historical achievements and customs. 'Cultural heritage' could therefore be seen as a 'push' factor as well as a 'pull' factor that motivates them to travel to the country of origin. Being proud of the homeland, the *Viet kieu* may also be interested in showing their children who live and grow up outside of Vietnam a rich Vietnamese culture and history on a return trip to Vietnam.

What emerges from this reading is that in the process of adapting to a new environment, the above factors are vital for *Viet kieu* continuity, development and even existence since they are viewed as contributing to holding the family together and maintaining continuity of lineage. They also play a crucial role in maintaining happiness and in contributing to the successes of the *Viet kieu*. The importance placed on marriage in *Viet kieu* travel indicates an attachment to roots, as marrying a traditional Vietnamese may guarantee the preservation of Vietnamese cultural tradition and the maintenance of family continuity. The *Viet kieu* are primarily motivated by a desire to experience culture, history and customs. They pursue the benefits of family togetherness and prefer to visit friends and relatives and places where they have originated from and where they have established and re-established their familial and kin relationships. Their common goal is to promote kinship and travel to Vietnam is an important part in achieving this goal. Furthermore, previous research by the authors reveals that return trips are frequently viewed as a moment in which the person returning is measured up for changes, for success and for bringing back rewards to their country (Nguyen and King 1998, 2002). In visiting their perceived homeland, they are forced to confront a changed Vietnam. In the period since they left, changes in Vietnamese society, economy and politics have been marked. In making trips, the *Viet kieu* are forced to confront these new realities and contemporary historical narratives.

This stated, being immersed in the home culture but seeking to maintain it nevertheless, the *Viet kieu* become 'guardians' rather than 'shapers'. They may exhibit such characteristics in common with other minorities who share the struggle to retain identity in the midst of a new and enveloping culture. Largely removed from the movement and flux of the home culture, their cultural forms become fixed and may experience a lag factor. Rules and norms may become an end in themselves rather than a means to an end and the attitude to such concerns is predominantly acquiescent rather than engaging.

One of the striking characteristics of the *Viet kieu* is how its members create communal spaces of belonging based on the perceived reproduction of traditions. The desire to maintain Vietnamese identity in the host country appears to be paramount for the *Viet kieu* and has emerged from a position of 'in-betweenness' where the relations between 'here' and 'there' need to be negotiated and redefined. This clearly relates to the creation of a diasporic space of cultural relations and a transnational culture between locations. What is suggested here is a shift from a culture of roots to a culture of the host country in a way that engages with the ideas of the original identities. The *Viet kieu* are in the process of building an adapted culture. This said, adaptation does not necessarily involve a shift towards Western, mainstream Australian values, but a shift towards a new set of behaviours brought about to allow an adaption to the new environment.

*Viet kieu* travel creates and maintains a shared sense of common origins and assists in the establishment and re-establishment of kin and social networks. The maintenance of social ties and attachments to familial and ancestral places creates a sense of complete Vietnamese identity. *Viet kieu* adapted culture may be defined by a deeply felt sense of identity and belonging. These factors combined with an attachment to traditional culture and to the homeland may help to predict *Viet kieu* travel behaviour and particularly to their ancestral home country.

## Comparing travellers and non-travellers

It is instructive to compare those who travel and those who do not with a view to gaining insights into the relationship between culture and motivations. The two groups, *Viet kieu* travellers (i.e. those who have travelled to Vietnam previously) and non-travellers (those who have not travelled to Vietnam yet) were found to differ in terms of identity, happiness and feelings towards Vietnam or Australia as their home. The former show closer contact with their family, friends and relatives in Vietnam compared to the latter. The incidence of feeling *Vietnamese-ness* and of viewing Vietnam as the home country is higher in the case of travellers, while non-travellers identify more closely with feeling an *Australian-ness*, of Australia as the home country and of having a happy life in Australia. Though differences in cultural attitudes toward travel do exist, a trip to Vietnam is nevertheless generally viewed as desirable and attainable by both groups,

The two groups exhibit a number of differences. The difference in religious beliefs between the two groups is of interest (see Table 11.1). This raises a question of whether travellers are more concerned about religious beliefs than non-travellers; and if this is the case, whether the maintenance of family religion is one of the main drivers for *Viet kieu* travel. More than twice as many *Viet kieu* travellers than non-travellers have lived in Australia for an extended period (17–26 years). The travellers therefore represent the first wave of migration after the reunion of Vietnam; that is, a more highly educated and more established, middle class group than those that followed. They quite clearly express a need to retain their identity and meaning through regular contact with the homeland and/or with other members of the Vietnamese diaspora.

*Table 11.1* Demographic characteristics of travellers and non-travellers

| Respondent characteristics | Travellers (171 or 39%) | Non-travellers (264 or 61%) | Significance* |
|---|---|---|---|
| Gender: | | | 0.88 |
| Male | 19.8 | 31.3 | |
| Female | 19.3 | 29.7 | |
| Age | | | 0.00* |
| 18–19 | 0.5 | 1.6 | |
| 20–29 | 12.6 | 23.4 | |
| 30–39 | 9.4 | 17.5 | |
| 40–49 | 9.7 | 16.8 | |
| 50–59 | 5.1 | 1.4 | |
| 60 and over | 1.8 | 0.2 | |
| Family structure | | | 0.00* |
| Single, living alone | 7.6 | 25.3 | |
| Single, with children | 2.8 | 2.3 | |
| Single, with extended family | 8.0 | 14.9 | |
| Married, with spouse | 6.7 | 10.3 | |
| Married, with spouse and children | 11.7 | 6.7 | |
| Married, with extended family | 2.3 | 1.4 | |
| Religion | | | 0.00* |
| Christian | 8.0 | 6.7 | |
| Buddhist | 21.1 | 24.8 | |
| None | 9.9 | 29.4 | |
| Qualification | | | 0.00* |
| Primary school | 1.8 | 1.6 | |
| High school | 12.4 | 17.9 | |
| Vocational school | 6.2 | 26.2 | |
| University degree | 14.3 | 13.3 | |
| Post-graduate degree | 3.4 | 1.8 | 0.00* |
| Occupation | | | |
| Manager/administrator | 1.4 | 1.6 | |
| Professional | 5.7 | 4.4 | |
| Trades – and salesperson | 8.5 | 4.6 | |
| Workers | 11.5 | 25.1 | |
| Retired | 1.8 | 0.2 | 0.00* |
| Student | 5.5 | 7.1 | |
| Unemployed | 2.3 | 5.1 | |
| Other | 2.3 | 12.9 | |
| Income | | | 0.00* |
| Below A$ 10,000 | 7.8 | 11.0 | |
| A$10,000–19,000 | 5.3 | 8.3 | |
| A$20,000–29,999 | 6.9 | 25.7 | |
| A$30,000–39,999 | 4.8 | 9.0 | |
| A$40,000–49,999 | 8.0 | 4.1 | |
| Over A$50,000 | 6.2 | 2.8 | |
| Length of residence | | | 0.00* |
| 1995–2001 (1–6 years) | 5.7 | 21.1 | |
| 1985–1994 (7–16 years) | 23.0 | 35.4 | |
| 1975–1984 (17–26 years) | 10.3 | 4.4 | |
| Migrant category | | | 0.00* |
| A refugee | 23.0 | 25.7 | |
| Family re-union migrant | 11.3 | 28.7 | |
| Overseas student | 4.8 | 3.2 | |
| Tourist | 0.0 | 2.8 | |

Source: survey results 2001.
*Chi-square significance at a level of 95%.

Over 50 per cent of travellers are repeat visitors which suggests that the *Viet kieu* have potential as a sustainable visitor market segment for Vietnam. Two-thirds of the traveller respondents reported that their first visit to Vietnam involved visiting family, friends and/or relatives, with a third citing holiday as the main purpose for travel. These proportions change in the case of repeat visitors, with nearly half claiming to visit family, friends and/or relatives, 42 per cent travelling for a holiday and 8 per cent on business. Both groups agree that they would travel to Vietnam to 'feel the warmth and love of the Vietnamese people', because such travel would provide them with a 'feeling of self-confidence, certainty and strength'. The association between prestige and overseas travel may also influence *Viet kieu* travel motives. Particular attitudes towards the importance of prestige, value for money and of showing holiday affordability attached to non-travellers may have the effect of delaying a trip to Vietnam by diverting it to other destinations or to spending money on alternative purchases. There are, of course, other factors that may prompt such postponement of travel and suggest the need for further examination.

The similarities and differences between travellers and non-travellers should be viewed as a function of the social and cultural complexities of the *Viet kieu*. The former responded based on their travel experience, while the latter answered hypothetically. The former grounded their responses in reality, the latter in imagined feelings. Travellers attributed travel behaviour to the pursuit of traditional cultural ends, while non-travellers referred to the need for show, exhibition and superstition in making travel decisions. The most significant difference between the two groups reflected travel as a symbol of success for the *Viet kieu* non-travellers than as a means to realize more practical, pragmatic ends.

## A conceptualization of travel in the diaspora

From the preceding discussion, it is clear that, in the case of the *Viet kieu*, migrant cultural adaptation and consumer behaviour through travel and tourism are related. These ideas may be integrated in basic model form (Figure 11.1).

Cultural adaptation theory explains how previous cultural background continues to influence the structure, function and values of migrant families in the host society. It offers a framework for studying consumption behaviour generally, but more specifically

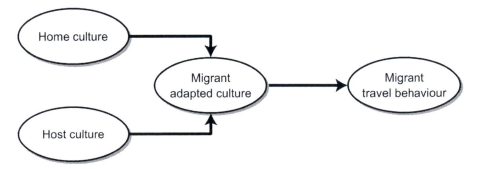

*Figure 11.1* The relationship between adapted culture and migrant travel behaviour.

how migrants' travel patterns are also an outcome of adjustments to host culture. In the case of cultural background, Crompton (1981) notes that one of the derived motives for travel may be to reinforce family ties and to enhance kinship. In this case, cultural norms and values from the home country are shown to have great resonance for societies such as the Vietnamese, which are characterized by close-knit nuclear and extended family structures. Conversely, the proposed framework also reflects the assembled evidence that points to the way in which migrant groups have built separate identities in a new cultural context and hence how they may adopt a new perspective on the significance and meaning of travel to different places. Thus, as values, identity, goals and expectations change, migrant consumption behaviour is transformed. Moreover, travel motives, patterns, expectations and experiences also change.

In the case of the *Viet kieu*, adapted culture is essential to gaining a fuller understanding of why and how people travel back to Vietnam. In a more abstract sense we would contend that it has further relevance. Although it is well-suited to the *Viet kieu* case, it may be well applied to other diasporas, albeit their historio-cultural conditions and hence outcomes may be more complicated (see also Figure 11.2). Furthermore, this model implies an essentially linear trajectory with the final implied outcome being

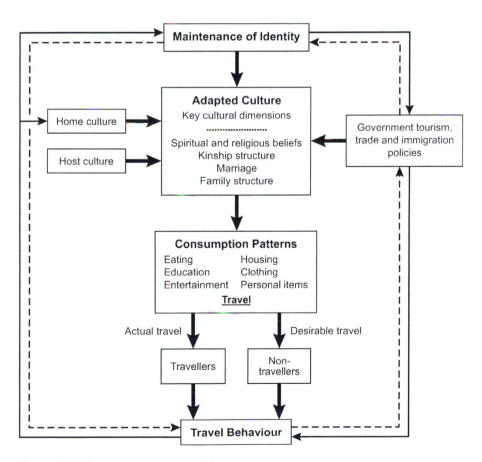

*Figure 11.2* Migrant consumption model.

the modification of migrant travel behaviour. We would note that there may, in fact, be feedbacks within this system. Home and host culture are in constant flux. Migrant travel behaviour has the potential to influence home and host culture thereby mediating new forms of adapted culture and subsequent rounds of migrant travel behaviour. One intriguing question for future research is the embeddedness and resistance of different home and host cultures associated with different diasporas and how their travel patterns unfold.

## Development of the migrant consumption model

Notwithstanding, this basic approach provides the theoretical possibility of developing the framework of a migrant consumption model (see Figure 11.2). This model proposes that travel behaviour is a consequence of migrant adapted culture, which is in turn attributable to three elements: the desire to preserve Vietnamese identity in the diaspora; internal forces associated with cultural adaptation of individual travellers to the adopted country; and external forces associated with tourism, immigration and government policy development.

The major dependent variable is travel behaviour. The preservation of Vietnamese identity is the 'push factor'. Causality between components is indicated by the use of arrows to indicate the direction of likely influence. The model also addresses behavioural variables that should have an impact on *Viet kieu* cultural values. Solid lines show hierarchically how determinants cause or influence other determinants. The model shows the relationships between the desire to maintain identity, cultural adapted dimensions and travel behaviour that in turn has an impact on the preservation of cultural identity. Kinship and marriage are included in the conceptual framework because they constitute a central concept in traditional Vietnamese social structures as an extension of family structures. Some relationships were generated from the semi-structured interviews that provided additional information about the relationship between travel behaviour and *Viet kieu* culture, norms and behaviour.

## Conclusion

The *Viet kieu* in Australia are a group seeking to maintain links with the past and the homeland as their culture undergoes a process of adaptation. Active articulation of their culture helps them to maintain many Vietnamese cultural values in the face of apparently perplexing external demands. Simultaneously, as they adjust to these demands, they acquire new cultural elements to mediate a new identity. The cultural past buffers against the upheavals of ongoing social change and offers direction for their future.

*Viet kieu* in Australia are a key diasporic group through which to gaze on wider relationships between diasporas and tourism. In a more abstract sense, *Viet kieu* travel functions mainly in family and cultural spaces. These spaces, settings and structures for travel contrast sharply with those articulated in Western readings of diaspora tourism. These characteristics are best seen in the context of the Confucian understanding of relationships and obligations. This 'Eastern' worldview involves expectations, norms and imperatives which are markedly different from a 'Western' worldview. Moreover, the assumption that travellers operate from within a context of free choice is incorrect when applied to the *Viet kieu*. Apart from emotional and cultural needs, a range of practical and moral issues also exert an influence. Practical reasons include marriage and business, whereas moral reasons include family and ancestral commitments and the desire to maintain a traditional Vietnamese culture and identity. A clear picture emerges of *Viet kieu* travellers as comfortably settled in Australia yet still calling Vietnam 'home' and quite consciously and deliberately maintaining a Vietnamese cultural identity. A greater ambivalence is evident in the case of non-travellers. Their thoughts concerning Vietnam are wistful, romantic and nostalgic, fed by memories and unchecked by reality. This ambivalence, together with the need to treat travel as a signifier of success, suggests a less settled group, more uncertain concerning its identity and less successful in integrating the two worlds of meaning. Travellers come across as being more secure and settled than non-travellers, recognizing and incorporating the ongoing context with Vietnam and/or the diaspora as integral to their adaptation, identity and meaning.

Thus, existing theories of travel consumer behaviour are predominantly Euro-centric in their presuppositions and, we would contend, not readily applicable to the consumption patterns of Asian migrants. Additional account is clearly required of Asian travel contexts, in particular factors such as the strength of group identity and belonging, as well as the high importance given to status, the need to save face, the custom of gift-giving and the importance of family ties and decision making. In the case of non-European diasporas, future research must investigate how such aspects also manifest themselves in migrant travel within the adopted country as well as travel to their related diasporic communities located in places other than the former homeland.

# References

Baker, S. and Kennedy, F. (1994) 'Death by nostalgia: a diagnosis of context specific cases', *Advances in Consumer Research* 21: 380–7.

Belk, R.W (1988) 'Possessions and the extended self', *Journal of Consumer Research* 15 (September): 139–53.

Blaine, T.W., Mohammed, G., Ruppel, F. and Var, T. (1995) 'US demand for Vietnam tourism', *Annals of Tourism Research* 22(4): 934–6.

Bureau of Immigration, Multicultural and Population Research (BIMPR) (1996) *Australia's Population Trends and Prospects 1995*, Canberra: Australian Government Publishing Service.

Chambers, I. (1994) *Migration, Culture, Identity*, New York: Routledge.

Chase, M. and Shaw, C. (1989) 'The dimensions of nostalgia', in C. Shaw and M. Chase (eds) *The Imagined Past: History and Nostalgia*, Manchester: Manchester University Press.

Clifford, J. (1997) *Routes*, Cambridge, MA: Harvard University Press.

Cohen, R. (1997) *Global Diasporas*, London: UCL Press.

Cox, D. (1980) *Migration and Integration in the Australian Context: An Introduction for the Helping Professions*, Melbourne: Department of Social Studies, University of Melbourne.

Crompton, J. (1981) 'Dimensions of the social group role in pleasure vacations', *Annals of Tourism Research* 8 (4): 550–67.

Davis, F. (1979) *A Yearning For Yesterday: A Sociology of Nostalgia*, London: Collier Macmillan.

Haraven, T. and Langenbach, R. (1981) 'Living places, work places and historical identity', in

M. Binney and D. Lowenthal (eds) *Our Past Before Us: Why do We Save It?*, London: Temple Smith.

Hiebert, M. (1993) *Vietnam Notebook,* Hong Kong: Review Publishing Company Limited.

Holak, S. and Havlena, W. (1992) 'Nostalgia: an exploratory study of themes and emotions in the nostalgic experience', in J. Sherry and B. Stemthal (eds) *Advances in Consumer Research*, vol. 19, Chicago, IL: Association for Consumer Research, pp. 380–7.

Holbrook, M.B. and Schindler, R.M. (1991) 'Echoes of the dear departed past: some work in progress on nostalgia', in R. Holman and R. Soloman (eds) *Advances in Consumer Research*, vol. 18, Chicago, IL: Association for Consumer Research, pp. 330–3.

Holbrook, M.B. and Schindler, R.M. (1994) 'Age, sex and attitude toward the past as predictors of consumer's aesthetic tastes for cultural products', *Journal of Marketing Research* 31(August): 412–22.

Kamptner, L. (1989) 'Personal possessions and their meaning in old age', in S. Spacapan and S. Oskamp (eds) *The Social Psychology of Ageing: Claremont Symposium on Applied Social Psychology*, London: Sage Publications.

Kaplan, H. (1987) 'The psychopathology of nostalgia', *Psychoanalytical Review* 74(4): 465–86.

Karnow, S. (1983) *Vietnam: A History*, New York: The Viking Press.

Kasinitz, P. and Hillyard, D. (1995) 'The old timer's tale', *Journal of Contemporary Ethnography* 24(2): 139–64.

King B. (1994) 'What is ethnic tourism? An Australian perspective', *Tourism Management* 15(3): 173–76.

Laenen, M. (1989) 'Looking for the future through the past', in D. Uzzell (ed.) *Heritage Interpretation,* vol. 1, London: Belhaven Press.

Muzny, C.C. (1985) 'The Vietnamese in Oklahoma City: a study of ethnic change', Doctoral dissertation, Ann Arbor: University of Michigan.

Nguyen, D.L. (1987) 'Cross-cultural adjustment of the Vietnamese in the United States', in Truong Buu Lam (ed.) *Borrowings and Adaptations in Vietnamese Culture*, Southeast Asia Paper No. 25, Centre for Southeast Asian Studies, School of Hawaiian, Asian and Pacific Studies, Hawaii: University of Hawaii at Manoa.

Nguyen, T.H. (1996) 'Ethnic Vietnamese travel from Australia to Vietnam', unpublished Master of Business in Tourism Thesis, Faculty of Business, Melbourne: Victoria University.

Nguyen, T.H. (2002) 'Travel behaviour and its cultural context: the case of the Vietnamese community in Australia', paper presented at the Tourism Outlook Conference, organized by Australian Tourism Research Institute (ATRI) and Cooperative Research Centre (CRC) for Sustainable Tourism Australia, Sydney.

Nguyen, T.H. and King, B.E.M. (1998) 'Migrant homecomings: *Viet kieu* attitudes towards travelling back to Vietnam', *Pacific Tourism Review* 1(1): 349–61.

Nguyen, T.H. and King, B.E.M. (2002) 'Migrant communities and tourism consumption: the case of the Vietnamese in Australia', in C.M. Hall and A. Williams (eds) *Tourism and Migration: New Relationships between Production and Consumption*, Dordrecht: Kluwer.

Nguyen, T.H., King, B.E.M. and Turner, L. (2002) 'Travel behaviour and migrant culture: the Vietnamese in Australia', *Asia-Pacific Tourism Association (APTA) Conference 2002 Proceedings*, Dalian, China.

Nguyen, X.T. (1994) *Vietnamese Studies in a Multicultural World*, Melbourne: Vietnamese Language and Culture Publications.

Nien-Chu Kiang (1995) 'Bicultural strengths and struggles of Southeast Asian Americans in school', in A. Darder (ed.) *Culture and Difference: Critical Perspective on the Bicultural Experience in the United States*, Westport, CT: Bergin and Garvey.

Prevot, H. (1993) *Social Policies for the Integration of Immigrants; The Changing Course of International Migration*, Paris: OECD.

Stern, B. (1992) 'Historical and personal nostalgia in advertising text: the *fin de siecle* effect', *Journal of Advertising* 31(4): 11–22.

Strauth, G. and Turner, B. (1988) 'Nostalgia, postmodernism and the critique of mass culture', *Theory, Culture and Society* 5: 509–26.

Te, H.D. (1962) 'Vietnamese cultural patterns and values as expressed in proverbs', Doctoral dissertation, Ann Arbor: University of Michigan.

Thomas, M. (1996) Place, Memory and Identity in the Vietnamese Diaspora, unpublished PhD Thesis, Canberra: Australian National University.

Thomas, M. (1999) *Dreams in the Shadows, Vietnamese-Australian Lives in Transition*, Sydney: Allen and Unwin.

Urry, J. (1995) *Consuming Places*, London: Routledge.

Urry, J. (2000) *Sociology Beyond Societies, Mobilities for the Twenty-first Century*, London: Routledge.

Viviani, N. (1984) *The Long Journey – Vietnamese Migration and Settlement in Australia*, Melbourne: Melbourne University Press.

Viviani, N. (1996) *The Indochinese in Australia 1975–1995 – from Burnt Boat to Barbecues*, Melbourne: Oxford University Press.

# 12 Mobilizing *Hrvatsko*

## Tourism and politics in the Croatian diaspora

*Sean Carter*

## Introduction

It has been argued by Mitchell (1997) that studies of transnationalism often fall into one of two traps; either they are too literal, or conversely too liminal. Literal accounts provide empirical evidence of transnational formations, but do so from a viewpoint which maintains a fixed theoretical conception of borders and movement. Liminal accounts, on the other hand, are too focused on the potential progressiveness of transnational practices and fail to recognize the realities 'on the ground'. This account of the relationship between Croatia, the Croatian diaspora (in particular, Croatian communities in North America) and tourism seeks to provide both literal and liminal perspectives of the ways in which the diaspora is accessed and mobilized, negotiated and remade through the medium of tourism.

In a literal sense, there are particular networks and institutions through which those in the Croatian diaspora can be accessed, including: *Hrvatska Matica Iseljenika* (Croatian Heritage Foundation, the prime organization in Croatia for building relationships between Croatia and the diaspora); various government ministries (Tourism, Emigration and Immigration); and several diaspora organizations (Croatian Fraternal Union, Croatian Catholic Union, Croatian-American travel agents). In addition to tracing these networks, it is also necessary to understand the liminal elements that are involved; that is, the ways in which *Hrvatsko*, or a *sense* of Croatian-ness is mobilized. The chapter is a based on an extended period of fieldwork in Pittsburgh in April 1999 and shorter periods of fieldwork in Zagreb in November 1998 and August 2000. Thus, while there are empirical elements to this chapter, it is also exploratory and suggestive, focusing less on the specifics of the nexus between Croatian diaspora and Croatian tourism and more on some of the broader ways in which diaspora might interact with tourism geographies. A general description of Croatia and the Croatian diaspora prefaces the discussion. The second section provides an overview of the various roles played by the diaspora during the recent conflicts in the Balkans and in the post-war period, focusing particularly on the 'travellings' of the diaspora during this time. The third section looks in more detail at a Croatian institution which plays a key role both in homeland–diaspora relations and in organizing visits to Croatia, the Croatian Heritage Foundation (or *Matica*).

## *Hrvatska Iseljenika*: **Croatia and its diaspora**

The Croatian declaration of independence heralded, in the words of the former President, Franjo Tudjman, the realization of a 'thousand year dream' (Tanner 1997). This alluded to the existence of the mediaeval state of Croatia established in AD 914 and a short-lived Kingdom of Croatia in the 1070s, which quickly fell under Hungarian rule. The following 900 years saw the territory today known as Croatia subsumed within other political formations, variously Hungary, the Habsburg empire, the Ottoman empire, the (first) Kingdom of Yugoslavia and finally, the (second) Socialist Republic of Yugoslavia (Figure 12.1), interspersed with the Nazi-supported *Ustasha* regime which nominally led the 'Independent State of Croatia' between 1941 and 1945 (*Ustasha* translates literally as 'upriser' and was the chosen name of the fascists in Croatia). Beginning in the early 1800s, the political, economic and cultural marginalization of Croats within these territorial formations led to a steady flow of emigrants, particularly to the Americas. Already by the 1830s, there were small Croatian communities in California, New Mexico and Louisiana. From the 1880s onwards, these smaller communities would be joined by mass Croatian emigration to the industrializing cities of the East, such as Cleveland, New York, Chicago and, in particular, Pittsburgh. This era of mass migration lasted until the start of the First World War. Irrespective of a plethora of methodological difficulties in determining the extent of migration, Croatian historians estimate that by this time 600,000 to 1 million Croats had emigrated to the USA (cf. Prpic 1971). There has been little emigration to the USA since 1914, but Croatian emigration has continued to other locations, notably Canada, Argentina, Germany and Australia (see Carter 2002 for a fuller discussion). Although it is impossible to provide any accurate or meaningful data on the size of the Croatian diaspora today, information from various official Croatian sources (collated in the Croatian Almanac 1998) suggests that only around two-thirds of those that identify themselves as Croatian actually reside within the borders of the six Republics of the former Yugoslavia. The largest 'host' countries are identified as Germany (450,000), Australia (300,000) and the USA (545,000).

This relatively large diaspora played a significant role in the independence movement in Croatia, both throughout the twentieth century and in the armed struggle for independence in the 1990s. Historically, elements of the Croatian diaspora have been among the most nationalistic and fiercely pro-independence sections of 'Croatian' society. This is, of course, not surprising, for at least two main reasons: first, as Anderson (1998) has shown, nationalism is often nurtured in exile; and second, those most opposed to Croatian rule by 'outside' powers are those that are either most likely to choose to leave Croatia, or to be forced to leave. For example, at the end of World War Two, large numbers of nationalist Croats (who had been involved in the Fascist Independent State of Croatia) fled from the incoming socialist Partisans. Similarly, throughout the 1960s, 1970s and 1980s, opponents of Tito's regime sought refuge elsewhere.

The large majority of those in the Croatian diaspora, however, did not hold the extreme beliefs of the *Ustasha*, but nevertheless, sympathy for the cause of Croatian independence (or at the least, greater autonomy) was widespread. This manifested itself in a number of ways, such as the founding of political organizations in diaspora communities which lobbied host governments; humanitarian funds set up to provide financial assistance to 'victims of oppression' in the homeland; and in a number of

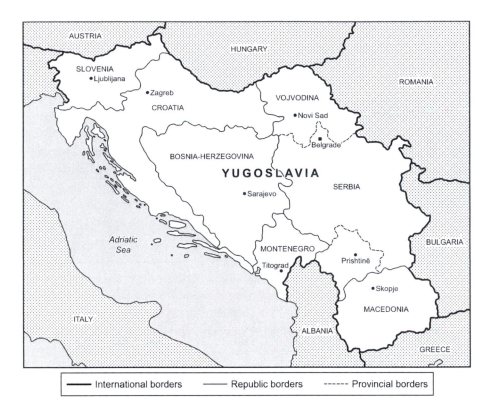

*Figure 12.1* The Republics of the Former Yugoslavia, 1991.
Source: adapted from Meier (1999).

extreme cases, acts of terrorism, such as the hi-jacking of a TWA flight in the USA. In the next section some of the ways in which the Croatian diaspora was involved in the politics of Croatia in the 1990s and, in particular, the relationship between these political acts and tourist practices, will be discussed in more detail.

## The homeland war and diaspora travellings

Franjo Tudjman became the first elected President of Croatia after elections while Croatia was still a constituent Republic of Yugoslavia. He came to power on an unashamedly pro-independence platform and via the most expensive and sophisticated election campaign in Croatian history. This was, in no small part, due to the connections that Tudjman had established in the diaspora. His supporters offered both finance and election know-how. In the conflicts that were to follow, the diaspora remained just as involved. In the most extreme of cases, this involvement took the form of individuals in the diaspora returning to fight on the front lines. In one case, a Canadian-Croatian returnee, Gojko Susak, became Tudjman's minister in charge of the armed forces. For the most part, however, diaspora support was more banal and mundane, although following Billig (1991), that should not necessarily be understood to mean harmless or insignificant. Activities in the diaspora included: fundraising for humanitarian causes (in *some* cases code for buying weapons); lobbying host governments in order to influence foreign policy; and PR and media campaigns aimed at influencing public opinion.

Diaspora engagement has not ended with the cessation of armed conflict in the region. Indeed, if at all, there is perceived to be a greater role for the diaspora in the post-war era than previously. Croatia has experienced and continues to undergo a number of 'transitions' simultaneously. In political terms, for instance, these have included such aspects as: the move from a constituent republic of a socialist state to an independent state with a developing parliamentary democracy; a programme of reforms designed to transform the economy from a form of socialist state planning to a market-oriented capitalist system; and, a society trying to come to terms with a history of oppression and recent civil war. Thus, the challenges facing Croatia are huge and Croatia not only has to deal with the economic impacts of transition (which has affected economies throughout central and eastern Europe), but also must cope with reconstruction following extensive war damage. In these circumstances, the Croatian diaspora and Croatian tourism take on particular significance. Prior to the dissolution of Yugoslavia, foreign tourism made up a vital element in the acquisition of foreign currency. Almost all of the receipts from foreign tourist were generated within Croatia, but given the federated nature of the Yugoslav state, these receipts were distributed amongst all the Republics. This re-distribution of capital amongst the Republics has been highlighted as one of the underlying causes of tension between the Republics which led to eventual collapse of Yugoslavia (Bicanic 1995; Woodward 1995). In particular, the Slovenes and the Croats resented the fact that there was a significant flow of money from their economies to Belgrade and the Serbian Republic. During the war itself, the lucrative Dalmatian coast was one of the most fought over pieces of territory – not for any strategic military reason, nor for reasons of 'ethnic cleansing', but for more petulantly destructive reasons because Croatia depended on the tourist receipts from locations such as Dubrovnik.

In recent years, there has been a growing call for states in central and eastern Europe, more generally, to develop their tourist industries as key features in their post-socialist transitions (see, for example, Hall 1998 and Williams and Baláž 2000). More specifically, the need for Croatia to revive its formerly lucrative tourist trade in order to aid its post-war reconstruction has been recognized (Gosar 2000). The settlement reached in the Dayton Accord of 1995 provided the basis for Croatia to achieve this (Figure 12.2). According to Jordan (2000: 525), it meant that Croatia had '65 per cent of the former Yugoslavia's tourism capacity (expressed in number of beds) and 81 per cent of total tourist nights spent by foreign tourists'. However, significant problems have remained for the Croatian tourist industry. In the aftermath of civil war receipts have failed to recover to their pre-war levels; the infrastructure of the tourism industry remains poor; and Croatia still suffers from a negative image amongst the majority of holiday-makers (Šerović 2001). Croatia has, therefore, sought investment from the diaspora both to achieve its economic aims, more generally, as well as to aid its re-development of the tourist infrastructure.

Engagement between the diaspora and the homeland has thus continued in the post-war years, although, at least from the perspective of some of those in the diaspora, the present and future role of the diaspora needs reconsideration. Indeed, this was one of

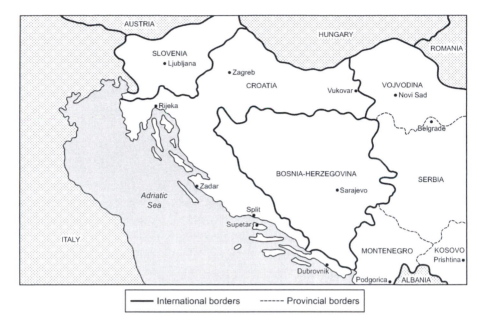

*Figure 12.2* Post-Dayton Croatia.
Source: adapted from Glenny (1996).

the recurring themes of the Association for Croatian Studies symposium held in Chicago in April 1999. More explicit guidance on the path forward has been offered by the Croatian Fraternal Unit (CFU). Established in Pittsburgh in 1894 under the name of the 'National Croatian Society', the CFU is the largest formally constituted organization in the diaspora with over 100,000 members throughout the USA and Canada. Principally the CFU is a mutual insurance company, but it also fulfils political, cultural and social functions in the Croatian diaspora. To this end, the President of the CFU, Bernard Luketich, highlighted public relations, tourism and investment as three particular roles of the Croatian diaspora in the post-war era. Elaborating on the latter two, he said,

> We need to work on helping tourism in Croatia, because even previously tourists used to go to Croatia from all over the United States and Canada, but now, first thing, there's a problem with air flights. We don't have a direct flight from Pittsburgh to Zagreb, or from New York to Zagreb, you know. That's a problem. Croatia's got to do something in that respect; and investments in Croatia, that's always part of our involvement, to help people from this country invest in Croatia, basically I think that's what it's all about.
>
> (Interview with Bernard Luketich, 21 April 1999)

In short, the Croatian diaspora played (and continues to do so) a number of important political and economic roles in the periods of conflict, transition and post-war recovery. Many of these practices involved 'travel' in some sense of the word, as well as displaying characteristics that we might call touristic. In the following section I want to make the connection between these political acts and tourist practices more explicit, by detailing the diasporic travellings of the Croatian diaspora during the 1990s. At the most extreme end of these travellings are those members of the diaspora who 'returned' to Croatia during the 'homeland war' to join fighting units. The term 'returned' is used in a flexible sense here akin to the diaspora literature (cf. Cohen 1997), as, although many had dreamt about, discussed and perhaps even agitated for return, for some, this was their first time in Croatia, having been born outside it in the diaspora.

### Politics and tourism in focus: extreme forms of 'volunteer' and 'adventure' tourism?

There is little in the way of quantitative evidence of the phenomenon of 'returning' to fight, but there are well-known individual cases, particularly from those locations in the diaspora of more recent emigrants, such as Toronto, Cleveland, Chicago and Australia. At the same time, there is some evidence that individuals from the diaspora were involved in attempts to illegally smuggle arms into Croatia. Shain (1999), for example, notes that in the summer of 1991 a number of Croatian-Americans were arrested in Miami and charged under the Arms Export Control Act. It is believed they were organizing a shipment of weapons to Croatia worth millions of dollars. One would hesitate to designate these kinds of activity as touristic, but speculatively, there are connections between such travellings and the risk elements of adventure and backpacker tourism (for example, see Elstrud 2001). The history of warfare suggests, for example, that young men are often drawn by the thrill and excitement of combat, in the same way that people today are increasingly attracted to extreme sports and adventure tourism, with an emphasis on (potential) danger (Cloke and Perkins 1998; Elstrud 2001). Numerous accounts from those involved in war speak of the reality and authenticity of such scenarios, which again echoes with sentiments expressed about adventure and backpacking tourism and with the search for the 'authentic' more generally in tourist experiences (May 1996; Ning 1999). There is insufficient space here to explore these ideas in any detail, but there are some useful connections to be made between war experiences and motivation on the one hand and extreme and adventure tourism on the other.

Together with those who 'returned' in order to fight on the front lines, others travelled to Croatia to assist in the war in other ways. One such case was a Croatian catholic priest from Pittsburgh, who, during the course of the war, visited Croatia to take mass with soldiers on the front line. The office of the Croatian Information Centre in Zagreb, set up to provide the global media with the Croatian side of events, was largely staffed by short-term returnees from the diaspora (Croatian Information Centre 1998), who, by and large, possessed better language skills and understood the workings of the modern news media more fully. In the latter stages of the conflict, Croatian-Americans worked as interpreters alongside UN peacekeeping forces in Bosnia. Again, one would hesitate to describe these as tourist practices, but there are certainly connections between these kinds of activities and the kinds undertaken by 'volunteer tourists'. Volunteer tourism has recently begun to be analysed as a distinct niche market within the tourism industry (Kottler 1997; Wearing and Neil 2000; Wearing 2001). Wearing (2001: 1) argues that such travellers are 'seeking a tourist experience that is mutually beneficial, that will not only contribute to their personal development but also positively and directly to the social, natural and/or economic environments in which they participate'. To this we might want to include contributions made to the 'political environment', although this is perhaps a much more subjective concept.

In the immediate post-war period, the Croatian government hoped that large numbers would return from the diaspora, for a number of reasons: first and foremost, they would bring capital and entrepreneurial know-how; and second, they would bring experience of democratic culture. In addition, the Croatian state saw itself as having a moral responsibility towards those in the diaspora, a commitment which was enshrined in the constitution. Article 10 [Citizens Abroad] of the new constitution reads,

> (1) The Republic of Croatia protects the rights and interests of its citizens living or staying abroad and promotes their links with the homeland. (2) Parts of the Croatian nation in other states are guaranteed special concern and protection by the Republic of Croatia.
>
> (Republic of Croatia 1990)

Therefore, the government set about making return more attractive by setting up a designated ministry for returnees (since disbanded) and putting in place a package of measures, such as tax, investment and property incentives, to make returning a more attractive proposition. Despite this, there has been little in the way of return migration – perhaps less than 1,000 from the USA for example (Sunič 1999). Throughout the lifetime of the second Yugoslavia, there was a strong expectation amongst those in diaspora that they would return should Croatia ever gain independence. However, when this became a reality, those who had imagined returning did not. When questioned on this issue, Bernard Luketich, the President of the CFU, responded,

> President Tudjman, in all of his addresses here, invites people to come back, tells people to come back to Croatia, but you know that's pretty hard. Someone comes here, they learn this American way of life, their families are here, it's not easy for them to go back to Croatia. It's nice to visit Croatia, go for a month, three weeks and then come back to the United States.
>
> (Interview with Bernard Luketich, 21 April 1999)

One of the consequences of the lack of permanent return, therefore, has been an increase in long-stay visits in Croatia by those in the diaspora. There is nothing in the way of official statistics to quantify these visits, but anecdotal and other evidence does suggest that this type of visit has increased. For example, although Luketich bemoans the lack of flights from Pittsburgh and New York to Zagreb, new routes from Zagreb and Chicago and Toronto have been introduced since the end of the war. Similarly, a Croatian consulate has recently opened in Chicago, partly to enable the processing of visas to take place at the point of demand. Consular days are also held in Pittsburgh for the same reason. Finally, on an anecdotal level, my fieldwork discussions with those in the diaspora in Pittsburgh, and more especially in Chicago, revealed that visits are much more frequent now than before. This is, in many cases, directly due to Croatian independence. For many in the diaspora, return visits during the communist era were simply not an option; many had left illegally, or as enemies of the state and return was practically impossible. Independence and the cessation of hostilities afforded the first opportunity for many to visit their friends and family in the homeland. There is now an established literature on VFR tourism (see Boyne *et al.* 2002; Jackson 1990; Morrisson 1995) and in many senses, much diaspora tourism in the Croatian context is an extension of this. The example of Croatia reminds us, however, that such trips not only rely on transport and financial determinants, but are also politically contingent.

Thus there are a range of travellings which were undertaken by the Croatian diaspora in North America during the upheavals in the homeland during the 1990s. Some of these travellings were in response to these events, others were a constitutive part of those events. The travellings I have described thus far are rather speculative. In the next section I provide a more detailed case study of one particular institution, the Croatian Heritage Foundation, a cornerstone of the connections between Croatia and the diaspora and one involved in organizing a range of more obvious 'tourist' activities.

## The 'Croatian Heritage Foundation' as a tourism mediator

*Hrvatska Matica Iseljenika* (Croatian Heritage Foundation – abbreviated to *Matica*) was founded in 1951 and is 'responsible for cultural, educational, athletic, publishing and information programmes for the Croatian diaspora. As a spiritual meeting-place of the Croatian nation spread throughout the world, the Foundation seeks to achieve, through its programmes, its key goal of preservation and development of the cultural identity of our people in the countries of their residence following examples of others who have shared a similar predicament' (HMI 1998: 91). There is no direct translation for the word *Matica*, but the closest approximations are 'matrix', or 'mother-bee'. The official English name of the organization translates it as 'homeland' and underscores its key link between the diaspora and the homeland, such that it occupies a prominent position in the cultural life of Croatia.

*Matica* is non-governmental, although it relies on state funding and the appointment of the head of *Matica* is generally seen as a political appointment. One of the key roles of *Matica* is publishing. It produces a monthly magazine for the diaspora, as well as a weekly supplement for the foreign edition of the Croatian daily *Vecernji List*. More significantly here, *Matica* also organizes around 100 different projects each year which aim to connect the homeland and the diaspora. Many of these are cultural exchange projects, where, for example, traditional Croatian music groups from the diaspora tour Croatia, or vice versa. Touring exhibitions of Croatian art have been organized, as well as international sporting events for young people from the diaspora and the homeland. Throughout *Matica's* programme there is a strong emphasis on youth, particularly in the series of events which form the regular core of *Matica's* activities, the summer schools. The brochure for this programme is entitled *Hrvatska za Mlade, Mladi za Hrvatsku* – Croatia for Youth, Youth for Croatia. For the most part, this programme offers a number of educational courses aimed at young people from the diaspora, which involve visiting Croatia for an extended stay (anything from 2 weeks to 3 months) to experience the homeland and to learn the history, culture and language of Croatia. At the most extreme end, *Matica* encourages participation in a degree programme in Croatology based at the University of Zagreb, which lasts for four years. More realistically, *Matica* and the University have devised a shorter programme (one or two semesters) to enable the study of Croatian language and literature as part of a degree programme based elsewhere.

The most popular programmes, however, are those which take place in the summer and are for a shorter period of time. These courses combine educational aims and leisure, with time being made available for socializing, sightseeing and activities. These courses also tend to be held in tourist locations, rather than in Zagreb. For example, both the Summer School of Croatian Language and the School of Croatian Folklore take place on islands in the Adriatic and last for just 2 weeks. These courses are clearly intended as both an educational and a tourist experience. The promotional literature for the School of Language and Culture, for example, explains,

> the summer school is profiled according to the students' needs and, apart from the intensive school program, there is time for recreation and entertainment. We have chosen Supetar . . . because it is one of the outstanding summer resorts in Croatia.
>
> (HMI 1998: 36)

The daily schedule begins with classes in the morning, 'leisure time' in the afternoon (swimming, tennis, basketball, sailing) and entertainment in the evening (concerts, dance, socializing). In addition to these two programmes, there is also a designated programme for children (aged 9–14), or effectively the little summer school of Croatian language and culture.

Aside from these educational tourism packages, *Matica* also organizes and promotes volunteer holidays. The most popular of these is the annual 'Task Force' project. These projects usually seek to protect and restore historical or natural environments damaged or destroyed during the homeland war and last for around three weeks. As with the educational packages, there is also a leisure element with organized activities and excursions. *Matica* also promotes the work of *Suncokret*, a non-governmental humanitarian organization, which runs centres to support the psycho-social needs of direct victims of the war. Each summer, *Suncokret* invites short-term volunteers to work alongside permanent staff in the delivery of its services. Although these voluntary positions are not restricted to those in the diaspora, *Matica's* promotion is targeted at diaspora volunteers.

*Matica* is perhaps the key actor in the organized tourist practices of the Croatian diaspora, but it is not the only one. There are, for example, a number of travel agents run by Croatians in the diaspora specializing in Croatian holidays, clearly marketed at the diaspora. Moreover, group trips are regularly organized by individual lodges within the Croatian Fraternal Union and advertised in their weekly newspaper, *Zajednicar*. What is particularly interesting about *Matica*, however, is the way in which the 'political', broadly conceived, is married to tourism practices. The production and consumption of *Matica's* tourist experiences constitutes a popular geopolitical practice, which is qualitatively different from the concept of VFR tourism within the diaspora tourism niche.

*Politics, practices and the mediation of diaspora tourism episodes:*
*beyond ethnic reunion*

Given the political edge to the tourism practices associated with *Matica*, how are we to make sense of the political elements of diaspora tourism? This is a question which has rarely been touched upon in the existing literature. The kinds of tourist practices highlighted do have connections with some of this concepts developed in this literature, but only tangentially. There has been a recent surge in interest in the relationship between tourism and migration for example (Hall and Williams 2002; Kang and Page 2000) and in particular the geographies of VFR tourism (Boyne *et al.* 2002) and the rise of 'ethnic tourism' (King 1994; Lew and Wong 2002; Nguyen and King 2002). The dominant motivation of tourists within these accounts is best described as 'ethnic reunion' where 'motivation commonly derives from a sense of belonging to or identifying with a way of life that has been left behind. This sense of lost 'roots' is a potent influence for travel' (King 1994: 174). Subsequent studies have been successful in describing and analysing such tourist patterns, but there has been little potential paid to other kinds of motivation and on differentiating between different kinds of diaspora tourist. The evidence from the travellings of the Croatian diaspora show that diaspora tourism is not always based on the ethnic reunion model, nor on some vague search for ethnic roots. Rather, it can be seen as a political act, as argued in more depth below. There are two elements to this politics-tourism couplet. The first relates to political motivations on the 'production' side of the tourist market, the second to political motivations on the consumption side. There are a number of accounts which touch on some of these issues. Krakover and Karplus (2002: 105), for example, highlight the Israeli state as a key actor in tourism and for reasons that go beyond the economic. They argue that 'the State and other semi-state agencies . . . substitute for and play the role of a very powerful 'friend or relative' in supporting the potential immigrants', with the intention of persuading Jewish tourists to become resident citizens. In this reading, a tourist is clearly seen as a potential immigrant (see Ch. 8).

In relation to the second point, there are a number of accounts which stress the political motivations of the tourist, particularly in relation to eco-tourism and volunteer tourism as examples (Wearing and Neil 2000; Wearing 2001; Wight 2001). There are also cases of more explicit political tourism, such as the kinds of 'solidarity tours' that are often run by left-wing groups to Cuba, for example. Hollander (1981: vii) coined this 'political pilgrimage', the 'reverential tour of politically appealing countries' Additionally, the similarities between pilgrimage and tourism have become well established within tourism research (see, for example, Smith 1992). There are also debates as to the extent to which international tourism is a force for understanding between cultures and nations and thereby can contribute towards world peace (for a review, see Tomljenovic and Faulkner 2000). Finally, there is a growing literature on the connections between conflict and tourism (Lennon and Foley 2000; Smith 1998). However, there have been few connections made between diaspora and ethnic tourism and political tourism.

## Conclusions

An interrogation into the tourist practices associated with *Matica* begins to show the significance of the political dimension. In particular, tourist experiences associated with *Matica* constitute a banal practice of belonging. On one level, taking an educational or volunteering holiday in the homeland of your parents or grandparents is rather unremarkable and benign. One approach is to dismiss these practices as irrelevant, throwaway, insignificant; examples of nothing more than 'dime-store ethnicity' (Stein and Hill 1977). However, I would contend that it is these small differences between, for example, holidaying in Croatia as opposed to elsewhere, which maintain effective diaspora identities. Despite the general acceptance of the idea of nations as imagined communities, there is an absence of knowledge 'about how these imagined communities have been perceived and integrated into people's everyday lives' (Linde-Laursen 1995: 1124). In a rare example of how we might understand the relationship between mundane differences and defining characteristics, Linde-Laursen (1995) analyses the importance of washing the dishes, which in Sweden is done to the right, in Denmark to the left. This reaffirms the assertion of Billig (1995: 6) that nations are, primarily, mundane and banal such that,

> daily, they are reproduced as nations and their citizenry as nationals . . . For such daily reproduction to occur, one might hypothesize that a whole complex of beliefs, assumptions, habits, representations and practices must also be reproduced.

Diasporic affinity can not, perhaps, be directly equated to national identities, but there is a strong case for reasoning that the reproduction of such associations might occur in similar ways. For example, Billig argues that the appeals of national leaders for support in times of crises is only possible because of the ways that banal national identities are reproduced daily. In turn, tourism can be seen as a banal and mundane practice, which, nevertheless, has certain political consequences. It is clear that holidays are usually seen as a break from routine and habit. However, holidaying itself is a routine and a habit and the choice of location is not usually thought of in political terms. In other words, the *process* of *choosing* a holiday destination is seen as banal and mundane, but banal in the sense that Billig invokes, rather than in the sense of the *choice* itself being unimportant or unremarkable.

There are two main ways in which 'small differences' are key in Croatian diaspora life. First, that 'small details build a sense of "us" as not like "them"' (Thrift 2000: 384); small differences make Croatian-Americans (however tenuous that connection) distinct from Polish-Americans, for example. Second, the distinctive sense of ethnic identity that this helps to constitute *can*, in *some* circumstances, be mobilized by 'diaspora leaders' to produce certain desired ends in the same way that national leaders are able to mobilize national feelings. For example, a number of staff whom I interviewed at the Croatian Information Centre in Zagreb had first visited Croatia as part of a *Matica* project, which eventually led to a long-term stay in Croatia working with or as part of 'nation-building' institutions.

Perhaps even more significant is the political dimension of the production side. *Matica* is a non-profit organization. It does not seek to attract diaspora tourists for economic reasons; after all, the state funds the activities of *Matica*. The projects that *Matica* organizes are not about ethnic reunion, in the sense which that term is used within the extant ethnic tourism literature. These activities are not about building bridges between family members, but rather, between those in the diaspora and the homeland. In its activities brochure, for example, *Matica* states:

All the CHF programmes endeavour to accurately depict or alone create representative examples of national culture which inspire the Croats and their descendents abroad in their efforts to protect, develop and promote the Croatian ethnic and cultural identity in the multicultural societies to which the Croatian people migrated to throughout history.

(HMI 1998: 91)

In other words, *Matica* is a nation-building institution. Tourism offers an institutional setting for the delivery of political aspirations and wider nation-building projects. *Matica* seeks to promote Croatia, principally to the diaspora and to maintain and protect their identification with Croatia. *Matica* seeks to extend the 'imagined community' that is the Croatian nation. The wider implications of this analysis are that it is necessary to differentiate types of diaspora tourism. Some Croatian diaspora tourist practices fall quite neatly into the VFR typology and to a lesser degree, ethnic reunion, whereas others, such as *Matica* experiences, do not. Furthermore, we need to be much more aware of the non-economic motivations of institutional actors. The work of Krakover and Karplus (2002) on the role of the Israeli state in tourist practices is one good example of this. From this discussion it has become clear that in the Croatian context, *Matica* is involved in a series of practices which seek not only to provide tourist experiences, but also to strengthen and sustain the national community. *Hrvatsko*, or a sense of Croatian-ness, is mobilized in two different senses of the word. Croatians in the diaspora are made mobile; they are encouraged to visit the land of their ancestors. At least as important, however, is their sense of mobilization as members of *Hrvatska Iseljenika*, the wider Croatia.

## References

Anderson, B. (1998) *The Spectre of Comparisons*, London: Verso.

Bicanic, I. (1995) 'The economic causes and consequences of new state formation during the transition', *East European Politics and Societies* 9: 2–21.

Billig, M. (1995) *Banal Nationalism*, London: Sage.

Boyne, S., Carswell, F. and Hall, D.R. (2002) 'Reconceptualising VFR tourism: friends, relations and migration in a domestic context', in C.M. Hall and A.M. Williams (eds) *Tourism and Migration. New Relationships between Production and Consumption*, Dordrecht: Kluwer.

Carter, S. (2002) 'The geopolitics of diaspora: Croatian community and identity in the United States', unpublished PhD manuscript, University of Bristol.

Cloke, P. and Perkins, H.C. (1998) '"Cracking the awesome foursome": representations of adventure tourism in New Zealand', *Environment and Planning D: Society and Space* 16(2): 185–218.

Cohen, R. (1997) *Global Diasporas*, London: Routledge.

*Croatian Almanac* (1998) Chicago: Croatian Franciscan Publications.

Croatian Information Centre (1998) 'Interview with Croatian Information Centre staff', unpublished transcript, S. Carter, Zagreb, 10 November.

Elstrud, T. (2001) 'Risk creation in travelling: backpacker adventure narration', *Annals of Tourism Research* 28(3): 597–617.

Glenny, M. (1996) *The Fall of Yugoslavia*, 3rd edn, London: Penguin.

Gosar, A. (2000) 'The recovering of tourism in the Balkans', *Tourism Recreation Research* 25(2): 23–34.

Hall, C.M. and Williams, A.M. (eds) (2002) *Tourism and Migration. New Relationships between Production and Consumption*, Dordrecht: Kluwer.

Hall, D.R. (1998) 'Tourism development and sustainability issues in Central and South-Eastern Europe', *Tourism Management* 19(4): 423–31.

Hollander, P. (1981) *Political Pilgrims: Travels of Western Intellectuals to the Soviet Union, China and Cuba 1928–1978*, Oxford: Oxford University Press.

Hrvatska Matica Iseljenika (1998) *Croatia for Youth: Youth for Croatia*, Zagreb: HMI.

Jackson, R.T. (1990) 'VFR tourism: is it underestimated?', *Journal of Tourism Studies* 1(2): 10–17.

Jordan, P. (2000) 'Restructuring Croatia's coastal resorts: change, sustainable development and the incorporation of rural hinterlands', *Journal of Sustainable Tourism* 8(6): 525–39.

Kang, S.K.-M. and Page, S.J. (2000) 'Tourism, migration and emigration: travel patterns of Korean-New Zealanders in the 1990s', *Tourism Geographies* 2(1): 50–65.

King, B. (1994) 'What is ethnic tourism? An Australian perspective', *Tourism Management* 15(3): 173–76.

Kottler, J.A. (1997) *Travel That Can Change Your Life: How to Create a Transformative Experience*, San Francisco: Jossey-Bass.

Krakover, S. and Karplus, Y. (2002) 'Potential immigrants: the interface between tourism and immigration in Israel', in C.M. Hall and A.M. Williams (eds) *Tourism and Migration. New Relationships between Production and Consumption*, Dordrecht: Kluwer.

Lennon, J. and Foley, M. (2000) *Dark Tourism: The Attraction of Death and Disaster*, London: Continuum.

Lew, A.A. and Wong, A. (2002). 'Tourism and the Chinese Diaspora', in C.M. Hall, and A.M. Williams (eds) *Tourism and Migration. New Relationships between Production and Consumption*, Dordrecht: Kluwer.

Linde-Laursen, A. (1995) 'Small differences – large issues: the making and remaking of a national border', *South Atlantic Quarterly* 94(4): 1123–44.

Luketich, B. (1999) 'Interview with Bernard Luketich', unpublished transcript, Pittsburgh: S. Carter, 21 April.

May, J. (1996) 'In search of authenticity off and on the beaten track', *Environment and Planning D: Society and Space* 14(6): 709–36.

Meier, V. (1999) *Yugoslavia. A History of its Demise*, London: Routledge.

Mitchell, K. (1997) 'Transnational discourse: bringing geography back in', *Antipode* 29(2): 101–14.

Morrisson, A.M. (1995) 'The VFR market: desperately seeking respect', *Journal of Tourism Management* 6(1): 2–5.

Nguyen, T.H. and King, B. (2002) 'Migrant communities and tourism consumption', in C.M. Hall and A.M. Williams (eds) *Tourism and Migration. New Relationships between Production and Consumption*, Dordrecht: Kluwer.

Ning, N. (1999) 'Rethinking authenticity in tourism experience', *Annals of Tourism Research* 26(2): 349–70.

Prpic, G. (1971) *The Croatian Immigrants in America*, New York: Philosophical Library.

Republic of Croatia (1990) *The Constitution of the Republic of Croatia*. Online. Available: http://www.usud.hr/html/standard_legal_refernces.html (accessed 13 July 2001).

Šerović, S. (2001) 'IVth biennial conference: tourism in Croatia on the threshold of the 21st century', *International Journal of Tourism Research* 3: 153–55.

Shain, Y. (1999) *Marketing the American Creed Abroad: Diasporas in the US and their Homelands*, Cambridge: Cambridge University Press.

Smith, V.L. (1992) 'Introduction: the quest in guest', *Annals of Tourism Research* 19(1): 1–17.

Smith, V.L. (1998) 'War and tourism: an American ethnography', *Annals of Tourism Research* 25(1): 202–27.

Stein, H. and Hill, R. (1977) *The Ethnic Imperative*, University Park, PA: Penn State University Press.

Sunič, T. (1999) 'Great expectations and small returns. . .? Immigration, emigration and migrations of Croats over the last ten years', paper presented at the Association for Croatian Studies symposium 'Croatian Diaspora in the USA on the Eve of the Third Millennium', St Xavier University, Chicago, 17 April.

Tanner, M. (1997) *Croatia: A Nation Forged in War*, London: Yale University Press.

Thrift, N. (2000) 'It's the little things', in K. Dodds and D. Atkinson (eds) *Geopolitical Traditions: A Century of Geopolitical Thought*, London: Routledge.

Tomljenovic, R. and Faulkner, B. (2000) 'Tourism, inter-cultural understanding and world peace, in B. Faulkner, G. Moscardo and E. Laws (eds) *Tourism in the 21st century: Reflections on Experience*, London: Continuum.

Wearing, S. (2001) *Volunteer Tourism: Experiences that Make a Difference*, Wallingford: CABI Publishing.

Wearing, S. and Neil, J. (2000) 'Refiguring self and identity through volunteer tourism', *Society and Leisure* 23(2): 389–419.

Wight, P. (2001) 'Ecotourists: not a homogeneous market segment', in D.B. Weaver (ed.) *The Encyclopaedia of Ecotourism*, Wallingford: CABI Publishing.

Williams, A.M and Baláž, V. (2000) *Tourism in Transition: Economic Change in Central Europe*, London: I.B.Tauris.

Woodward, S. (1995) *Balkan Tragedy: Chaos and Dissolution after the Cold War*, Washington, DC: Brookings Institute.

# 13 Sojourners, *guanxi* and clan associations

## Social capital and overseas Chinese tourism to China

*Alan A. Lew and Alan Wong*

## Introduction

Tourism is not only one of the more instrumental tools for enhancing social capital, but is likely to become even more important in a world where international migration for economic, political and other reasons has created diasporic populations from virtually every country and culture on the planet. Robert Putnam (1995a) has been credited with popularizing the concept of social capital, which he defined as a usable resource created by open, collective and cooperative networks built on relationships of trust 'that enable participants to act together more effectively to pursue shared objectives' (Putnam 1995b: 665). Frances Fukuyama defined social capital as 'the cultural propensity for people to seek solutions by establishing horizontal links that are outside the government or the state and organized by civil society itself' (from Fukuyama 1995, cited in *Association Management* 2002: 75). Set in a postmodern or post-Fordist conceptual framework, social capital resources are not reduced by usage, but are instead strengthened and enhanced by greater levels of member participation (Ostrom 2000). Unlike traditional forms of economic capital, human capital, or cultural capital (all of which relate to attributes of individuals), social capital is situated in the quality of relationships and is not easily quantifiable or measured (Mohan and Mohan 2002). Friendship and goodwill are examples of this. They are best created through face-to-face interactions and they become resources when 'mobilized to facilitate action' (Adler and Kwon 2002).

Tourism is instrumental in enhancing social capital by bringing people together in face-to-face interactions that can, in properly structured circumstances, lead to mutually beneficial relationships. Belief in this aspect of tourism underlies support for sustainable tourism approaches and ecotourism product developments, as well as broader assertions of tourism as a force for intercultural understanding and global peace-making (cf. IIPTT 2002). Unfortunately, few tourist experiences actually achieve the goal of creating social capital, even if the capital is as amorphous as understanding and peace.

What is missing from the intercultural communication and global peace scenarios is a broader embedding of tourism as a component of a larger system of social capital institution building. For most tourism, such institutions do not exist. Diasporas, however, can provide an institutional framework within which tourism facilitated social capital can be realized. Tourism brings hosts and guests into face-to-face relationships, which are enhanced by the diasporic bond between the two groups.

Social capital is created when this common bond is mobilized into more significant relationships of action. Not all diaspora cultures may be conducive to building social capital in this way. The culture of the overseas Chinese, however, has developed norms of behaviour and created civil institutions that may represent the best example available of how tourism can be a key component in enhancing social capital relationships to the benefit of both diasporic populations and their homelands of origin.

## Guanxi *and Confucian social capital*

Motivations for maintaining ties to a geographically and historically distant homeland are many. For the Chinese there is a racial identity that ties them to China and separates them from other racial groups in their adopted lands. While race alone can be a superficial basis for establishing identity (Chang 1997), in the Chinese case, Han Chinese ethnic culture is so closely tied to the Chinese race, that it transcends a good portion of the great diversity of origin and life experiences among overseas Chinese. Chen (2002) reflected on being ethnically Chinese, yet raised in an English-speaking environment,

> Perhaps more so than any other race, being Chinese carries with it expectations beyond the physical. It's a complete package: linguistic, historical, psychological as well as physical. To be Chinese and not speak the language fluently, well, the mind boggles.
>
> (Chen 2002: 1)

Chineseness can be denied, but it cannot be escaped (Ang 2001). Especially for ethnic Chinese living far from societies that are predominantly Chinese, a trip to China can allow an emersion in racial (if not fully ethnic) sameness that is only possible in a few locations outside of east Asia. For some it can be a reaffirming experience strengthening cultural identity and providing personal meaning in life. For most it can at least address a curiosity of what it means to be Chinese.

Part of what it means to be Chinese is to carry the legacy of a long history of traditional values and obligations that are centred on the family and extended to community and other relationships. These relationships form the basis of a formal social capital network, which has long supported migrations of Chinese overseas and has helped to maintain their relationships with their homeland. The Chinese concept of *guanxi*, which pervades most of the cultures of east Asia, demonstrates the depth of significance that social capital can play in a complex network of human obligations and face-saving sensitivities (King 1994; Nguyen and King 1998). *Guanxi* has often been considered a major difference between Eastern and Western social order, philosophy and world view (Haley *et al.* 1998). *New York Times* journalist Fox Butterfield illustrated this point:

> I began to appreciate how differently Chinese order their mental universe than do Westerners. We tend to see people as individuals: we make some distinctions, of course, between those we know and those we don't. But basically we have one code of manners for all . . . Chinese, on the other hand, instinctively divide people

into those with whom they have a fixed relationship, a connection, what the Chinese call *guan-xi* and those they don't. The connections operate like a series of invisible threads, tying Chinese to each other with far greater tensile strength than mere friendship in the West would do. *Guan-xi* have created a social magnetic field in which all Chinese move, keenly aware of those people with whom they have connections and those they don't . . . In a broader sense, *guan-xi* also help explain how a nation of one billion people coheres.

(Butterfield 1983: 74–5, cited in King 1994: 110)

*Guanxi* relationships and their associated reciprocal obligations are often used within Chinese society to leverage social resources for personal and group advantages. At the same time, they can lead to nepotism, favouritism, corruption, group oppression and limits on one's freedom of behaviour. The latter can become a motivation for leaving home so as to more freely interact in a world of 'strangers' (Fei 1967; De Glopper 1978). *Guanxi* has also been credited with allowing early Chinese entrepreneurs to succeed in places where others could not because it substituted for a weak legal and commercial system (Backman 1999: 225). The relatively recent popularity of 'social capital' (cf. Adler and Kwon 2002; Mohan and Mohan 2002) in some ways reflects a Western re-discovery of a pervasive Eastern tradition.

Travel and tourism back to China may be conceptualized as a traditional part of the *guanxi* social capital system among overseas ethnic Chinese (Lew and Wong 2002). Chinese merchants have a long history of travelling abroad, either on the Silk Road through Central Asia (to ancient Greece and Rome) or by the 'Porcelain Route' by sea through South and Southeast Asia (Pan 1990; Poston *et al.* 1994). In 1841, the British forced China to open its doors to international trade through the first Opium War and impoverished Chinese labourers emigrated in droves to Southeast Asia and the rest of the world (Lew 1995). Unlike their merchant forefathers, most of these coolie labourers were of impoverished rural peasant origin and most were 'sojourners' whose ultimate goal was to return to China after they made their fortunes overseas (Wang 1991; Brogger 2000).

The sojourner form of temporary migration predominated among the overseas Chinese who had left China in the nineteenth and early twentieth century, prior to the establishment of the People's Republic of China in 1949. After 1949, the sojourner model was transformed into an existential tourism relationship between overseas ethnic Chinese and their China homeland (Cohen 1979). Existential travel could be considered a postmodern form of sojourning; that is, one that is more flexible and less essential than the permanent return of the sojourner, yet still allows the traveller to adhere to expected norms of behaviour within the traditional overseas Chinese society (Lew and Wong 2004).

In addition to the topophilic attachment to China that many Chinese feel (Tuan 1974), travel back to the homeland (whether to the home village or to China in general) was a way of meeting the basic *guanxi* requirements of an ethical Chinese. Chinese ethics, based on Confucianism and Taoism, are primarily focused on relationships, with family relations being paramount (Haley *et al.* 1998). Most traditional Chinese would agree with Putnam's statement that 'the most fundamental form of social capital is the family' (1995a: 73; see also Bubolz 2001). Chinese culture has codified the structure of family relationships through the social philosophy of *Kung Fu Zi*

(Confucius, 551–479 BCE) and in Taoism, both of which established clear *guanxi* loyalties and obligations in father–son, husband–wife and older sibling–younger sibling relationships. Chinese ethics even extend into the grave, as children have a filial duty to regularly pay respects and remembrances to both immediate and more distant ancestors, including tending to gravesites. Related to this is a strong tradition of genealogical record keeping, which dates back in China some 3,000 years. Most Chinese villages maintained detailed genealogical records that followed the male line of village members back to the legendary periods of Chinese history (Lim 2000).

The ultimate goal in defining and regulating relationships is the maintenance of social harmony. Networks extending beyond the family unit typically use *guanxi* to incorporate and extend family-type relationship duties. Thus, the village, the township, the county, the province and the country of China become incorporated into a single family-*guanxi* ethic. Business relationships with non-Chinese may also take on similar roles and patterns, though only after a well-established personal rapport is established. Relationships that fall outside of this family-*guanxi* ethical realm are generally considered insignificant and may even be subject to unethical (or a-ethical) treatment (Haley *et al.* 1998).

Maintaining *guanxi* relationships from afar was a challenge for Chinese sojourners in the nineteenth and early twentieth centuries. However, they were able to apply traditional Confucian norms to their special circumstances, which allowed them to lead an ethical life outside of China. Woon (1989) identified the following as paramount values among the earlier sojourner migrant population (emphasis added):

1   The importance of the extended family and a feeling of insecurity in a place without the extended family reference group;
2   Filial piety pressures to return home to care for elderly and ancestral graves and acceptance by villagers upon returning home, despite having left;
3   The presence of an open, class society in southern China, allowing upward socioeconomic mobility; and
4   Increased prestige among fellow expatriate sojourners through donations and *home visits prior to retirement.*

The last point shows that existential travel back to China became an integral part of the more prosperous sojourner's life. Many overseas Chinese returned to China on a frequent basis prior to the Communist revolution in 1949. These traditional sojourner values, including existential travel, continue to influence contemporary overseas Chinese, even though the sojourner model is rare today. Existential tourism back to China, rather than return migration, is the predominant model among the vast majority of ethnic Chinese living outside of China today.

## Contemporary travel back to China

For overseas Chinese travelling back to China, expectations in visitor behaviour vary considerably between earlier immigrants and later generations and between first visits and subsequent visits. Overseas Chinese who migrated during or prior to the 1950s are expected to participate in filial piety rights and *guanxi* obligations to a greater degree, especially as part of their first visit back to the home village. This often includes providing red envelopes (*hung baos*) with money to all relatives, which could include an entire village, providing roast pigs for the ancestral grave visit, hiring lion dancers and setting off fire crackers for the house and grave visits and providing a feast for the extended village family. Of course, the more elaborate these festivities are, the higher the prestige of/for the visitor.

While some traditional overseas Chinese will specifically travel to their home village for special ancestral ceremonial days (Lim 2000), others will include such events as only one part of a broader range of considerations in their travel planning and motivations. Either way, for these overseas Chinese, this aspect of filial piety continues to serve as a motivational element in the decision to travel back to China.

Overseas Chinese migrants are welcome in their home villages because they have a known blood relationship and because villagers view them as wealthy relatives who can benefit the ancestral community. Oxfeld (2001) emphasized the role of 'face' and money in driving both overseas Chinese tourism motivations and village reactions to visiting overseas relations. These can lead to beneficial outcomes, but can also lead to self aggrandizement and manipulation when abused. Meetings with local government and school officials centre on talks of charitable donations and economic investments and can seem to overly heighten the visitor's potential contributions. Donations to schools, in particular, can considerably enhance a visitor's social standing in the community.

Subsequent visits after the first one are generally less elaborate, although a range similar activities are typically included. Visits by second and later generations of overseas Chinese, as well as those who have been disconnected from China for a long time and have lost knowledge of proper behaviour, are far less elaborate. For these 'bananas' (yellow on the outside and white on the inside) only a few of these elements may be included, such as a meeting with local government or school officials and a red envelopes for only the very closest relatives.

Thus, Confucian values and norms of behaviour have evolved into sojourner values, which in turn have been modified into more diluted, but nevertheless present, contemporary overseas ethnic Chinese values (Wong 1997). They are widely recognized among overseas Chinese and practised by many, though clearly not all. For those that do practise these values to some degree, tourism has become an essential means of realizing or meeting one's *guanxi* obligations.

In a survey of overseas ethnic Chinese visitors to Hong Kong, Lew and Wong (2004) found that home village connections were about twice as common among ethnic Chinese residing in more distant, non-Asian countries (59 per cent to 86 per cent for Australia, the USA and Western Europe) than for those residing closer to China (38 per cent to 42 per cent for Malaysia, Taiwan and Singapore). Furthermore, the survey found that overseas Chinese averaged 53 per cent of visitors from the six surveyed countries/regions overall, with every country having a larger proportion of ethnic Chinese visitors to Hong Kong than was present in their general populations. These data support anecdotal and case study evidence that overseas ethnic Chinese still value maintaining close ties to their home village region. They also suggest that existential travel to China is a significant part of China's tourism market and plays an important role in local economic development for many areas, especially in southern coastal China.

## An institutionalized network of ethnicity

A key component of Putnam's (1995a) concept of social capital is the role of voluntary associations in expanding the strength found in family-based systems of social capital to larger social groups and communities. Because migration often fragments and disperses family units, immigrant Chinese modified their traditional family-based relationship values to develop a system of *guanxi* support through formal voluntary associations. Such voluntary associations are widespread within overseas Chinese communities where they often became a form of extended family, or extended village, for immigrant Chinese. These associations played a vital role in enabling, supporting and maintaining migrant ties to their home village areas, including existential tourism back to China.

When most Chinese migrate overseas they follow a long established 'network of ethnicity' (Mitchell 2000). Most ethnic-based networks of this type are voluntary and have evolved as part of the informal economy (Sassen 1989). Earlier migrant cohorts create a structure which inculcates and influences the experience of newer immigrants (Castles and Miller 1998). In this way they create social capital both informally and formally (Portes 1994). The overseas ethnic Chinese, in particular, have developed a highly formalized system of social capital through a variety of voluntary social associations that provide support for new immigrants, while at the same time strengthening the immigrants' ties to China. Although there are many forms of overseas ethnic Chinese voluntary/social associations, these can be summarized into three types (based on Lim 2000):

1    *Lineage, clan or surname associations.* These could be based on actual blood relations, with members coming from a paternal lineage region where all villagers are related through male lines. More likely, however, they are surname associations that generally welcome overseas Chinese who share a common surname, but typically are associated with a geographic region.
2    *Geographical, place and dialect associations.* Geographical associations generally range from those based on a province, such as Fujian or Guandong, to a city or county region. They rarely extend to townships below the county level and they are often closely associated with dialect associations, which also tend to be associated with political boundaries in China. Southern China, for example, contains at least major 200 dialect groups, many of which are mutually unintelligible (Seng 2002). In recent decades, dialect groups have weakened considerably as Mandarin (*putonghua*) has come to be the lingua franca dialect of China.
3    *Special interest associations* including *trade, guild and business associations* and *culture and sports associations.* Business associations include Chinese chambers of commerce, as well as more specific trade groups, such as coffee or rubber producers in Southeast Asia. Cultural and sporting interests include Chinese music and opera associations, poetry and calligraphy groups and martial arts clubs. All of these special interest groups can also be based on geographical or dialect regions and may be closely tied to associations of those types.

Versions of the first two of these association types replicate the village social organization typical of southern China. They offer opportunities to develop *guanxi* relationships and to reaffirm Chinese traditions. They also work to maintain relationships between migrants and their home village area. As such, overseas Chinese voluntary associations are among the longest continuing civil institutional structures created largely for building and using social capital through tourism. Though their organization is based largely on traditional Chinese social values, they can also offer lessons for more modern efforts to create structures that both strengthen social capital and develop the existential tourism market segment.

In addition to these overseas-based institutions of social capital, there is also a range of organizations in China that serve similar roles. The most prominent among these are the Overseas Chinese Affairs Offices, which were established immediately after the founding of the People's Republic of China (PRC) in October 1949 (Huang 2000). From the start, their objectives were to implement policies passed by the Communist Party of China (CPC) related to overseas Chinese. In the 1950s, these included the 'Method of Setting Up Education and Schools by Overseas Chinese,' and the 'Favoured Treatment to Overseas Chinese and Compatriots for Investment in National Companies' policy. (The term 'compatriots' is used in China to refer to Chinese residing in Hong Kong, Macau and Taiwan. For the purposes of this chapter, compatriots are included in the definition of overseas Chinese.) The second major task of local and provincial Overseas Chinese Affairs Offices was to strengthen relationships between overseas Chinese and China (Douw 2000).

From the 1950s to 1970s these offices fought a major propaganda battle with similar governmental bodies on the island of Taiwan in efforts to win the political support of overseas Chinese. Tourism played a significant role in these efforts as Taiwan would provide almost free trips to overseas Chinese from around the world to visit the island and China would use visits by overseas Chinese to display the great successes of Maoist communism (Lew 1987). China's Cultural Revolution (mid-1960s to mid-1970s), however, was a low point in China's relations with overseas Chinese as they and their relatives in China were accused of being rich landlords and farmers who exploited the poor (Huang 2000). By the early 1980s, however, virtually all of these individuals were rehabilitated and confiscated lands were returned.

Overseas Chinese Affairs Offices today exist at every level of government in China, from the national to the sub-county township, though local level offices only exist in major immigrant source regions. Examples of their work include publishing magazines and websites with news items of interest to overseas ethnic Chinese, assisting overseas Chinese in establishing claims to ancestral village property and helping overseas Chinese find their home village if ties have not been well maintained (Lew and Wong 2003; cf. GOC 2002). Overseas Chinese Affairs Offices also organize trips abroad, primarily to Southeast Asia and North America, to encourage home visits and investments by members of overseas Chinese voluntary associations.

Various other groups, both overseas and in mainland China, are involved in overseas Chinese-related affairs and most promote some form of existential tourism. These include associations of 'returned overseas Chinese', consisting of Chinese who were born and raised for at least part of their lives in a foreign country and then returned to China (cf. XOCF 2002), private business associations (like chambers of commerce) (cf. CTPW 2002), some museums (such as the Overseas Chinese Museum in Xiamen City, Fujian Province), educational institutions from local schools to major universities that have developed relationships with, and research interests in, overseas Chinese.

These other organizations are especially prominent in Fujian and Guandong Provinces in southern coastal China.

Thus, there exists a wide range of institutions that build upon and support the traditional Chinese cultural values of overseas ethnic Chinese and which continually beckon them to return home as existential tourists. These are clearly institutions that work to build social capital, which members can then tap into to achieve a variety of personal objectives. Many of these organizations consciously work to build and utilize social capital networks by expanding the existential tourism roles of overseas Chinese into ones of philanthropic benefactor and economic investor. The issue of social capital becomes less clear in these instances as these organizations are mostly Chinese government sponsored and clearly have ulterior (economic and political) motives that can damage the sense of trust that is at the core of pure social capital building institutions. While originally intending to strengthen interpersonal relationships, traditional Chinese *guanxi* cultural values seem to have fostered a range of institutions that may actually make it more difficult to manoeuvre the complex minefield of trust, obligation and face in seeking an existential relationship with one's home village or homeland.

### Building and using social capital through tourism

The complexity of governmental and quasi-governmental offices and associations that work to develop overseas Chinese social capital have had quite significant impacts on areas where large numbers of overseas Chinese are derived. At the local level, overseas ethnic Chinese have made major contributions to enhancing the livelihood of their home village regions (Lew and Wong 2003). In Taishan County (Guandong Province), one of the more prominent source areas of overseas ethnic Chinese to North America, the strength of the social capital built upon traditional Chinese values, maintained through voluntary associations and developed by overseas Chinese Affairs Offices, can be seen in the level of donations recorded by the Taishan Overseas Chinese Affairs Office from 1978 to 1998. During this period, there were some 16,550 projects amounting to US$14 million (these figures do not include remittances sent to individual family numbers) (Huang 2000). Major areas of donations included: 577 school related projects; 40 medical related projects; 282 technology, culture and sports projects; and the constructions of 19 residential building blocks, 118 bridges and 1,040 km of roads.

These numbers, however, do not show how the social capital of overseas Chinese relationships has been transformed from existential tourism to philanthropy and investment. In a study that examined one aspect of this process, Lew and Wong (2003) looked a the contents of magazines for overseas Chinese readers published by township and county level Overseas Chinese Affairs Offices in southern China. Figure 13.1 shows the model of social capital formation which they used to explain the number and types of articles that these magazines included. They found that the first step, and the dominant area of emphasis, in the process was that of creating a foundation built on a sense of shared, common origin. Once this was done, face-to-face contacts through existential tourism, sometimes connected with philanthropic donations, was the crucial next step toward building supportive relationships between overseas Chinese and their homeland relations. In describing traditional business relations in a coastal Taiwan community, De Glopper (1978: 297) noted that,

> The very first thing to say about the structure of business relations in Lukang is that one does not do business with people one does not know. No one deals with strangers.

Although Chinese society has changed some over the past couple of decades, the importance of face-to-face interactions is still high. When properly structured, tourism can allow interpersonal relationships to be created upon which social capital can evolve to benefit the home village community. This is also the type of activity which voluntary overseas Chinese associations have facilitated. While economic investment was found to be a major goal of the Overseas Chinese Affairs Offices, only a very few articles in the magazine publications actually covered this topic explicitly because it represents a level of commitment that can only be secured after the more basic levels of *guanxi* relationships have been firmly established in face-to-face meetings brought about by existential tourism.

The importance of face-to-face relationships in the building of these social capital resources is what made travel and tourism a significant part of the lifestyle of early sojourners and this pattern continues to be important today among more recent overseas migrants. Driven by the remnants of *filial piety* and supported by the ongoing activities of overseas voluntary organizations and government agencies within China, overseas Chinese tourism today has become a crucial link in building social capital and moving it from a sense of existential camaraderie among co-ethnics to a form of economic and social development.

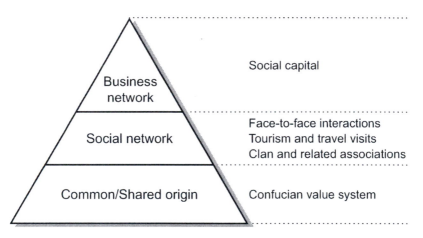

*Figure 13.1* A model of social capital building through tourism.

## Conclusions

There are at least 60 million ethnic Chinese who reside outside of the People's Republic of China (Poston *et al.* 1994; Ma 2002). That number has grown considerably since China opened its doors in 1978 (cf. WHF 2002). Furthermore, overseas Chinese tend to be among the wealthiest ethnic populations in the world (*The Economist* 1993). Together, overseas Chinese and compatriot Chinese account for close to 80 per cent of China's total foreign direct investment (FDI) (Tefft 1994; Cheong 2000).

Clearly overseas and compatriot Chinese comprise a valuable resource which mainland China has turned into a form of social capital that has been tapped into in a variety of ways to enhance the country's economy and improve the lives of the Chinese people.

The shrinking time-space economy of the 1990s with economic globalization, the expansion of the Internet and cheaper and more accessible travel will likely transform the traditional sojourner and existential tourist experience and could create new means of building social capital for communities in China and elsewhere through both 'real' tourism and future worlds of 'virtual' tourism. Some argue that the creation of virtual communities may make face-to-face interactions, such as those which tourism enables, less significant (Rich 1999). With the rapid growth of the Internet in China (Greenspan 2002), there is probably a significant role that virtual travel can play in helping to maintain social capital relationships, but only after these have been established through interpersonal contact. Tourism remains an essential catalyst for the transformation of relations into usable social capital.

Given the large and ever increasing, numbers of migrants worldwide (Zlotnik 1998), the Chinese experience could serve as a model for how developing countries can maximize the social capital of their global offspring by encouraging and developing these new spaces of tourism. Both tourism and social capital have often been looked upon as economic development tools for less developed areas (Wilson 1997; World Bank 2002), though it has also been argued that an over-emphasis on social capital diverts the responsibility for development from governments to the voluntary sector (Mohan and Mohan 2002).

Social capital in China was born at least 2,500 years ago; it was used among overseas Chinese to structure their society when no orderly government structures existed (in early Southeast Asia and in ethnic Chinatown enclaves around the world); and it continues to play a role today in structuring existential tourism and economic development efforts in many parts of China (Lew and Wong 2003b). Diasporic populations need and use existential tourism to create an existential wholeness in their lives overseas. And as Figure 13.1 shows, it is only through existential tourism that the homeland can overcome the geographic space of diaspora and, through face-to-face interactions, convert shared ethnicity into the social actions leading to an enhanced quality of life for all.

## References

Adler, P.S. and Kwon, Seok-Woo (2002) 'Social capital: prospects for a new concept', *Academy of Management Review* 27 (1): 17–41.

Ang, I. (2001) *On Not Speaking Chinese: Living between Asia and the West*, London: Routledge.

*Association Management* (2002) 'Associations: fostering Fukuyama's social capital', *Association Management* 54(6): 75.

Backman, M. (1999) *Asian Eclipse: Exposing the Dark Side of Business in Asia*, Singapore: Wiley (Asia).

Brogger, B. (2000) 'Singapore Huiguan members' donations and investments in Qiaoxiang areas: reasons, problems and rewards', in C. Huang, G. Zhuang and T. Kyoko (eds) *New Studies on Chinese Overseas and China*, Leiden, Netherlands: International Institute for Asian Studies.

Bubolz, M. (2001) 'Family as a source, user and builder of social capital', *Journal of Socio-Economics* 30(2): 129–31.

Butterfield, F. (1983) *China: Alive in the Bitter Sea*, London: Coronet Books.

Castels, S. and Miller, M. (1998) *The Age of Migration*, New York: The Guilford Press.

Chang, Weining C. (1997) 'Ethnic identify of overseas Chinese', in Zhang Guotu (ed.) *Ethnic Chinese at the Turn of the Centuries*, vol. 2, Fuzhou, China: Fujian Peoples' Publishing Co.

Chen, Ee Wen (2002) 'No place like home', *South China Morning Post* LVIII (172, 24 June): *Sunday Morning Post*, Features Section, p. 1.

Cheong, Young-Rok (2000) 'Mode of entry of overseas Chinese foreign direct investment in China', in C. Huang, G. Zhuang and T. Kyoko (eds) *New Studies on Chinese Overseas and China*, Leiden, Netherlands: International Institute for Asian Studies.

Chinese Taishan People Website (CTPW) (2002) *Taishan International Web*. HuaFa Computer Technology Inc.: Taishanshi, Guangdong, Online. Available: http://www.taishan.com/english (accessed 2 February 2002).

Cohen, E. (1979) 'A phenomenology of tourist experiences', *Sociology* 13: 179–201.

De Glopper, D.R. (1978) 'Doing business in Lukang', in A.P. Wolf (ed.) *Studies in Chinese Society*, Stanford, CA: Stanford University Press.

Douw, L. (2000) 'Diasporas and transnational institution-building: some research questions', in C. Huang, G. Zhuang and T. Kyoko (eds) *New Studies on Chinese Overseas and China*, Leiden, Netherlands: International Institute for Asian Studies.

*Economist, The* (1993) 'China's diaspora turns homeward: overseas Chinese investment in China', *The Economist,* US version, 329 (no.7839, 27 November): 33–4.

Fei Hsiao-t'ung (1967) *Hsiang-tu Chung-kuo (Peasant China)*, Taipei: Lu-chou Sh'u-pan-she.

Fukuyama, F. (1995) *Trust: The Social Virtues and the Creation of Prosperity*, Harmondsworth: Penguin.

Greenspan, R. (2002) 'China pulls ahead of Japan', *InternetNews.com – ISP News* (22 April). Online. Available: http://www.internetnews.com/isp-news/article.php/1013841 (accessed 7 June 2002).

Guandong Overseas Chinese Affairs Office (GOC) (2002) *Guangdong Overseas Chinese Affairs Homepage*, Guangzhou: GOC. Online. Available: http://www.gdoverseaschn.com.cn (accessed 2 February 2003).

Haley, G.T., Tan, Chin Tiong and Haley, Usah C.V. (1998) *New Asian Emperors: The Overseas Chinese, their Strategies and Competitive Advantages*, Oxford: Butterworth Heinemann.

Huang Ren Fu. (2000) *Taishan Past and Present 500 Years*, Macau: Macau Publishing (in Chinese).

International Institute for Peace Through Tourism (IIPTT) (2002) *International Institute for Peace Through Tourism*, Stowe, VT: IIPTT. Online. Available: http://www.iipt.org (accessed 2 February 2003).

King, Ambrose Yeo-chi (1994) 'Kuan-hsi and network building: a sociological interpretation', in Tu Wei-ming (ed.) *The Living Tree: The Changing Meaning of Being Chinese Today*, Stanford, CA: Stanford University Press.

Lew, A.A. (1987) 'The history, policies and social impact of international tourism in the People's Republic of China', *Asian Profile* 15(2): 117–28.

Lew, A.A. (1995) 'Overseas Chinese and compatriots in China's tourism development', in A.A. Lew and L. Yu (eds) *Tourism in China: Geographical, Political and Economic Perspectives*, Boulder, CO: Westview Press.

Lew, A.A. and Wong, A. (2002) 'Tourism and the Chinese diaspora', in C.M. Hall and A.M. Williams (eds) *Tourism and Migration: New Relationships Between Production and Consumption*, Amsterdam: Kluwer.

Lew, A.A. and Wong, A. (2003) 'News from the motherland: a content analysis of existential tourism magazines in China', *Tourism Culture and Communication* 4(2): 83–94.

Lew, A.A. and Wong, A. (2004) 'Existential tourism and the homeland seduction: the overseas Chinese experience', in C.L. Cartier and A.A. Lew (eds) *The Seduction of Place: Geographical Perspectives on Globalization and Touristed Landscapes*, London: Routledge.

Lim, P.P.H. (2000) 'Genealogy and tradition among the Chinese of Malaysia and Singapore', in C. Huang, G. Zhuang and T. Kyoko (eds) *New Studies on Chinese Overseas and China*, Leiden, Netherlands: International Institute for Asian Studies.

Ma, L.J.C. (2002) 'Space, place and transnationalism in the Chinese Diaspora', *The Centre for China Urban and Regional Studies Occasional Paper Series* No. 20, Hong Kong: Polytechnic University.

Mohan, G. and Mohan, J. (2002) 'Placing social capital', *Progress in Human Geography* 26(2): 191–210.

Mitchell, K. (2000) 'Networks of ethnicity', in E. Sheppard and T.J. Barnes (eds) *A Companion to Economic Geography*, Oxford: Blackwell.

Nguyen, T.H. and King, B.E.M. (1998) 'Migrant homecomings: Viet kieu attitudes toward travelling back to Vietnam', *Pacific Tourism Review* 1(1): 349–61.

Ostrom, E. (2000) 'Social capital: a fad or a fundamental concept?', in P. Dasgupta and I. Serageldin (eds) *Social Capital: A Multifaceted Perspective*, Washington, DC: World Bank.

Oxfeld, E. (2001) 'Imaginary homecomings: Chinese villagers, their overseas relations and social capital', *Journal of Socio-Economics* 30(2): 181–6.

Pan, L. (1990) *Sons of the Yellow Emperor: A History of the Chinese Diaspora*, Boston: Little, Brown.

Portes, A. (1994) 'The informal economy', in N.J. Smelser and R. Swedburg (eds) *The Handbook of Economic Sociology*, Princeton, NJ: Princeton University Press.

Poston, D.L. Jr, Michael Xinxiang Mao and Mei-yu Yu (1994) 'The global distribution of the overseas Chinese around 1990', *Population and Development Review* 20(3): 631–45.

Putnam, R. (1995a) 'Bowling alone: America's declining social capital', *Journal of Democracy* 6(1): 65–78. Online. Available: http://muse.jhu.edu/demo/journal_of_democracy/v006/putnam.html (accessed 27 June 2002).

Putnam, R. (1995b) 'Tuning in, tuning out: the strange dis-appearance of social capital in America', *Political Science and Politics* 28: 667–83.

Rich, R. (1999) 'American volunteerism, social capital and political culture', *Annals of the American Academy of Political and Social Science* 565: 15–34.

Sassen, S. (1989) 'New York's informal economy', in A. Portes, L. Castel and L.A. Benton (eds) *The Informal Economy: Studies in Advanced and Developing Countries,* Baltimore: Johns Hopkins University Press.

Seng, S. (2002) 'The diminishing Chinese dialects', *Chineseroots.com Newsletter*. Online. Available: cs@chineseroots.com to: allmembers@chineseroots.com (accessed 1 February 2002).

Tefft, S. (1994) 'The rootless Chinese: repatriates transform economy, yet endure persistent resentment', *Christian Science Monitor*, 30 March.

Tuan, Yi-Fu. (1974) *Topophilia: A Study of Environmental Perception, Attitudes and Values*, Englewood Cliffs, NJ: Prentice Hall.

Wang, Gungwu (1991) *China and the Chinese Overseas*, Singapore: Times Academic Press.

Wilson, P.A. (1997) 'Building social capital: a learning agenda for the twenty-first century', *Urban Studies* 34(5–6): 745–60.

Wong, H. (1997) 'The North American (US) Chinese experience', in Zhang Guotu (ed.) *Ethnic Chinese at the Turn of the Centuries*, vol. 2, Fuzhou, China: Fujian Peoples' Publishing Co.

Woon, Y.-F. (1989) 'Social change and continuity in South China: overseas Chinese and the Guan lineage of Kaiping County, 1949–87', *The China Quarterly* 118: 324–44.

World Bank (2002) *Social Capital for Development*, The World Bank Group: Washington,

DC. Online. Available: http://www.worldbank.org/poverty/scapital/index.htm (accessed 24 June 2002).

World Huaren Association (WHF) (2002) *Chinese Diaspora*, San Francisco: WHF. Online. Available: http://huaren.org/diaspora (accessed 2 February 2002).

Xaimen Overseas Chinese Fraternity (XOCF) (2002) *Xiamen Federation of Returned Overseas Chinese*, Xiamen (Amoy), China: XOCF. Online. Available: http://www.xmqs.xm.fj.cn (accessed 2 July 2002).

Zlotnik, H. (1998) 'International migration 1965–1996: an overview', *Population and Development Review* 24(3): 469–510.

# Part III

# Mobilizing diasporas for tourism

# 14 Diaspora, cultural capital and the production of tourism

## Lessons from enticing Jewish-Americans to Germany

*Tim Coles*

### Introduction: creating value from diasporic cultural capital

In a recent edition of the travel lifestyle magazine *Wanderlust*, an anonymous correspondent observed wryly that there had been a notable new development on the tourism scene in Afghanistan. Under the drôle headline 'Kyber passed out?', it was noted that:

> The inexorable spread of Irish theme bars has reached Kabul: The Irish Club opened on St. Patrick's Day (17 March). Owned by an Irish-Australian who's been working in Afghanistan for 11 years, the bar serves Guinness and Irish farmhouse breakfasts to hungry diplomats, journalists and NGO workers and also offers Aussie barbecues, internet access and guest rooms.
>
> (*Wanderlust* 2003: 8)

The report conceded that 'unconfirmed reports suggest that it may actually be quite good'. Reading between the lines, there is an apparent undercurrent of incredulity. Irish pubs are a well-known and a widely spread global phenomenon. Although many of them sell themselves on their self-perceived and self-confessed authenticity, often such claims are exaggerated and are accepted at face value as such. Broadly speaking, any customer about to enter one can be pretty sure of what s/he is about to encounter and of what will be presented for his or her edification. But an Irish pub hybridized with a strong Australian influence? And that melange of Irish and Australian cultural traits and iconography as an attraction and a tourist service in Afghanistan which, lest we should forget, is an Islamic country?

This short report precipitates serious conceptual issues in two respects. The first relates to how experiences and encounters are produced specifically for diaspora tourists. As McKercher and du Cros (2002: 8) argue more generally,

> cultural tourists want to consume a variety of cultural experiences. To facilitate this consumption, cultural heritage assets must be transformed into cultural tourism products. The transformation process actualizes the potential of the asset by converting it into something that the tourist can utilize. This transformation process, though abhorrent to some, is integral to the successful development and sustainable management of the cultural tourism product.

To date, supply-side accounts of diaspora-related tourism products, their appointment and their promotion have been largely overlooked in favour of accounts that concentrate on the diaspora tourist and the experiential (Bruner 1996; Timothy 1997; Hollinshead 1998; Epstein and Kheimets 2001; Stephenson 2002; Duval 2003). Part of the reason for the marginalization of the transformation process is, according to Klemm (2002: 85), that ethnic minority groups (and hence diasporic communities) are invariably viewed as elements of tourism products rather than market segments which purchase holidays. Promotional methods and selling techniques of major tour operators and chains of travel agents are not well adjusted to the needs of 'others' (British-Asians in Bradford), comprising as they do portfolios of products which were intended for much different, conventional 'British-European' markets. Klemm and Kelsey (2002) expose important differences in service delivery and the channels of distribution employed to access British South-Asians. Flight-only transactions with their 'own' communities are the core business for Asian agencies, while general ignorance prevails among the 'mainstream' British travel companies of the needs, preferences and behaviours of potential British-Asian travellers. Culturally appropriate marketing methods and messages are necessary (if not obvious) ways in which others' demand for holidays (even mass-market tourism) may be mobilized.

Thus, promotion of diaspora-related products and experiences has sporadically entered the radar of tourism research. Cohen *et al.* (2002, see also Ch. 8) have explored both the pedagogic construction of the Israel Experience offered to young Jews in the diaspora, as well as the importance of informal marketing in developing interest in and consumption of these tours (Cohen 1999). Participants recruited through personal contacts tended to be more satisfied with their tours than those who were the subject of conventional media. In contrast, Morgan *et al.* (2002: 70, cf. Ch. 15) view diaspora tourists as a viable and accessible market that can be reached by cost-effective, non-traditional methods such as database marketing and public relations as well as informal word-of-mouth.

What these accounts infer is that diasporas are still for the most part naïvely understood by producers and governors of tourism. Instead, the prevailing paradigm among diaspora tourists is perceived to be auto-production of highly individualized experiences, often involving visiting friends and relatives (Stephenson 2002; Duval 2003), sites of relevance to family history (Timothy 1997; Ioannides and Cohen Ioannides 2002; Nash 2002; Fowler 2003) and aided by travel guides to landscapes of cultural relevance (cf. Gruber 1999; also Ch. 6). Not only do producers not realize that diaspora tourists have different demands and behaviours, however subtle (Klemm 2002) from mainstream markets, but also that they are riven by internal cleavages and inconsistencies borne of their dynamism, the nature and negotiation of their roots and their routes and their continually unfolding identities (Safran 1991; Gilroy 1993; Clifford 1994, 1997; Cohen 1997; Mitchell 1997; Urry 2000; Braziel and Mannur 2003; see also Ch. 2). Ignorance of these conceptual advances also motivates such (false) perceptions and comments (as embodied in the report above) that diasporic tourism spaces, experiences and products can at times seem contrived and inauthentic (Hoelscher 1998; Friedmann 1999; Adams 2002). In fact, authenticity is a negotiated and highly personal concept and can be associated with unfolding cultures to reflect changing structural imperatives and independent agencies (Cohen 1997; Ning 1999; Markwick 2001).

Diasporas are complex entities, but from the perspective of tourism production they have the potential to deliver destinations important gains in visitor numbers and spending. For instance, VisitScotland estimates that 20 per cent of Americans visiting Scotland do so because they have Scottish roots and 10 per cent claim that genealogy is the prime motive for their journey (Fowler 2003). As Morgan *et al.* (2002: 70) observe, there are over 90 countries chasing less than one-third of the world's international tourists. Set in this context, diasporas are a more obvious, as yet largely untapped group from which potential increases can be secured. Scots ancestry is claimed by 28 million people worldwide. If only a fraction of those potential tourists are enticed to visit, this would represent an important additional increase in international demand and revenue. For this reason, they observe a new strategy for Scottish tourism, which identified the potential volume and value of business through marketing to diasporas, was launched in early 2002 (see also STB 2001; Fowler 2003).

According to Zukin (1990: 38), cultural capital is the various sorts of capital invested in culture and cultural industries in which 'symbolic' consumption practices provide a basis for capital accumulation (cf. Zukin 1995: Short and Kim 1999). This capital can take many forms; it may be emotional, creative, intellectual and/or financial. Van Hear (1998: 62) argues that the especially rich histories, narratives and experiences associated with migrant communities and their trajectories imbues them with a wealth of 'migratory cultural capital'. Given that the pathways taken and experienced by different migratory groups are almost inevitably going to be singular by nature, migratory cultural capital offers tourism producers, especially National Tourism Organizations, the basis for the sorts of unique selling propositions that are so sadly lacking from many destinations (Morgan *et al.* 2002; Coles 2003a). This is especially evident in the case encountered above. It is the Irishness combined with the Australianness of the iconography and cultural references; the very particular fusion of the two sets of cultural influences; and the specific geographical setting in which this combination is negotiated, that delivers the novelty, the unique selling proposition and, ultimately, the appeal of the attraction. Thus, as a second conceptual aspect to which the report alludes, it is pertinent to note that there is rich cultural capital imbued in diasporic communities and their members. The question arises as to how that cultural capital may be released most effectively and best value be derived from this erstwhile hidden asset.

This chapter investigates recent attempts made by the state tourism authority to attract Jewish-Americans and Jewish-(German-)Americans to Germany for their major vacations. It explores how, in the words of the *Deutsche Zentrale für Tourismus* (DZT, German National Tourist Board 2001a), 'Jewish citizens of the USA in general and the large number of German-Jewish Americans in particular are to be enticed to make joint visits to Germany and Israel'. The objectives are to chart the mechanics used by the state tourism authority and its agents to construct experiences and products for diasporas; and to assess what may be learned from the construction process, especially when read in the wider contexts of market development and contemporary and historical Jewish spaces. Where state-led activities have been explored, they have been largely introspective; that is, inspections of marketing campaigns and their associated products have paid little regard to the wider diaspora, its emergent and unfolding form and characteristics and its inevitable internal fractures and variations (Hollinshead 1998; Braziel and Mannur 2003; see Chs 1 and 2). Equally, discourses have isolated campaigns to attract diaspora tourists from other marketing campaigns simultaneously operating. As discussed below, this is an artificial position because, when reduced to its principal components, diaspora tourism is essentially a specialized form of (personal) 'cultural' and 'heritage' tourism (Timothy 1997). Practically all state tourism authorities in their tourism marketing and management programmes over the past two decades have focused on cultural tourism and heritage tourism in one form or another at one time or another. The relative strength and frequency of these other campaigns has the potential to obfuscate the efforts to stimulate diaspora tourism.

## Germany, the 'moving millions' and the Jewish diaspora in the USA

Migration to the USA, due to the country's lenient immigration policies, boosted the population by 23.5 million people from 1880 to the introduction of quotas in 1921 (Kraut 1982: 2). The contribution of Germans and Germany to demographic growth and the transatlantic migration phenomenon is not to be underestimated. According to Bade (1995: 511), approximately 5.5 million Germans migrated to the USA between 1816 and 1914. Many German-born migrants were Jewish, but reliable estimates of the true total are elusive. Other Jewish migrants passed through Germany on their way to the USA. For Jews from the Pale of Settlement in Poland, Russia and the Ukraine the most popular route to the USA was through eastern Prussia and middle Germany, to the ports of Hamburg and Bremen. Although options existed to travel across Austria-Hungary, fierce local anti-Semitism and state restrictions on shipping lines diverted many. Indeed, the HAPAG shipping company was so popular with Jews as a point of passage from Europe that a synagogue was established in its embarkation centre (Kraut 1982).

Once in the USA, Jewish immigrants' experiences were different to several other prominent groups (cf. Riis 1890). Whereas many immigrants, especially young and male and from southern Europe, were described by immigration officers as 'birds of passage' because of their high frequency of return trips home, Jewish-Americans seldom left. There are several complex reasons for this, not least that Jewish families migrated en masse unlike other migrants and that religious tolerance in the USA was in stark contrast to the anti-Semitism in many of their home countries. Although at first many Jewish leaders were concerned that the secularity of the USA would not be a place for orthodoxy to flourish, later migrations reflected a revised viewpoint. As Kraut (1982) notes, many migrant Jews were well-educated and highly-qualified artisans, tradespersons, merchants and professionals. Many of the long-established Jews to whom the USA had been kind were philanthropists, patrons of the arts and heavily involved in community life. Self-help organizations, especially for recent immigrants, helped to assimilate Jews into American urban society. Today the Jewish population of the USA is estimated to be over 6 million.

German migration to the USA continued in the 1920s, but not always at the elevated levels of the pre-war era due to changing economic, social and political conditions. The next major Jewish migration to the USA was precipitated by the Nazis' rise to power in the early 1930s. Between 1933 and 1945 imprecise estimates suggest that well over half a million refuges left German-speaking countries to other European states or overseas with the number of Jewish émigrés also over 500,000 (Bade 1995: 514).

Through often circuitous routes, America proved the most popular end destination, absorbing as it did 48 per cent of these German-speaking emigrants. From 1933 to 1941 over 104,000 Germans and Austrians entered the USA, more than 80 per cent of whom were Jewish and roughly 50 per cent of whom did not enter until after 1938 (Bade 1995: 514).

While Jewish migration from and through Germany characterizes European migration episodes before 1940, since the end of the Second World War, there have been subtle changes. The Jewish community in Germany has steadily increased, is the fastest growing community in Europe and has become the third largest in Europe (Webber 1997), comprising as it does 87,756 members and 82 communities with 70 synagogues and 27 Rabbis (DZT 2001b; Statistisches Bundesamt 2001). Although Germany has had curious immigration policies (see Bade 1995), this growth in the Jewish population has been fuelled in no small measure by an enduring commitment since 1950 to the 're-migration' of diaspora Germans from the Soviet Union and eastern Europe (Bade 1995; Münz and Ulrich 1998; Tress 1998; Spiegel 2002).

Most strikingly, there has been a relative shift in the Jewish population westwards in Germany. In part, this is a function of changing boundaries, statistical calculation and internal partition until 1989. However, excluding Berlin, the proportion of the Jewish population living in the east has declined. Today, the scattering of the Jewish population mimics the earlier distribution. North Rhine-Westphalia is home to the largest Jewish population (27.2 per cent), followed by Bavaria (17.5 per cent), Berlin (12.8 per cent) and Hessen (12.7 per cent) (Statistisches Bundesamt 2001). In the 1920s the Jewish population was greatest in Berlin (where 30 per cent lived in 1925), the Rhine Province of Prussia (10.3 per cent centred on Cologne, Bonn and Düsseldorf), Hessen-Nassau (9.4 per cent, today NE Hessen), Bavaria (8.7 per cent) and Lower Silesia (5.3 per cent), today in Poland (Statistisches Amt 1930).

It is impossible to review here the growth of the Jewish-American community, nor the richness and diversity of Jewish diasporic experiences (Boyarin and Boyarin 1993; Kugelmass 1993, 1994; Kotler-Berkowitz 1997; Webber 1997). With respect to the construction of tourism, there are salient points to reflect on. First, many Jewish-Americans belong to families that left Germany (and eastern Europe) a century ago. As third- or fourth-generation Jewish-Americans, many will not have a conscious memory of Germany; many may not have visited Germany before; and many will be reliant on the popular media (with its propensity for stereotyping and selectivity) for their constructions of Germany, German-ness, Germans and their attitudes to Jewry (cf. Gruber 2002). Second, the Jewish landscape and settlements of Germany which migrants left behind before the Holocaust – and even before 1914 – were different to contemporary Jewish-Germany. Although there are obvious similarities in settlement patterns, while favoured cities and artefacts from bygone times remain embedded in the landscapes, the profoundly evil Nazi regime expunged many Jewish families, spaces and commemorations. The Holocaust eliminated rural Jewish life in Germany; the redrafting of borders left many Jewish homes beyond the boundaries of (East) Germany in Poland and Czechoslovakia where German references were removed (Ther and Siljak 2001); and the same conditions did not exist in the GDR to foster the concerted rebuilding of Jewish communities after the Holocaust (cf. Gruber 2002). In light of these developments, this chapter proceeds to consider how the *DZT* has worked collaboratively to entice Jewish-Americans to Germany; endeavoured to capitalize on the cultural capital imbued in diaspora; and how its approach is positioned against its wider motives and ambitions for tourism promotion and governance.

## Tourism development in contemporary Germany

Throughout the 1990s, the German market has been characterized by modest expansion mainly from internal sources with significant spatial variations and realignments in both production and consumption (Schnell 1998; Coles 2003b). At the Federal level, the number of arrivals and overnight stays increased between 1992 and 1999 by 20.9 per cent and 4.8 per cent respectively as the benchmarks of 100 million visitors and 300 million overnight stays were exceeded for the first time in 1999 (Spörel 2000: 246). On one level, such results would appear to denote a respectable performance by the German tourism sector; on another, such headline figures hide several structural weaknesses and deficiencies which the *DZT* has attempted to address by a concerted marketing effort both domestically and overseas. The principal difficulty is that for the last quarter of a century Germany has been a net exporter of tourists and tourism-related capital (see Schnell 1998; BMWI 2000; Spörel 2000). Moreover, consumption in Germany has been dominated by domestic visitors. Throughout the 1990s the ratio of overnight stays of domestic visitors to foreign guests has remained virtually constant at 8:1 at the federal level (Spörel 2000: 249). This has been compounded by concerns of how to retain visitors; how to induce repeat visits especially among foreign guests; whether the requisite quality and standard of service is being delivered in the country; and, finally, how to incorporate the destinations and attractions of the former east with their implications for the restructuring of consumption and hence revenues to existing (western) businesses. Foreign visitors are perceived as pivotal to future growth because of their much higher *per capita* and *per diem* spending as well as their often longer length of stay and considerable market potential.

The *DZT*'s overseas marketing activities reflect these imperatives. In addition to several cooperation agreements with foreign partners to promote Germany and to enhance accessibility, the *DZT* has been pursuing a comprehensive programme of events and offers to attract foreign visitors. Five basic themes have been prevalent. Urban tourism contributes over 50 per cent of overnight stays to the tourism industry. To stimulate further consumption of urban environments with their strong cultural and heritage offers the *DZT* has been vigorously pursuing its existing *German City Experience* campaign in which 72 cities are featured (DZT 2001b: 16). As a partner, the *DZT* is also involved in the similar Magic Cities (Berlin, Dresden, Düsseldorf, Frankfurt am Main, Hamburg, Hanover, Cologne, Munich, Stuttgart) and Historic Highlights (Augsburg, Bonn, Bremen, Erfurt, Freiburg, Heidelberg, Lübeck, Münster, Potsdam, Regensburg, Rostock, Trier, Würzburg) initiatives run by independent consortia. Beyond this mainstream segment, German tourism governors are keen to access more specific niches such as the youth market (15–34 years), the culinary tourist, and to capitalize on the excellence of health and spa tourism. Finally, in a scheme called 'Holidays in Harmony with Nature' the *DZT* is aiming to capitalize on the country's green credentials and commitment to sustainable tourism by attracting tourists with environmental interests.

Alongside these broad campaigns the *DZT* has organized a series of theme years, including: EXPO 2000 in Hanover; Bach Year 2000 centred on Leipzig; Gutenburg

Year 2000; Romanesque Year 2001; and both Romanticism on the Rhine (from Lake Constance to Holland) and Hermann Hesse year for 2002 (in Baden-Württemburg). The choice of the historical personalities, narratives, movements, locations, environments is deliberately intended to spread the benefits of tourism, but more importantly to attract visitors from the key foreign markets. Six states – the Netherlands, USA, UK, Italy, Switzerland and France – were responsible for over 51 per cent of overseas overnight stays in Germany in 2000. Although the Netherlands was the most important source, the USA was a close second generating 5 million stays, (or 12.7 per cent of the total) and progressively closing the gap in the late 1990s. Visitors from the USA are viewed as a key market because they contributed €1.744billion (9.1 per cent) of all direct tourist spending in 2000. As high as this figure may appear, this market is considered to have future growth potential because it was the fourth highest total contributor to German tourism revenue in 2000, per diem rates were lower than the other leading contributors and European destinations comprised only 18 per cent of the American holiday market (DZT 2001b).

To increase awareness of 'Germany the brand', the *DZT* has presented its main products in the USA with subtle, yet significant modifications. A dedicated version of the *DZT* website has addressed the particular needs of the American market. In terms of principal themes offered to the American audience in 2000, the standard menu of the *German City Experience*, *Culinary Germany* and '*Wellness*' (health and spa) was supplemented by a pilot scheme *Germany – Land of Music* (with Bavaria and Saxony), *Gay Germany* and its 'Jewish Traveller' promotion. Both these latter initiatives, like the youth strand, were (and are) intended to portray Germany as a progressive, liberal, tolerant society that has emerged from under the dark clouds of shame that descended on it half a century ago.

### Presenting Germany as a destination for Jewish-Americans

In the words of the DZT in its guidebook *Germany for the Jewish Traveller* (2000: 4),

> for travellers from all over the world, a visit to Germany is exciting and enormously fulfilling. For Jewish travellers – and particularly Jewish-Americans – it is a country which offers a spectacular; fascinating, poignant and thought-provoking kaleidoscope of experiences.

Happy to endorse the same volume, in his foreword the President of the Central Council of the Jewish Communities in Germany, Ignatz Bubis, argued that a visit to Germany is crucial to members of the diaspora because such a trip allows,

> visitors to pay homage to the memory of those who perished at the hands of a regime whose crimes will serve as a reminder to never let such atrocities happen again. It shows that Germany has learned from its history and that there is a new democratic country, which has been built in the last 50 years.

> (DZT 2000: 1)

Indeed, the contribution of such visits to reaffirming aspects of diaspora identity and belief systems is reflected by the appeal to diaspora loyalties and long held beliefs of victimization (Cohen 1997; Webber 1997). Invoking the words of Joseph Greenblum, a scholar writing in the journal *Judaism*, the guide notes that,

> visits by Jews to Germany 'symbolize the failure of the Nazis to erase Jewish memory . . . That failure would be powerfully demonstrated by a visit to sites of Jewish significance in the very heartland of what was once the Nazi empire . . . Such pilgrimages by Jews would recognize and support the "other Germany" – its accomplishments in reclaiming Jewish history and its seriousness in coming to terms with the past and with itself'.
>
> (DZT 2000: 2–3)

Diaspora mechanisms were used to re-launch the guidebook. The first version was published in 1997; a second edition (on which this campaign is based) was launched by a letter from the Head of the DZT in New York to 3,600 rabbis in the USA, presenting the programme to them. Forty-five destinations are offered to the Jewish traveller in the guidebook. Its style is very reminiscent of the guides for independent Jewish travellers published by Ruth Gruber (1999), but without her definitive hallmark of perspicacious detail. The destinations range from the main contemporary and historical homes of German Jewry (Berlin, Frankfurt, Munich, Cologne, Hamburg) to smaller towns, many of which had former vibrant communities (Affaltrach, Ichenhausen, Gröbzig, Wörlitz). Attractions range from communities and synagogues – past and present– to artefacts and reminders: museums, townscapes and fabric and, more chillingly, sites of Nazi atrocity and genocide. Intended primarily at the independent traveller, the guide delivers only information on places that could have significance to the Jewish tourist.

No further information on accommodation or transport is provided, albeit these may be found on the DZT's website. This is also a portal to German destination regions as well as the guide's virtual source. For the less adventurous traveller, a dedicated, structured set of tour packages aimed at delivering Jewish-American travellers doubly enriching spiritual experiences has also recently been launched. The *Germany and Israel Tours* variously appeal to the emotional loyalties of diaspora Jews who have prayed to be 'Next Year in Jerusalem', as well as to their desires to learn more about Jewish heritage in Germany and the Holocaust.

A four-stakeholder partnership has been responsible for delivering the *Germany and Israel Tours*; the *DZT* has united with its Israeli counterpart as well as the private sector partners Lufthansa and TAL Tours. The latter is a specialist operator in tours to Israel and runs the commercial management of the scheme. Included in the basic portfolio of *Germany and Israel Tours* are six main packages of varying degree of flexibility and prescription, independence and structured touring (Box 14.1).

The tours range in price from US$2,995 to US$3,795. Superior, business-class accommodation is offered, transfers are included, as well as a range of other features and options. Within Germany the portfolio of packages is based on three main tourism gateways, each of which is extolled for its Jewish credentials: Berlin, Frankfurt am Main and Munich. These are viewed as starting points from which to branch out via formal (and informal) excursions to closeby cities and attractions in most cases with strong Jewish connections. For instance, visitors to Berlin are reminded of the Sachsenhausen concentration camp nearby as well as Dresden and Leipzig, both of which had strong Jewish communities before 1933 and which are a day excursion

---

**Box 14.1** The basic components of Germany and Israel tours offered in 2002

- *Contrasts and similarities of Germany and Israel* – A 13-night holiday, based on 3 nights in Frankfurt am Main, 4 nights in Munich and 6 nights in Jerusalem.
- *Past and present of Germany and Israel* – A 13-night holiday, based on 2 nights in Frankfurt am Main, 1 night in Dresden, 4 nights elsewhere in Germany and 6 nights in Jerusalem.
- *City packages* – Comprising a minimum 4-night stay in Germany and Israel respectively, with flight, car hire and choice of hotels.
- *Life of Anne Frank tour* – A structured, guide-led tour of 14 nights, including: Frankfurt am Main, Amsterdam, Bergen-Belsen, Hanover, Rhine Valley, Worms, Mainz and Black Forest. Option of extending to Israel on completion of German leg.
- *Music tours of Germany and Israel* – Dedicated tour packages associated with TAL Tours' sponsorship of two main musical competitions.
- *Spas of Germany and Israel* – 14 nights in selected spas: 7 in Germany; 7 in Israel.

Source: TAL (2001).

---

away from the capital. The most notable exception is the case of Munich where the Neuschwanstein Schloss (the so-called 'fairytale castle') is picked out as one of the predominant tourism magnets with strong resonances for American audiences (as an inspiration for the Disney motifs). Beyond these three principal centres of Jewish life and heritage, several other cities are identified as worthy a potential visit, including: Augsburg, Cologne, Dresden, Düsseldorf, Freiburg, Hanover, Leipzig, Mainz. With the exception of Freiburg, each of these cities is also mentioned in the *Germany for the Jewish Traveller* guide book and hence they are logical choices also to feature within the *Germany and Israel Tours* portfolio (Table 14.1).

### Hybrid holidays for hybrid peoples?

*Germany for the Jewish Traveller* and *Germany and Israel Tours* represent hybridized forms of vacation themselves. Compromises, negotiation and mediation characterize their content, their production, their market position and their appeal. They strike an interstitial way in their cultural reference points between the contemporary and the historical; dedicated product development and reflections of generic marketing campaigns; the spiritual and the secular; epiphany and the familiar; and perhaps most of all, what is desired and what can be provided.

A strong case can be made that the places selected for the guidebook and the tours strike a balance between the Jewish landscapes of the past and the present. While major centres such as Berlin and Frankfurt faithfully reflect past and present realities, other parts of Germany are subject to a more selective treatment. Contemporary progress in

*Table 14.1* The commodification of towns and cities for Jewish-(German-)American travellers to Germany in context

| | Town as part of . . . | | | | |
|---|---|---|---|---|---|
| | Germany for the Jewish Traveller | Germany and Israel | Magical Cities | Historic Highlights | Germany WWW* |
| Augsburg | X | X | | X | X |
| Bamberg | X | | | | |
| Berlin | X | | X | | X |
| Bonn | X | | | X | X |
| Braunschweig | X | | | | |
| Cologne | X | X | X | | X |
| Dessau | X | | | | |
| Dresden | X | X | X | | X |
| Düsseldorf | X | X | X | | X |
| Erfurt | X | | | X | X |
| Essen | X | | | | |
| Frankfurt am Main | X | | X | | X |
| Freiburg | | X | | X | X |
| Halle | X | | | | |
| Hamburg | X | | X | | X |
| Hanover | | X | X | | X |
| Koblenz | | X | | | X |
| Leipzig | X | X | | | X |
| Lübeck | X | | | X | X |
| Magdeburg | X | | | | X |
| Mainz | X | X | | | X |
| Munich | X | | X | | X |
| Nuremberg | X | | | | X |
| Potsdam | | X | | X | X |
| Regensburg | X | | | X | X |
| Rothenburg (o–d–T) | X | | | | |
| Saarbrücken | X | | | | X |
| Speyer | X | X | | | |
| Stuttgart | X | | X | | X |
| Trier | X | | | X | |
| Weimar | X | | | | X |
| Wiesbaden | X | X | | | X |
| Worms | X | X | | | |

Source: abridged by author from above sources.
* featured on *Deutschland Tourismus*, the DZT's website.

North Rhine-Westphalia is not fully reflected, while Bavaria and the south appear to be over-represented with a level of detail not evident elsewhere. One can speculate about the reasons. Bavaria is Germany's principal destination region, the south was the American Zone of Occupation, it has one of the highest levels of development and North Rhine-Westphalia does not have the same strength of brand definition. Eastern destinations beyond Berlin give the impression of being under-represented relative to the contributions of their Jewish communities to German society, economy, culture and politics. However, they were not necessarily the most strongly populated in the past, they are not at present, they have been blighted by isolated cases of anti-Semitism and rightwing neo-Nazi politics (Barber 2001) and they have been subject to recent supply-side restructuring to bring them up to Western standards. Finally, the urban is stressed over the rural and the larger cities are afforded more attention. This raises issues of what should be commodified versus what actually can be. Many Jews lived in larger towns and cities before 1993; many also lived in smaller towns and villages and followed a rural life which was practically eliminated by the Nazis. Many of the smaller towns, especially in the east, do not have the attractions and/or the infrastructure to draw visitors.

Such considerations are important in light of the cost of the holidays and the commensurate level of expectation. By stressing the three metropolises of Berlin (the capital), Frankfurt ('Mainhattan') and Munich (gateway to the Alps, Italy and Austria), diaspora tourists are being offered cities they are likely to know of from their vantage points in North America. They are being presented cities of a size and scale with facilities and services which can be equated to North America in addition to their apparently strong Jewish credentials. In short, the familiarity and security one would expect at home is being blended with the anticipation, excitement and discovery of the unknown. Many cultural tourism (and hence, in this respect, diaspora tourism) activities are secondary features in trips for which the primary purpose is rest and relaxation, often in high quality environments (cf. McKercher and du Cros 2002).

When viewed in their wider settings, further dualities between mainstream destinations and specialized niche locations are exposed. The choice of locations in the diaspora programmes appears a compromise between the need to market generic key strengths and the desire to ensure that relevant diaspora attractions are incorporated (Table 14.1). The promotion of Germany to Jewish-Americans appears to offer another mechanism for increasing the size of the total audience exposed to Germany's major destinations and marketing campaigns. Crucially, this additional constituency is foreign, high spending and in a key target market, the USA. Put another way, the diaspora market would appear another audience for the major marketing priorities, albeit in a subtle repackaged manner. Urban tourism promotion is a key action area in the *DZT*'s promotional activities. Not all the potential Jewish diaspora locations and attractions are listed in their dedicated publicity material (see DZT 2000). However, those selected faithfully reproduce the priorities in other mainstream urban tourism campaigns through its website, 'Magic Cities and Historic Highlights'. Similarly, health tourism is directly implicated and the leitmotiv of a tolerant, liberal and atoned society is reinforced.

As a final observation, there appears a trade-off between one of quintessential features of the diaspora condition and the commercial realities of running a profitable tour operation. Intra-diaspora disparities are one of the defining features of diaspora and such differences exist within host states such that the Jewish-(German-)American community is not a single, entirely homogenous group. Such heterogeneity is to a degree acknowledged. A range of destinations throughout the country is offered (perhaps reflecting the variety of family origins). Furthermore, the rhetoric of the guidebook and the tours is to treat the diaspora traveller as an individual, as the recipient of a very particular tailor-made experience. The possibility certainly exists to fashion an experience relevant to the individual, but the reality is that there is a finite set of choices available. Further specialization beyond this diet is not evident unless the tourist is able to act independently to devise his or her own agenda. For operators, the plurality of diaspora is awkward. While they may wish to cater more deeply for difference, an apparent threshold appears to exist beyond which extended choices and ranges of destinations can make packages unwieldy, unworkable and erode profitability by incurring much higher costs of promotion and administration.

## Conclusion: linking supply with demand to unlock cultural capital

Diasporic groups present tourism research with several challenges of conceptualization and interpretation, not least in how to unravel and understand travel patterns, but also how travel to foreign territories is induced and produced, by whom and for whom. As Gruber's work (1999, 2002) demonstrates, many visits by Jewish-Americans to Europe are privately organized and are designed around familial spaces and references, which are supplemented with major mainstream destinations and attractions. Increasingly, state tourism authorities and private sector businesses have recognized the volume and value of diaspora tourism. This is not a uniquely German phenomenon; in Wales, Scotland, Ireland, England, Hungary, Austria, Poland, France and Spain diasporas have been identified as significant potential sources of demand (Gruber 2002; Morgan *et al.* 2002; Nash 2002; Fowler 2003). By the nature of their association with place, diaspora members are expected to be repeat visitors to 'home' and to spaces on their diasporic routes (see Gruber 2002).

If tourism producers and governors are serious about capitalizing on diasporic demand, dedicated strategic marketing approaches are required, not merely the repackaging and selling of existing products. Though the temptation may be to capitalize on economies of scale and scope by adapting existing formulae to address such niches, tourism producers must avoid repackaging extant products, events and experiences with a diaspora façade. Otherwise, they risk just adding to the increasing congestion in heritage and cultural tourism markets (McKercher and du Cros 2002). Appeals to diasporic tourists will be largely unheard among the multiple, competing messages of other heritage and cultural tourism products and experiences.

Rather, if the value imbued in diasporic cultural capital is to be released, not only is more market research required, but also a far deeper appreciation of the potential market is required than some of the simple constructs that are currently used. Production must acknowledge the complex nature of diasporas and it must connect more eruditely with experiential, demand-side discourse. Product design must recognize the resources available to deliver diasporic experiences and it must understand how space and place are valorized by diasporas and how evaluation varies within particular communities. Diasporic groups tend to be reduced to 'essentialized', singular, homogenous entities. As Gruber (2002: 126) notes of the 'widely varied target audiences' that encompass Jewish tourists, 'what works for some targets may deeply offend or alienate others'. Ioannides and Cohen Ioannides (2002) hint that one possible first step is to audit tangible cultural resources, notwithstanding the importance of symbolic commemoration where references to Jewish life have been eliminated (cf. Gruber 1999, 2002). Stereotypical appeals to 'home' and kinship underestimate the variety and relevance of other types of spaces in the roots and routes of the diaspora and the individual diaspora member. They also mask an ignorance of what is available for commodification, what type of visitor may be attracted and what alternative types of products and experiences may be assembled and expected. As the products and places presented to Jewish-(German-)Americans demonstrate, there are different types of diaspora tourist with different expectations, motivations and requirements; some are committed to intensely personal voyages of discovery, while others embed their search for their Jewish heritage as part of their vacation, their escape from work and the everyday world. Some initial steps have been taken. For instance, the Scottish Tourist Board (2001: 4) has recognized three types of diaspora visitor: family historians, Scots aficionados and homecomers. Furthermore, they correctly note, 'each of these segments has distinct levels of knowledge, emotional attachment and motivation and must therefore be considered separately when formulating a marketing plan'. As McKercher and du Cros (2002) reflect, cultural tourism is a break first and foremost from the everyday world that is punctuated to varying degrees by visits to cultural sites; the level of engagement determines the type of cultural tourist. In order to inform production, more research is required on diaspora visitor types based on their likely engagement with diasporic themes and on their reaction to the vacations presented to, and played out by, them.

Finally, there are limits to the commodification process and these limits can impact on diasporic identities. For instance, the full heterogeneity within diasporas may frustrate attempts to obtain more detailed market intelligence on them and the niches within them. Although a more nuanced understanding of diasporic identities is necessary, from a practical perspective it is almost impossible to explore the fullest extent of variations as part of the preparation of products. Hence, compromises, such as the identification of three broad groups of diasporic visitor to Scotland, will be made. However, production must acknowledge the mutual implications between tourism and diasporic identities, as well as the dilemmas and more abstract outcomes that selective commodification inevitably generates. Producers should recognize their possible role – however influential – as mediators of diasporic identities through the post-event valorization of experiences. Just as cultural heritage managers, they act as *de facto* gatekeepers of resources that contribute to identity formation (Ioannides 1998; McKercher and du Cros 2002; Timothy and Boyd 2003). Diasporic identities are not static; they are constantly evolving as experiences, encounters and performances are contemplated, evaluated and acted upon. Diasporic tourism products deliver identity-shaping stimuli through cognitive feedbacks from experiences. Political upheavals, boundary changes, mass migrations and a century of European history have taken their toll on what heritage can be gazed upon, the jurisdiction of contemporary tourism governors over historical diaspora-generating spaces, what can be packaged and delivered to diaspora tourists and what is available for their appreciation, interpretation and self-reflection. Similarly, the nature of places bundled together and presented to the diaspora tourist will be a function of the information the producer has on the proposed consumer and how this is acted upon. Just as tourism through visiting friends and relatives in the diaspora may mediate, qualify and/or reinforce diasporic identities through the social relations played out during visits (Stephenson 2002; Duval 2003), structured tourism products to sites of diasporic relevance also have the ability to negotiate and moderate identities. Ironically, therefore, through the compromises of production, more complex identities may be mediated which in the future producers will have to try to address to reinvent diaspora tourism products.

# References

Adams, J. (2002) 'The "Bavarianization" of German Texas: ethnic tourism development in Neu Braunfels', paper presented at the Annual Conference of the Association of American Geographers, Los Angeles, March.

Bade, F.J. (1995) 'From emigration to immigration: the German experience in the nineteenth and twentieth centuries', *Central European History* 28(4): 507–35.

Barber, B.J. (2001) *Jihad vs. McWorld. Terrorism's Challenge to Democracy*, New York: Ballantyne Books.

Boyarin, D. and Boyarin, J. (1993) 'Diaspora: generational ground of Jewish identity', *Critical Enquiry* 19(4): 693–725.

Braziel, J.E. and Mannur, A. (eds) (2003) *Theorizing Diaspora: A Reader*, Malden, MA: Blackwell Publishing.

Bruner, E.M. (1996) 'Tourism in Ghana: the representation of slavery and the return of the black diaspora', *American Anthropologist* 98: 290–304.

Bundesministerium für Wirtschaft und Technologie (BMWI 2000) *Tourismus in Deutschland*, Magdeburg: Gebr. Garloff GmbH, 2e Auflage.

Clifford, J. (1994) 'Diasporas', *Cultural Anthropology* 9: 302–38.

Clifford, J. (1997) *Routes: Travel and Translation in the Late Twentieth Century*, Cambridge, MA: Harvard University Press.

Cohen, E.H. (1988) 'Authenticity and commoditization in tourism', *Annals of Tourism Research* 15: 371–86.

Cohen, E.H. (1999) 'Informal marketing of Israel Experience Educational Tours', *Journal of Travel Research* 37(3): 238–43.

Cohen, E.H., Ifergan, M. and Cohen, E. (2002) 'A new paradigm in guiding: the madrich as a role model', *Annals of Tourism Research* 29(4): 919–32.

Cohen, R. (1997) *Global Diasporas: An Introduction*, London: Routledge.

Coles, T.E. (2003a) 'Urban tourism, place promotion and economic restructuring: the case of post-socialist Leipzig', *Tourism Geographies* 5(2): 190–219.

Coles, T.E. (2003b) 'The emergent tourism industry in eastern Germany a decade after unification', *Tourism Management* 24(2): 217–26.

Deutsche Zentrale für Tourismus (undated, *c.*2000) *Germany for the Jewish Traveler*, New York: Selbstverlag. Online. Available: http://www.deutschland-tourismus.de (accessed 28 August 2001).

Deutsche Zentrale für Tourismus (2001a) *US-tour operator is arranging trips to Germany for American Jews*, Frankfurt am Main: DZT Press Release (24 January). Online. Available: http://www.deutschland-tourismus.de (accessed 16 October 2001).

Deutsche Zentrale für Tourismus (2001b) *Marketing Report 2000*, Frankfurt am Main: Selbstverlag. Online. Available: http://www.deutschland-tourismus.de (accessed 17 April 2002).

Duval, D. (2003) 'When hosts become guests: return visits and diasporic identities in a Commonwealth eastern Caribbean community', *Current Issues in Tourism* (forthcoming).

Epstein, A.D. and Kheimets, N.G. (2001) 'Looking for Pontius Pilate's footprints near the Western Wall: Russian Jewish tourists in Jerusalem', *Tourism, Culture and Communication* 3(1): 37–56.

Fowler, S. (2003) 'Ancestral tourism', *Insights* March: D31–D36.

Friedmann, J. (1999) 'The hybridization of roots and the abhorrence of the bush', in M. Featherstone and S. Lash (eds) *Spaces of Culture: City–Nation–World*, London: Sage.

Gilroy, P. (1993) *The Black Atlantic: Modernity and Double Consciousness*, London: Verso.

Gruber, R.E. (1999) *Jewish Heritage Travel: A Guide to East-Central Europe*, Northvale, NJ: Jason Aronson.

Gruber, R.E. (2002) *Virtually Jewish: Reinventing Jewish Culture in Europe*, Berkeley, CA: University of California Press.

Hollinshead, K. (1998) 'Tourism and the restless peoples: a dialectical inspection of Bhabha's halfway populations', *Tourism, Culture and Communication* 1(1): 49–77.

Hoelscher, S. (1998) 'Tourism, ethnic memory and other-directed place', *Ecumene* 5(4): 369–98.

Ioannides, D. (1998) 'Tour operators: the gatekeepers of tourism', in D. Ioannides and K.G. Debbage (eds) *The Economic Geography of the Tourist Industry*, London: Routledge.

Ioannides, D. and Cohen Ioannides, M.W. (2002) 'Pilgrimages of nostalgia: patterns of Jewish travel in the United States', *Tourism Recreation Research* 27(2): 17–25.

Klemm, M. (2002) 'Tourism and ethnic minorities in Bradford: the invisible segment', *Journal of Travel Research* 41: 85–91.

Klemm, M. and Kelsey, S.J. (2002) 'Catering for a minority? Ethnic groups and the British travel industry', paper presented at Tourism Research 2002 – An Interdisciplinary Conference in Wales, Cardiff, September.

Kotler-Berkowitz, L. (1997) 'Ethnic cohesion and division among American Jews: the role of mass-level and organizational politics', *Ethnic and Racial Studies* 20(4): 797–829.

Kraut, A. (1982) *The Huddled Masses: The Immigrant in American Society, 1880–1921*, Wheeling, IL: Harlan Davidson.

Kugelmass, J. (1993) 'The rites of the tribe: the meaning of Poland for American Jewish tourists', in J. Kugelmass (ed.) *YIVO Annual*, vol. 21, *Going Home*, Evanston, IL: Northwestern University Press.

Kugelmass, J. (1994) 'Why we go to Poland: Holocaust tourism as secular ritual', in J.E. Young (ed.) *The Art of Memory: Holocaust Memorials in History*, Munich: Prestel.

Markwick, M.C. (2001) 'Tourism and the development of handicraft production in the Maltese islands', *Tourism Geographies* 3(1): 29–51.

McKercher, B. and du Cros, H. (2002) *Cultural Tourism: The Partnership between Tourism and Cultural Heritage Management*, Binghampton, NY: Haworth Press.

Mitchell, K. (1997) 'Different diasporas and the hype of hybridity', *Environment and Planning D: Society and Space* 15: 533–53.

Morgan, N. Pritchard, A. and Pride, R. (2002) 'Marketing to the Welsh diaspora: the appeal of hiraeth and homecoming', *Journal of Vacation Marketing* 9(1): 69–80.

Münz, R. and Ulrich, R. (1998) 'Germany and its immigrants: a socio-demographic analysis', *Journal of Ethnic and Migration Studies* 24(1): 25–56.

Nash, C. (2002) 'Genealogical identities', *Environment and Planning D: Society and Space* 20(1): 27–52.

Ning, W. (1999) 'Rethinking authenticity in tourism experience', *Annals of Tourism Research* 26(2): 349–70.

Riis, J. (1890) *How the Other Half Lives*, New York: Charles Scribners' Sons (reprinted 1997, Harmondsworth: Penguin).

Safran, W. (1991) 'Diasporas in modern societies: myths of homeland and return', *Diaspora* 1(1): 83–99.

Schnell, P. (1998) 'Germany: still a growing international deficit?', in A.M. Williams and G. Shaw (eds) *Tourism and Economic Development: European Experiences*, 3rd edn, Chichester: Wiley.

Scottish Tourist Board (STB) (2001) *Genealogy Tourism Strategy and Marketing Plan*, Edinburgh: STB.

Short, J.R. and Kim, Y.-H. (1999) *Globalization and the City*, Harlow: Longman.

Spiegel, Der (2002) *Die Flucht der Deutschen*, Hamburg: Spiegel Verlag Rudolf Augstein GmbH and Co. K.G. (Spiegel Special. Das Magazin zum Thema 2).

Spörel, U. (2000) '1999 – Rekordjahr im Deutschen Inlandstourismus. Ergebnisse der Beherbungsstatistik', *Wirtschaft und Statistik* 2000(4): 245–52.

Statistisches Amt (1930) *Statistisches Jahrbuch für das Deutsche Reich 1930*, Berlin: Selbstverlag.

Statistisches Bundesamt (2001) *Statistisches Jahrbuch 2001*, Wiesbaden: Selbstverlag.

Stephenson, M. (2002) 'Travelling to the ancestral homelands: the aspirations and experiences of a UK Caribbean community', *Current Issues in Tourism* 5(5): 378–425.

TAL Tours, German National Tourist Board, Israel Tourist Board, Lufthansa (undated, c.2001) *Germany and Israel, 2001–2002*, Valley Stream, NY: TAL Tours.

Ther, P. and Siljak, A. (2001) *Redrawing Nations. Ethnic Cleansing in East-Central Europe, 1944–1948*, Lanham, MD: Rowman and Littlefield.

Timothy, D.J. (1997) 'Tourism and the personal heritage experience', *Annals of Tourism Research* 24(3): 751–54.

Timothy, D.J. and Boyd, S.W. (2003) *Heritage Tourism*, Harlow: Prentice Hall.

Tress, M. (1998) 'Welfare state type, labour markets and refugees: a comparison of Jews from the Former Soviet Union in the United States and the Federal Republic of Germany', *Ethnic and Racial Studies* 21(1): 116–37.

Urry, J. (2000) *Sociology Beyond Societies: Mobilities for the Twenty-First Century*, London: Routledge.

Van Hear, N. (1998) *New Diasporas: The Mass Exodus, Dispersal and Regrouping of Migrant Communities*, London: UCL Press.

Wanderlust (2003) 'Kyber passed out?', *Wanderlust: Passion for Travel* 58 (June/July): 8.

Webber, J. (1997) 'Jews and Judaism in contemporary Europe: religion or ethnic group?', *Ethnic and Racial Studies* 20: 257–79.

Zukin, S. (1990) 'Socio-spatial prototypes of a new organization of consumption: the role of real cultural capital', *Sociology* 24: 37–56.

Zukin, S. (1995) *The Cultures of Cities*, Cambridge, MA: Blackwell.

# 15 Mae'n Bryd I ddod Adref – It's Time to Come Home

## Exploring the contested emotional geographies of Wales

*Nigel Morgan and Annette Pritchard*

## Introduction

The question of identity, of how we define ourselves in relation to others and society, has become increasingly important in today's sharply transitional times (Hall 1996, 1997). Paradoxically, as national identities have become globally threatened, ethnicity, which implies an active cultural definition, has emerged as a key marker of self and others. Indeed, ethnicity has become a crucial source of identity, defining how we see ourselves 'within the possible range of culturally constructed selves' (Osborne 2002: 160) and is as much a marker of difference as of similarity; that is, setting us apart as well as binding us together. This elevation of ethnicity to centre stage has seen a concomitant rise in the importance of the concept of diaspora in social, economic, political and cultural discourses. Any diaspora is a complex and multi-layered phenomenon but all are characterized by a desire to endure as a distinct collective despite spatial dispersement and such hyphenated communities invest considerable psychological and social energy into maintaining expressions of their identity. Understanding any diaspora, thus involves not merely understanding the physical migrations of a dispersed people, but also the kinematics of cultures, stories, myths and imaginings. In this way, tracing 'the sense of being a diaspora . . . is not about charting the movement of a group in the past so much as [understanding] how a group continuously seeks to invent itself' (Parsons 2000: 2). Such continuous (re)inventions are journeys of being and becoming which invoke and merge the mythologies of the new, promised land and the cherished, sacralized memories of the homeland. Thus, diasporic communities continually reconnect with an otherness which originally represented self, articulating narratives of 'others' as 'self' and 'self' as 'others' through explorations of travel, reminiscence and home.

In this chapter, we explore how differences between identities are marked out and defined through the lens of diaspora tourism. We examine homeland appeals to hybridized communities in migrated spaces, some of the narratives that give credence to a particular identity and explore the ways in which place and culture are entwined. This is achieved by examining which customized excursions were offered to the Welsh diaspora in the Wales Tourist Board's (WTB) *Homecoming 2000 – Hiraeth 2000* campaign video – *Maen Bryd I ddod Adref – It's Time to Come Home.* Through such video analysis, we scrutinize which experiences occupy centre stage, which are hidden; which cultural icons are given precedence, which are ignored; what is invoked to epitomise the home country and what is set aside by the agencies of cultural marketization. As a visual text, video 'can . . . be seen as part of the dominant ideology of a society, reproducing and enhancing its preferred images while appearing to present entirely accurate representation' (Crawshaw and Urry 1997: 182). Yet, of course, such texts privilege particular definitions of cultural representation and identity and certain versions of ethnic place accounts. The privileged story elevated in the promotional material produced for this WTB tourism-marketing campaign is contributing to defining and redefining Wales and the Welsh identity for diasporic consumption.

At the same time, it is clear that, like any ethnicity, Welshness revolves around the relationship between identity and subjectivity, between defining self and other, or inter-subjectivity, and in defining the wider identity of Welshness in terms of cultural, social and political solidarity. In saying this, we would question the notion that identity must involve an essentialist idea such as Welshness, something that is immutably, naturally and essentially simply existing. Welsh states of mind are contested entities and:

> Faultlines are drawn over the question of what constitutes the identity and heritage of Wales and what qualifies to be recognized as Welsh heritage. These are profoundly political questions, which have considerable import for the ways in which appeals to national feeling are mediated in the public sphere.
>
> (Dicks 2000: 62)

Definitions of ethnicity and nationhood are always problematic and multifaceted. McCrone *et al.* (1995: 45) have commented how,

> ethnicity . . . becomes a form of rhetoric read off a dominant white culture which is highly implicit. Hence there is a black but no white consciousness, female but no male . . . Scottish but little English . . . ethnicity helps to define the periphery to the centre rather than the other way round.

Defining Welsh identity is very difficult (Bowie 1993), although scholars have established three basic, but conflicting, strands to Welsh identities: *Y Fro Gymraeg*, Welsh Wales and British Wales (Osmond 1988). Both *Y Fro Gymraeg* and Welsh Wales have, as Gruffudd (1994: 33) argues, their own 'imagined communities' and 'imagined landscapes'. The former is wrapped up with those rural myths and narratives which have become inextricably linked with the discourse of Welsh nationalism and fostered a Welsh *Y Fro Gymraeg* identity based on land (gwlad) and language (iaith), embodied in the folk (gwerin) (Morgan 1983; Dicks 2000). Here, the 'real' Wales is to be found in the unsullied mountains, in song and nature and in pure but poor communities. The second strand of identity is Welsh Wales, or 'valleys' Wales. In contrast, this identity is industrialized and urbanized, working class and anglophone, essentially labourist in viewpoint and modern in approach (Smith 1999). Rarely is Welshness understood to be truly urban or multicultural, with metropolitan Wales equated with 'British Wales', the third and hitherto most peripheral strand of Welsh identity (Gruffudd 1994).

Of course, the hyphenated Welsh communities' own narratives, images and myths are as central to the production of the Welsh diaspora experience as those which emerge from Wales itself, particularly if one agrees that Wales 'has long existed not as a distinct *nation-state* . . . so much as a *state of mind*' (Parsons 2000: 11). While recent political developments (namely devolution and the creation of the National Assembly for Wales – NAFW) are changing this status, Wales' 'state of mind' remains vital. This is even more so, given that appeals to diasporic tourism can be seen not only to be a response to government initiatives but also as enriching and rewarding experiences which culminate in an enhanced sense of identity, history and community. Thus, those scripts and representations privileged in tourism promotional material aimed at the Welsh diaspora are arguably central to the continuous remaking of Wales as a state and a state of mind.

### *Production, signs and reception as cultural analysis*

This study analyses the WTB's campaign video as auto-ethnography; in other words, as a 'text a culture has produced about itself' (Dorst 1989: 4). The Welsh *Homecoming 2000 – Hiraeth 2000* (hereafter *Hiraeth 2000*) campaign was produced in Wales to appeal to fragmented Welsh communities around the globe. We draw on approaches from cultural geography, post-colonial studies, critical and cultural studies and particularly cultural history to scrutinize the *Hiraeth 2000* video as a cultural text, as a site of cultural production which is the culmination of both social interaction and individual experiences (Pink 2001). As such, our chapter examines the video as a cultural artefact and discusses the systems of production and signification that 'gave rise to the artefact and from which it derives its meanings' (Ashplant and Smyth 2001: 5–6). Production is concerned with authorship, mode of publication and contemporary, historical and cultural context. Signification focuses on the conventions within which cultural texts are produced (language, performative dimensions of delivery, genres, etc.). To complete the analysis, the reception of the artefact should also be investigated – that is, how contemporaries and those who come later interact with the cultural artefact. This chapter of necessity, however, focuses on the production and signification phases of the *Hiraeth 2000* video.

### The production phase – Wales' *Hiraeth 2000* initiative

In contrast to the activities of, for example, Israel, Ireland, Scotland and some Caribbean destinations (Morgan *et al.* 2002), the first campaign targeting the Welsh diaspora, the *Hiraeth 2000* initiative, was launched to synergize with the UK's Millennium 2000 celebrations. The WTB stimulated and marketed a programme of events under this umbrella title to project 'Wales as a country which has a view of the past with an eye on the future' (WTB 1998a). This initiative was entirely consistent with the new NAFW strategic plan, which identified as a key priority the need to 'raise the international profile and influence of Wales and establish it as a first class place to live, study, visit and do business' (NAFW 2000: 1).

Launched in November 1998, *Hiraeth 2000* had four key elements: the harnessing of residents' visiting friends and relatives (VFR) connections; a dedicated, interactive website; the formalization and exploitation of links to Welsh expatriate communities across the world; and PR-led events and activities. Communications imagery for *Hiraeth 2000* was chosen with care to reflect a sense of place, positioning Wales as a vibrant, forward-looking destination with a hint of nostalgia attractive to the target segments. While ancillary printed material was important, harnessing VFR connections was central to the *Hiraeth 2000* campaign. Media advertising and PR events raised awareness of the campaign in Wales to encourage residents to invite relatives and friends living away from the country to visit in 2000. Those who nominated their friends and relatives received invitation packs for forwarding overseas and those who were nominated for packs were entered into a 'Homecoming Database' which finally totalled 11,000 names and addresses. These 'Homecoming Packs' included an emotive video underlining key recent developments in Wales, together with a brochure and a personalized letter from the then First Minister of Wales, Alun Michael. Significantly, while *Hiraeth 2000* was a WTB initiative, the video was produced in conjunction with the Welsh (economic) Development Agency (WDA).

There were two primary markets for the WTB campaign. One group was composed of émigrés – those people from Wales who had relocated for career or business reasons and who could visit home to see friends and relatives. VFR tourism is a significant sector of the Welsh market and when the *Hiraeth 2000* initiative was planned it accounted for 2.5 million trips and £70 million of spend by UK tourists and 207,000 trips and £48 million of spend by overseas visitors (WTB 1998a). The other group comprised descendants from Welsh lineage who may never have sought to experience the country of their forebears. The 'hyphenated Welsh' can be found around the globe, especially in Canada, Australia, South Africa and Patagonia. The key market, however, was undoubtedly the USA, home to the biggest and most active Welsh diaspora community. Florida and California have the largest Welsh-American populations, while Delaware, New York, Massachusetts, Pennsylvania, Ohio, Virginia, Minnesota, Wisconsin and the Carolinas also have significant Welsh communities (Owen 1999).

The Welsh (defined as those claiming Welsh ancestry) nevertheless account for only 1 per cent of the US population (Owen 1999). Only 55,000 Welsh emigrated to the USA from 1890 to 1939, compared with over a million English, over a million Irish and half a million Scots (Olsen 1994). Nevertheless, 'there is scarcely a family in Wales . . . which does not have its American and Canadian dimension' (Williams 1985: 179–80). Moreover, there has been much recent interest in Americans re-discovering their Welsh heritage. The 1990 US Census revealed over a million Americans reported Welsh ancestry – an increase of 22 per cent over the 1980 Census. According to the National Welsh American Foundation (NWAF), this growth was unlikely to reflect an increase in 'new' Welsh-Americans, but more likely an increase in those individuals recognizing their ancestry (WTB 1998a). This interest was further reflected in the successful NAWF *Census 2000: Count us in America* campaign which gained recognition for Welsh-Americans as an ethnic group in the US Census.

This growing diasporic interest in Wales and its language has been fostered by both collective and personal means. Welsh-Americans, through various avenues (including local, regional and national societies, the Internet and personal home pages): '[a]re, it seems, fighting back against the WASP categorization and are seeking to establish themselves as individuals with a distinctive *Celtic* ancestry' (Parsons 2000: 170). This sense of identity takes many forms. Many collective organizations such as the Welsh *Gymanfa Ganu* Association seek to 'preserve, develop and promote our Welsh religious and cultural heritage . . . and traditions' (www.wrigga.org). Similarly, the NWAF aims to 'provide a link between Welsh-Americans in the USA and the Welsh in the home-land who share a common interest in their culture, heritage and the promotion of the Welsh language' (www.Wales-USA.org). On a more individual level, there has also been a huge growth in online Welsh language learners and in Welsh language Internet chat rooms (Parsons 2000). Equally, personal home pages are instructive reflections of Welsh-Americans' perceptions and constructions of self. However, these may 'oftentimes reveal a curious preoccupation with an imagined land half buried in the mists of a Celtic twilight' (Parsons 2000: 18–19), they are significant in the creation of narratives and representations and they are helping to (re-)shape Welsh identity from beyond Wales. Research undertaken for the WTB at the 1997 *Gymanfa Ganu* (a peripatetic celebration of Welsh culture held annually in North America) to inform the *Hiraeth 2000* initiative reflected some of these preoccupations. It indicated that the appeal of Wales for Welsh Americans lies in: family/cultural affinity; history and heritage; scenery and outdoors activities. Thus, formal tracing of their roots is a secondary motivation for these visitors who instead prefer '. . . to feel and experience the local mood and ambience of their ancestral home' (WTB 1998a).

### The **Hiraeth 2000** *video as a cultural artefact*

The starting point for any advertising campaign is normally the commissioning agency's creative brief. These are revealing documents because, although the bodies 'that represent tourism . . . can only provide structures into which our imaginative practice enters' (Crouch 1999: 4), these structures are created by powerful agencies operating in a world of top-down strategies, harnessing and reflecting the agendas of dominant political interests (Hewison 1987). This particular campaign had to balance the ongoing overseas positioning of 'Wales the tourism destination' as a land of nature and legend, against a need to convey a sense of the vibrant, newly devolved Wales as a place for investment and economic development. The creative brief very clearly outlined the requirements for a campaign that portrayed these two faces of Wales. It had to depict the country in 'a very nostalgic light', reflecting a creative treatment which needed '. . . to consider the past but project the contemporary view of Wales and seek to reflect the growing confidence in the future without alienating the traditional views of the target segment' (WTB 1998b).

The campaign thus had to appeal to diasporic nostalgia without reinforcing any outdated, negative views the target market might have of Wales. It also had to display the passion, friendliness and welcome of Welsh people and the tourism products Wales has to offer. Significantly, it also expressly stipulated that the campaign synergize with the three key principles which underpinned the UK government's approach to the millennium: namely, that cultural enrichment should be available to all; the desire to create a society which encourages inclusivity and at the same time respects diversity; the requirement that regeneration and renewal should be vital and measurable forces in society (WTB 1998b). Less obviously but more critically, the campaign was one strand of the NAFW's wider drive to position Wales internationally. An advertising agency from Cardiff (the capital) was appointed to produce the video for the WTB, the WDA and the other campaign sponsors. This is noteworthy as previous WTB marketing campaigns had been created by London-based agencies. This campaign was unusual, therefore, in the sense that it was authored within Wales in its entirety.

The creative brief pointed to which versions of identity were privileged in this campaign as it stipulated the need to project a confident, contemporary and positive identity. Such versions of identity are hugely important, given privileged tourism stories' potential contributions to marginalized communities' assertions of self; versions of identity that then define place and imagined communities (Anderson 1991). Privileged tourism stories are only one element in the production (and contestation) of place myths (Shields 1991). However, as Massey (1994) has discussed, there are social ramifications of top-down promotion of particular placed heritage identities. Such identities often fail to allow for a progressive sense of place, which enables the telling of conflicting stories (in this case differing versions of Welsh identities), which in turn reflect cultural diversity and invite more nuanced accounts of place. In this instance, the WTB and significantly, the WDA, in responding to the wider political agenda of the NAFW administration, became engaged in a process of cultural marketization, offering particular definitions of cultural representation and identity and particular versions of ethnic place accounts – here the meaning of Wales and what it means to be Welsh. Ironically, while the creative brief expressly referred to the demands of the UK government (WTB 1998b), as we will see below, it is difficult to discern where these influenced the final product. By contrast, the influence of the NAFW administration on the campaign and the accompanying video is much more apparent, even though it was never specifically articulated in the creative brief.

### Constructing a British Wales?

Turning to the campaign video, we must ask precisely which tourism stories does it privilege? Appeals to the emotions are key as the video invites émigrés and the ancestral Welsh to their homeland, not only to (re-)discover old connections, but also to acquaint themselves with contemporary Wales. Such a campaign is consistent with the trend in tourism towards a 'new Romanticism consumed as emotions and spiritually rather than for more utilitarian purposes' (Prentice 2001: 8). As such, it was given additional piquancy because the potential visitors received the video at the invitation of their family and friends. The message was clear: someone in the 'old' country, thinks enough of you to want you to come home. At the same time, the political agenda was made explicit in the accompanying letter from the then First Minister of Wales, Alun Michael:

> Someone special is waiting to welcome you home to Wales. That's why you are receiving this video – to remind you of what you're missing and why it's time to come home for a visit.
>
> (WTB 2000)

The invitation to return transcends individual relationships and there was a sense that the nation itself was waiting to welcome diaspora Welsh home for the millennium. As Alun Michael's letter continued, the millennium:

> is a once in a lifetime occasion. Welsh people will be coming from all over the world to be here, knowing that family and friends and the whole of Wales are waiting to welcome them home . . . and I'm sure you will want to see them and Wales once more.
>
> (WTB 2000)

This appeal is firmly grounded in the Welsh concept of *Hiraeth*. A literal translation is very difficult. *Hiraeth* encapsulates a range of emotions associated with the Welsh *mamwlad*, or motherland. Vital to the Welsh psyche, it is an imperative connecting people with home, articulating a range of emotions, including longing, yearning, homesickness, nostalgia and grief (Collins Spurrell 1991). The video's opening expresses in both words and metaphors how the *Hiraeth* Welsh émigrés feel for their motherland. The video opens to a bright sky and fades into a shot of a woman in a light, airy living space. The camera then pans into a scene of metropolitan America, moving upriver and framing a bridge in the background. These scenes, with the bright, intense light convey a sense of realization dawning as the magnitude of the migration event becomes clear. Émigrés' voices explain how 'It was an absolute, total wrench, for the family and myself'. The sense of loss and the upheaval is powerful: 'I didn't really think about it until I was on the plane and you think, urrgh . . . I'm going half way around the world'.

These voices of people far from home reinforce the specialness of Wales, one that perhaps is best appreciated through the act of leaving. The image of the US city fades and the camera shifts to a shot of the South Wales heritage coast. Then we return to the light modern living space where we see a family. A woman is preparing food, the son is playing with a toy car and the father makes a telephone call. As the man dials from his telephone, images of red, green and white streamers are seen connecting around the globe to a night-time view of Wales from space. The outline of the country is clearly visible as are its urban centres, north and south. The camera follows the coloured streamers as they enter a cottage beside a stream in twilight, rural landscape. Lights illuminate the cottage's windows and the scene shifts to the homely interior setting, with red painted walls.

There follows a brief telephone conversation between David (the émigré) and his parents. Obviously emotional, David asks his father (Gwyn) 'so how's things in good old Wales then?' This provides the cue for the video to shift from the private world of David's family to the public world of Wales. Adopting a bird's-eye approach, the camera sweeps over romantic, empty landscapes of misty, ethereal castles, over the Isle of Anglesey ('Mother of Wales') and across mountains, coastlines and valleys. The beauty, perfection and sacred nature of the Welsh landscape is emphasized and articulated by the migrant voices directing the viewer's gaze to the 'perfect proportions of sky and land'. Those who have left are able to re-evaluate their relationship with Wales so that 'you suddenly see the place through completely different eyes'. They tell us that when you are away from the country, you even miss the things you didn't realize you would such as 'the seasons' and 'even the weather', 'the frosty mornings' and 'the snow on the mountains'. Above all, you miss home, the warmth and the emotion. As one voice echoes, 'it was just nice to get back to the warmth and the feeling that we all have from being back home'. This Wales deserves celebration and interpretation by skilled craftspeople (not 'just' laypeople) as the viewer is told: 'if I were an artist I could explain it better'. Similarly, the accompanying soundtrack in this segment reinforces the sacred nature of Wales as it echoes the choral Latin chants of the Mass, culminating in an audible 'Gloria'.

Gwyn's response to his son's question is, 'You know son, just the same, nothing changes much around here'. However, the video suggests that 'old Wales' has very definitely changed into vibrant, contemporary 'new Wales'. Earlier we saw the landscape of 'old Wales' as spectacular, but passive and uninhabited. Now we encounter 'new Wales', a vibrant, populated landscape, home to bustling cosmopolitan cityscapes and adventurous activity-oriented playscapes. Beginning with a shot of sunrises over a sports stadium and a switch of music to a modern beat, we see scenes of a modern cityscape with glass and marble buildings. The modern music then reaches a crescendo, signalling the next video segment as there follows a series of fast-paced shots of opera concerts, sports events, fireworks and business activities, including shots of Cardiff International Airport. Interestingly, the cosmopolitan cityscapes are exclusively drawn from Cardiff. Significantly, now, the émigré voices are silenced and no longer direct the viewer's gaze in this segment – perhaps because they lack experience of the 'new Wales'. Instead the voices leading us here are those of people who have returned to Wales or those who live there.

The returnees comment on the biggest recent change: the achievement of partial self-government for the first time in modern Welsh history. The video reminds the viewer that contemporary Wales is very different to pre-1999 Wales. As one returnee comments: 'I would love to have been here when we elected ourselves an assembly'. Wales is seen to have not only changed politically, but to have become a place worthy of outside attention – 'suddenly it's there on the map and people are talking about it' – so much so that, as the video demonstrates, beautiful celebrities from beyond Wales are quite happy to be seen and photographed here. Presumably in the past, 'old Wales' was a place off the map, a marginalized space not worthy of discussion. As if to cement this perspective, an authoritative voice from the business world tells the viewer Wales has: '. . . got everything I need' against a visual collage of computer keyboards, hi-tech infrastructure and aeroplanes emerging from hangers. Political, cultural and economic liberation are seen to fuse in this brave new devolved Wales which expatriates can enjoy as soon as they cross the border and return home: 'when we got onto the Severn Bridge and when we saw the Red Dragon [the welcome to Wales road sign] . . . I loved it'.

Other cues reinforce the video's message that Wales is an exuberant, vibrant, modern destination that now has 'wish you were here' appeal. The accompanying soundtrack is fast paced and builds to a crescendo. Similarly, the performance qualities of this section of the video are strong: images of city, culture and playscapes interact to entice the viewer. We see a cricketer acknowledging spectators, pop stars reaching out to crowds who wave in collective response, a conductor in evening dress orchestrating an audience and fireworks exploding across the sky. As this segment closes, a guitarist's final flourish signals the end of the rock music sound track and the final shot of all is of a golfer bowing to camera.

With the video shifting back to the private world of Gwyn, David and Jack (David's young son), the obvious sense of excitement builds as an announcement is soon to be made. Jack tells his grandparents 'Dad says to tell you . . . it's time to come home'. The surprise and delight of the soon-to-be reunited family are matched by the background music, which rises in volume and intensity. A male, Welsh-accented voice-over reminds the viewer: 'Wales. [pause] It's what you remember and so much more and, as the millennium approaches, think about this: [pause] maybe it's time to come home.' It is in this final segment of the video that the people of Wales now make their most significant appearance. They become the focus and face of the story as the anticipation of homecoming builds. A sense of waiting for a decision, for a signal to act is conveyed in many of the scenes as boys wait for the whistle in rugby and swimmers prepare to run into the sea. Suddenly, an explosion of fireworks signals the move to action – the rugby players (smiling and shouting) begin running and the swimmers jump into the sea. Throughout these scenes, the accompanying music exhorts the viewer to answer the call to come home – with 'yeah, yeah, yeah, yeah' being the only audible lyrics. Throughout this final segment, the people featured are passionate, active, living and loving life to the full, whether eating, playing or revelling in the rain.

### Silent voices: your Wales, our Wales, my Wales?

Until now, we have read the obvious meanings in the video. However, images have no fixed or single meanings and whilst mass media texts such as this attempt to generate only one meaning, there are many meanings to the video as different people will use their own subjective knowledge to interpret it (Pink 2001). Moreover, in discourse analysis, it is incumbent upon the researcher '. . . to read against the grain of the text, to look at silences or gaps, to make conjectures about alternative accounts which are excluded by omission' (Tonkiss 1998: 258). Thus, now we need to ask which voices are given precedence and credibility and which are silenced and marginalized in this construction of Wales?

While prominence is given to Wales' rural landscape, it would be a mistake to suggest that this video reflects and champions the *Y Fro Gymraeg* identity. Indeed, the Welsh language is completely absent from the video – appearing only as a bilingual title on the video sleeve. No bilingual signs welcome people to Wales; no Welsh language voices feature in the video. Yet whilst the Welsh language is muted, the Welsh landscape is celebrated. But it is an empty landscape, devoid of people, communities, even animals. It is merely a blank canvas for visitors at play. In this playscape, ancient castles and ruined abbeys can be seen, but living communities cannot be found. In fact, very little uniquely Welsh or Celtic imagery features. Its 'culturescapes' pay no reference to the *Eisteddfod* (Europe's biggest peripatetic cultural festival), the main forum for Welsh language cultural achievement. There is similarly no place here for the Welsh flag, one of the most potent symbols of nationhood. Instead the only red dragon featured is a soft toy seen in David and Catrin's house in America – a sentimental, commercialized and essentially neutered souvenir. The Welsh ethnoscape is silenced and the little that is depicted – the daffodils and the Welsh dresser in David's parents' cottage – is restricted to the private sphere of the home, rather than celebrated in the public arena. The lack of Welsh imagery is in direct contrast to other current Welsh overseas marketing campaigns, which heavily feature the language, cultural and Celtic traditions of Wales (Pritchard and Morgan 2001). Instead the video appears more reminiscent of the WTB's domestic UK marketing which makes virtually no reference to the uniquely Welsh dimensions of Wales (Pritchard and Morgan 2001).

Welsh Wales is similarly missing. No 'valleys' communities are seen, no reference is paid to the industrial heritage of Wales, or to its radical political tradition. No coalmines or slate quarries appear, although those remaining in Wales are popular visitor attractions. Additionally, notwithstanding the video celebrates a particular kind of Wales (British Wales), it does little to convey any sense of Welsh achievement or contribution to world life. There is no reference to Wales' political, economic, cultural or sporting heritage. Rugby (Wales' national sport) features but no stars of the past or the present figure, only young boys. Similarly, Wales has long been associated with the performing arts and, whilst this is a strong and enduring metaphor in the video, little is made of Wales' specific contribution in this field. There is limited reference to opera, but no Welsh performing artistes obviously feature. Instead this arena is dominated by international pop stars. An appearance by the Welsh band, the Manic Street Preachers does little to convey that Wales is a prolific producer of musical talent; instead Wales is constructed as a place that derives its celebrity status from hosting international stars from elsewhere. What remains is a construction and proclamation of British Wales – a metropolitan, modern but ultimately rootless and placeless Wales.

### Situating the production of the video

In addition to reading against the text, discourse analysis also seeks to reinterpret social practices by situating their meanings in broader historical and structural contexts. Thus, we must consider the political context for the production of this video to understand why it presents this particular view of British Wales – a previously tangential strand of Welsh identities. It was commissioned and produced by metropolitan Cardiff-based agencies and it juxtaposes the dramatic but empty rural playscapes of Wales with its populated cityscapes. Here we have binary opposites – representing Cardiff and its rural other. In analysing the video therefore, it is difficult to escape the reading that the Cardiff view of Welsh identity – British Wales – has been privileged over others. The NAFW (located in Cardiff Bay) has provided a new focus for Wales' identity. It is perhaps not surprising therefore – but very revealing – that the video privileges this new expression of nationhood over other identifiers. Indeed, First Minister Rhodri Morgan, in submission of evidence to the UK Parliament's Welsh Affairs Committee's investigation of 'Wales and the World', argued that the NAFW provides the foundation point for building Wales' profile abroad. He said: 'the Assembly gives us a start because nobody can say that there is no such thing as Wales', even though the nation has existed for much longer (NAFW 2000). Indeed, the First Minister has made it a personal priority to promote the Assembly and the interests of Wales abroad. Significantly, the NAFW, through its International Relations Unit, is working with key organizations in Wales (including the WTB, the WDA, the Arts Council for Wales and BBC Wales) to harness all available resources to develop marketing material which positions Wales as:

> a vibrant, outward-looking and ambitious country . . . [reflecting] pride in Wales' rich history and culture and, at the same time, seek to overcome a number of outdated and industrial images that have become associated with it.
>
> (NAFW 2000: 2)

As First Minister Morgan noted at the Committee, this is a complex task as:

> Half the problem really is in deciding how you merge the pride that we have in our linguistic, cultural, industrial and political history *that is Wales' past* with the image and confidence you have about your future.
>
> (WAC 2000: point 5, emphasis added)

In so expressly confining Wales' linguistic, cultural, industrial and political history to its past, we would ask what role this heritage has in Wales' future? This is particularly pertinent given that this heritage was not celebrated in the video. The political dominance of Labour – a party and philosophy which has expressly defined itself as modern and forward-looking and contrasted itself with what it regards as backward, traditional, rural Welsh-speaking Wales (Thomas 1999) – cannot be easily dismissed in this creation of place myths and identities. This is, however, an expression of 'New' rather than 'Old' Labour and modern Wales is constructed as a place for enterprise, consumerism, leisure and marketization. The valleys' identity, imbued with the radical 'Old' Labour tradition, is redundant (and hence not worthy of inclusion) in the metropolitan and consumption-oriented brave new world of a Cardiff-dominated Wales. In a similar vein, the rural Wales of *Y Fro Gymraeg* has been reduced to a mere adventure playground as opposed to living, breathing communities.

It is also interesting to note how this video, redolent of such New Labour views of Wales, directly contradicts the cherished memories and imaginings of Wales' diaspora. We have already alluded to some of their concerns above, but the way in which Wales' identities are discussed and problematized by its politicians is illuminating in this context. Whilst recognizing that 'we do need to tap into the Welsh diaspora' (WAC 2000: point 15), First Minister Morgan notes:

> There is a certain generation problem sometimes that the Wales of Gymanfa Ganu and the male voice choir is sometimes very high profile to people who left in the 1930s, 1940s, 1950s or 1960s or whatever and who are still going strong today . . . But then you are *into this problem,* is that the only image of Wales that you want to portray wonderful though it is? . . . What does it do for the image of Wales? . . . We have to be proud of that image . . . but you have to build on that and have something more modern and vibrant.
>
> (WAC 2000: point 17, emphasis added)

## Conclusion

In this chapter we have explored notions of identity and ethnicity and examined how these are continually contested and engaged in a process of becoming, rather than simply existing, in an essentialist form. We have chosen to spotlight this by focusing on the appeals of Wales' agencies of cultural marketization to the Welsh diaspora, concentrating specifically on the video, which accompanied the *Hiraeth 2000* initiative. We have elsewhere (Morgan *et al.* 2002) evaluated the marketing success of the initiative, which generated £21.2 million in revenue from a campaign budget of £276,684, a return on investment of 77:1. Here, however, we have stressed that, successful or not, such texts are much more than mere promotional vehicles but instead reflect the tensions surrounding definitions of self, identity and the debates over whose stories are given voice. We are all engaged in the making of places through the telling of stories and myths and the promotion of images, whether from within or outside those places. Although it has not been possible here, there is a clear need for work which explores the reception of such promotional texts and to investigate diasporic experiences of homeland and its appeals. In the case of Wales, interest in its heritage and ancestry is growing, particularly in America where its profile has risen dramatically in recent years. For so long the Celtic chameleons, the Welsh are shrugging off their penchant for blending in as they attempt to stand out. Collective organizations and perhaps even more importantly, cyber-Cymru are offering the hyphenated Welsh community the ability to make its own contribution to the making and remaking of Wales through the telling of stories, myths and imaginings (Parsons 2000). Their stories of language, culture, heritage and Celtic 'glamour', however, conflict with the stories told by agencies responsible for promoting Wales to the Welsh diaspora – for the Wales promoted in the video is a Wales stripped of what made (and continues to make) Wales Welsh.

The *Hiraeth 2000* campaign is the product of a 'reconfigured temporal-spatial consciousness' (Dicks 2000: 58), prompted in no small part by the emergence of a NAFW which is concerned with Wales' position on the world stage. It cannot be ignored that close parallels exist between the output of touristic and cultural agencies and political priorities. Here political definitions of place are constructing a Wales where much of what makes Wales Welsh has been problematized and confined to the past – something to be proud of but put aside in a move to embrace the Wales of today and tomorrow. We are presented with a metropolitan-Cardiff view of the new Wales, derived from the commissioning and the advertising agencies. In the process, the concerns of much of Wales – of sustaining vibrant rural and valleys communities, of threats to the Welsh language and of concerns of Cardiff's growing dominance – are silenced, not only in this video but in how prominent Welsh politicians conceptualize Wales in the world. Whilst alternative versions of Welsh identities exist and deserve articulation, in this instance, no one is giving them voice.

# References

Anderson, K. (1991) *Vancouver's Chinatown: Racial Discourse in Canada, 1875–1980*, Montreal: McGill-Queen's University Press.

Ashplant, T.G. and Smyth, G. (2001) 'In search of cultural history', in T.G. Ashplant and G. Smyth (eds) *Explorations in Cultural History*, London: Pluto Press.

Bowie, F. (1993) Wales from within: conflicting interpretations of Welsh identity in S. Macdonald (ed.) *Inside European Identities*, Oxford: Berg.

Collins Spurrell (1991) *Welsh Dictionary*, Glasgow: Harper Collins.

Crawshaw, C. and Urry, J. (1997) 'Tourism and the photographic eye', in C. Rojek and J. Urry (eds) *Touring Cultures Transformations of Travel and Theory,* London: Routledge.

Crouch, D. (1999) 'Introduction: encounters in leisure/tourism', in D. Crouch (ed.) *Leisure/ tourism Geographies: Practices and Geographical Knowledge*, London: Routledge.

Dicks, B. (2000) *Heritage, Place and Community*, Cardiff: University of Wales Press.

Dorst, J.D. (1989) *The Written Suburb: An American Site, an Ethnographic Dilemma*, Philadelphia, PA: University of Pennsylvania Press.

Gruffudd. P. (1994) 'Tradition, modernity and the countryside: the imaginary geography of rural Wales', *Contemporary Wales* 6: 33–48.

Hall, S. (1996) 'New ethnicities', in D. Morley and K.H. Chen (eds) *Stuart Hall: Critical Dialogues in Cultural Studies*, London: Routledge.

Hall, S. (ed.) (1997) *Representation: Cultural Representations and Signifying Practices*, London: Sage.

Hewison, R. (1987) *The Heritage Industry: Britain in a Climate of Decline*, London: Methuen.

Massey, D. (1994) 'Double articulation: a place in the world', in A. Bammer (ed.) *Displacements: Cultural Identities in Question*, Bloomington and Indianapolis: Indiana University Press.

McCrone, D., Morris, A. and Kiely, R. (1995) *Scotland – The Brand. The Making of Scottish Heritage*, Edinburgh: Edinburgh University Press.

Morgan, N.J. and Pritchard, A. (2001) 'Contextualising destination branding', in N.J. Morgan, A. Pritchard and R. Pride (eds) *Destination Branding: Creating the Unique Place Proposition*, Oxford: Butterworth Heinemann.

Morgan, N.J., Pritchard, A. and Pride, R. (2002) 'Marketing to the Welsh Diaspora: the appeal to hiraeth and homecoming', *Journal of Vacation Marketing* 9(1): 69–80.

Morgan, P. (1983) 'From a death to a view: the hunt for the Welsh past in the romantic period', in E. Hobsbawm and T. Ranger (eds) *The Invention of Tradition*, Cambridge: Cambridge University Press.

National Assembly For Wales (NAFW) (2000) Memorandum submitted by the NAFW to The Select Committee on Welsh Affairs, Minutes of Evidence, 2000. *Wales in its World Context: The Role of the UK Government in Promoting Wales Abroad*. Online. Available: www.parliament.the-stationery-office.co.uk/pa/cm199900/cmselect (accessed 1 November 2001).

National Welsh American Foundation (2002) *Mission Statement*. Online. Available: http://www.Wales-USA.org (accessed 2 June 2002).

Olsen, J.S. (1994) *The Ethnic Dimension in American History*, 2nd edn, New York: St Martin's Press.

Osborne, R. (2002) *Megawords*, London: Sage.

Osmond, J. (1988) *The Divided Kingdom*, London: Constable.

Owen, R. (1999) 'Home from home – with just a touch of hiraeth', *Western Mail Magazine*, 20 November: 5.

Parsons, W. (2000) 'From Beulah land to Cyber-Cymru', *Contemporary Wales* 13: 1–26.

Pink, S. (2001) *Doing Visual Ethnography*, London: Sage.

Prentice, R. (2001) 'Experiential cultural tourism: museums and the marketing of the new romanticism of evoked authenticity', *Museum Management and Curatorship* 19(1): 5–26.

Pritchard, A. and Morgan, N.J. (2001) 'Culture, identity and representation. Marketing Cymru or Wales?', *Tourism Management* 22: 167–79.

Rojek, C. and Urry, J. (1997) 'Transformations of travel and theory', in C. Rojek and J. Urry (eds) *Touring Cultures: Transformations of Travel and Theory*, London: Routledge.

Shields, R. (1991) *Places on the Margin: Alternative Geographies of Modernity*, London and New York: Routledge.

Smith, D. (1999) *Wales: A Question for History*, Bridgend: Poetry Wales Press.

Tonkiss, F. (1998) 'Analysing discourse', in C. Seale (ed.) *Researching Society and Culture*, London: Sage.

Thomas, A. (1999) 'Politics in Wales: a new era?', in D. Dunkerley and A. Thompson (eds) *Wales Today*, Cardiff: University of Wales Press.

Wales Tourist Board (1998a) *Homecoming 2000 – Hiraeth 2000: Draft Programme and Delivery Plan*, Cardiff: WTB.

Wales Tourist Board (1998b) *Homecoming 2000 – Hiraeth 2000: Creative Brief*, Cardiff: WTB.

Wales Tourist Board (2000) *Homecoming 2000 – Hiraeth 2000 Video Pack*, Cardiff: WTB.

Welsh Affairs Committee (2000) *Minutes of Evidence*. Examination of witnesses, questions 1–19, Monday 30 October 2000. Online. Available: www.parliament.the-stationery-office.co.uk/pa/cm199900/cmselect (accessed 25 October 2001).

The Welsh National Gymanfa Ganu Association homepage. Online. Available: http//www.wrigga.org (accessed 2 June 2002).

Williams, G.A. (1985) *When Was Wales?* Harmondsworth: Penguin.

# 16 India and the ambivalences of diaspora tourism

*Kevin Hannam*

## Introduction

The Indian diaspora has been the subject of a great deal of empirical research in recent years because of its diversity and complexity. This needs to be placed in the context of India's foreign policies and attempts by the Indian government to overtly engage with the Indian diaspora. In this chapter, it becomes evident that the idea of the diaspora being a 'hidden asset' in terms of its economic potential was not taken seriously by the Indian government until the late 1990s. In particular, this chapter analyses and evaluates India's recent tourism policies aimed at the Indian diaspora. Based upon documentary sources, the central argument is that India's relationship both with its diaspora and its tourism industry is at best ambivalent. This is in direct contrast to other states, not least those presented in this volume (see Chs 12 and 15), where there has been much more positive, enthusiastic and proactive engagement with their respective diasporas for reasons including income generation, place promotion and nation-building. There are several probable reasons for this. First of all, India has failed to engage with its diaspora historically, a feature compounded by the devolved structure of tourism development and governance, which largely prevents any strategic inter-national engagement. Furthermore, evidence suggests that second and third generations of the Indian diaspora are ambivalent about returning to India; they would rather utilize family networks than engage directly with the state and its attempts to reach out to them through marketing and promotions. Finally, the complexity of the regional, ethnic and religious divisions within and among the Indian diaspora largely frustrate a unified approach towards diaspora tourism.

## Defining India's diaspora

The Indian diaspora is estimated at between 14 and 20 million people worldwide, which is relatively small in comparison with other diasporas. However, the Indian diaspora has, nevertheless, been the subject of a great deal of empirical research in recent years because of its diversity and complexity. For example, the edited collections by Clarke *et al.* (1990), Vertovec (1991), Van der Veer (1995a) and Petievich (1999) all cover different aspects of the entire South Asian diaspora. In particular, Van der Veer highlights the considerable hostility in the USA towards recent Indian immigrants. After 1965 US immigration laws restricted migration to those with professional skills, business interests or sizeable amounts of capital to invest. However, interestingly, he argues that the Indian diaspora in the USA maintains a continuing interest in its roots; that is, the search for an elusive and largely mythical India. Meanwhile, in her research, Lessinger (1999) argues that Indian migrants in America construct new hybrid identities for themselves, combining as they do elements of Indian-ness and American-ness. She also notes that 'most immigrants are slightly defensive about having left India, for which all retain strong, if ambivalent, emotions' (Lessinger 1999: 20). In the UK context Bhachu (1999: 71) has recently researched the multiple migratory experiences of British Punjabi women. She argues that South Asian women's 'marriage and dowry patterns are, like their identities, continuously negotiated and determined not by their migration histories but powerfully filtered through by the codes of their local and national cultures and also by their class positions' (Bhachu 1999: 71). Moreover, 'those who have emigrated more than once possess very powerful communication networks which have been greatly facilitated and enhanced by global communications' (Bhachu 1999: 72).

This research has also demonstrated that the Indian diaspora is internally riven with complex and multifarious fractures and is hence extremely heterogeneous. Groups within the Indian diaspora have had a variety of different historical trajectories and have developed in widely divergent historical contexts in many parts of the world (Van der Veer 1995b). It is the fragmented nature of these contexts and experiences that complicates the idea of an Indian diaspora. The differences within the Indian diaspora derive from political and geographical origins (India, Pakistan and Bangladesh), often in combination with religious and ethnic schisms (Hindu, Muslim, Sikh, Buddhist, Christian, etc.). In turn this is overlain with both linguistic and economic divisions (Bhardwaj and Madhusudana Rao 1990). Moreover, the sense of national conscious-ness varies from region to region and it is often specific regions to which individual expatriate Indians remain attached (Lall 2001).

In spite of these complex fractures in Indian society and culture, it has been argued that the Indian migratory experience can actually lead to a stronger sense of identity on the margins of the host society. Thus,

> those who do not think of themselves as Indians before migration become Indians in the diaspora. The element of romanticization which is present in every nation-alism is even stronger among nostalgic migrants, who often form a rosy picture of the country they have left and are able to imagine the nation where it did not exist before.
>
> (Van der Veer 1995b: 7)

Each diaspora constitutes a multiple weaving of many disparate narratives of identity. However, at times these may come together in a confluence of narratives as the experience of diaspora is lived and re-lived. As Brah (1996: 196) has argued, 'diasporic identities are at once local and global. They are networks of transnational identifications encompassing "imagined" and "encountered" communities'.

Based upon previous research, five major groups of migrants within the Indian diaspora have been identified (Clarke *et al.* 1990). First, there are emigrants who left India under the British colonial system (Cohen 1997). For example, indentured labourers and migrants who travelled to Southeast Asia, Africa, the Caribbean and Polynesia up to 1930s. The second group comprises commercial migrants who left

India for Africa, Australia, Europe and the Americas just before and immediately after independence (*c.*1920–1960). Low and high skilled migrants who left India for short-term contract work in the Middle East from the 1970s onwards constitute the third group. Fourth, are the so called 'brain drain' migrants who left India for higher education and better jobs in the UK, the USA, Canada and, largely from the 1980s onwards, Australia. Finally, there have been migrants to other parts of the South Asian subcontinent (notably Pakistan and Bangladesh) in the aftermath of Independence. Some of these migrants have maintained their Indian citizenship, some of them have acquired the nationality of their host society, but most have maintained at least informal links with the country of their origin (Lall 2001).

Subsumed within these categories are two terms that are widely used in India in debates concerning the Indian diaspora: namely, NRIs (Non-Resident Indians) and PIOs (Persons of Indian Origin). Both terms represent functionalist attempts to categorize together for the sake of debate and discussion, in a neutral, almost politically correct manner, the multifarious groups who constitute the Indian diaspora. According to the government definition as detailed in the Foreign Exchange Regulations Act (1973), NRIs are defined as,

1   Indian citizens who stay abroad for employment or for carrying on a business or vocation or for any other purpose in circumstances indicating an indefinite period of stay outside India.
2   Indian citizens working abroad on assignment with foreign governments/ government agencies or international/regional agencies like the UNO (including its affiliates), the IMF, the World Bank, etc.
3   Officials of the central and the state government and public sector undertakings deputed abroad on temporary assignments or posted to their offices (e.g. Indian Diplomatic Mission) abroad.

Put simply, an NRI is an Indian citizen who holds an Indian passport and who lives abroad indefinitely. Being classified as an NRI, though, brings with it lots of special economic and bureaucratic incentives and privileges in India. Colloquially, however, the term NRI is often referred to in the pejorative sense as a 'Not Required Indian' (Dutt 2003; Ved 2003). Conversely, a PIO is officially defined as a person who has at any time held an Indian passport, or a descendant one of whose parents or grandparents held an Indian passport, or a person who has been a permanent resident of undivided India at any time. A partner of a citizen of India or of a person of Indian origin is also deemed to be of Indian origin even though he or she may be of non-Indian parentage. Notwithstanding, there are certain exceptions. For instance, nationals of Pakistan and Bangladesh are not able to claim PIO status for political reasons. Moreover, designation as a PIO does not offer the same range of benefits as being an NRI. More recently the Indian government has sought to redress the imbalance by introducing a range of limited inducements for PIOs. For example, the Persons of Indian Origin Card scheme launched at the end of March 1999 gave a small range of economic, educational and cultural benefits, along with a twenty year visa albeit for a substantial fee.

In spite of their technical differences and although the former has greater official significance to the Indian treasury, the terms NRI and PIO are largely used inter-changeably in parliamentary debates. However, neither NRIs or PIOs have any formal *de jure* political rights in India whatsoever. This is largely because of a deep-seated fear that nationals of other South Asian countries (in particular Pakistan and Bangladesh) would claim to be able to vote in Indian parliamentary and local elections (Bahadur Singh 1979; Lall 2001; Ministry of External Affairs 2001). Indeed, part of the Indian government's problem in attempting to regroup the Indian diaspora under one umbrella definition or another is the fact that Pakistani and Bangladeshi citizens are, in almost all cases, as much of 'Indian' origin as the migrants to Southeast Asia or Africa or the West Indies who left during the colonial period. The government of India has argued that, by default, this would allow Pakistanis and Bangladeshis simultaneously to hold an Indian passport. Based on geopolitical insecurities, the implication is that this would pose a security threat to India (Lall 2001). This is the key reason why the debate on dual nationality for members of the diaspora has stagnated.

## India's diaspora policy

When India gained independence in 1947, the Nehru government's foreign policy excluded the issue of expatriate Indians from policy formulation and actively encouraged the Indian diaspora to integrate into their host societies. According to Lall (2001: 41), 'In Nehru's eyes the expatriate Indians had forfeited their Indian citizenship and identity by moving abroad and did not need the support of their mother country'. Nehru had made expatriate Indians alien in a legal sense and this position had several drawbacks. It meant that India could not get involved when part of the diaspora was going through political, economic or social discrimination (Bahadur Singh 1979). Also, despite continuing informal ties between members of the diaspora and their families in the place of their origin, the diaspora was not encouraged to take part directly in the economic development of India.

Nehru's position reflected his idealistic and intellectual sympathies rather than realistic and practical interest in foreign affairs. Under Nehru, India's priorities had changed from being a de-territorialized, anti-colonial, nationalist movement which included all the Indians from around the world. Rather, there was a move towards a territorialized nation-state project with internal integration as the central priority. India was thus caught in a dilemma when it came to its diaspora. As Lall (2001: 76) notes, 'its foreign policy dictated independence from all foreign involvement and its focus on non-alignment and good relations with the developing nations excluded a specific policy towards that community'.

Rajiv Gandhi's politics in the 1980s, however, heralded a new era in which the potential of the NRI was actively discussed. In particular, two new categories of the Indian diaspora (temporary workers in the Middle East and the highly educated economic migrants in the West) brought about a renewed government interest in the diaspora. These new categories of NRIs made substantial money and maintained informal family ties in India. However, although India tried to facilitate opportunities for the remittances coming from the Gulf, it largely failed to open up the economy for any serious NRI investment beyond the family unit. Indeed, the idea of the diaspora as a 'hidden asset' in terms of its economic potential was not taken seriously by the Indian government until the 1990s. After the end of the Cold War, liberalization of the Indian economy brought with it fresh hope that the diaspora would invest heavily in the Indian economy. However, even after liberalization, there was little the government was prepared to do to establish a formal relationship with diaspora communities and members (Lall 2001). Successive governments remained suspicious of the latent power of the Indian diaspora. It was felt that the diaspora might attempt to corrupt the political system by buying votes in elections. Anti-globalization protestors also felt that there was a threat that the diaspora would corrupt Indian culture. Dr Rajhans, who spoke in a debate on the conferment of rights to People of Indian Origin in the *Lok Sabha* (Indian Parliament), on 25 July 1987 argued that,

> if they are given the right of getting elected here and are given the citizenship, the culture of this country will be changed. Everywhere we will hear pop music and see peep shows. God knows what else will be seen here. We will be finished and they will dominate. We will not be able to stop that situation.
>
> (Cited in Lall 2001: 167)

Furthermore, in many parts of India it was still felt that when NRIs come back to India, it was solely for the purpose of profit. Whatever profit was made would be taken back out of the country and not reinvested thereby reinforcing the economic and social inequalities in India.

The Indian government has never developed a meaningful relationship with its diaspora despite recent attempts at re-engagement. Ultimately, India has remained more concerned with the threat to security posed by its neighbours in South Asia than with embracing diaspora Indians and the economic, cultural and social opportunities they afford India. At present the ties between India and its diaspora are still mainly articulated through a global-local ligature manifested by transnational family relations and the private interests an individual expatriate may have in the locality, region or state of perceived origin. The actual dialogue between the central government and the diaspora as a whole has stagnated governed as it has been by the old ideals of non-interference and suspicion, despite rhetorical change (Lall 2001).

In spite of a long-standing orthodoxy of ambivalence towards the diaspora, there have been some more recent signs that the Indian government has recently begun to re-think its stance on the diaspora once again. In August 2000 the Indian government decided to appoint a High Level Committee on the Indian diaspora, headed by a current MP and former High Commissioner of India to the UK. Its terms of reference were to:

1   Review the status of Persons of Indian Origin (PIOs) and Non-Resident Indians (NRIs);
2   Study the characteristics of the Indian Diaspora;
3   Study the role the Indian diaspora may play in the development of India;
4   Examine the regime that governs the travel and investment problems faced by PIOs and NRIs; and
5   Recommend a broad but flexible policy framework for forging beneficial relationships.

(Ministry of External Affairs 2001)

The report was published in December 2001 and once again it emphasizes the economic investment potential of the diaspora. It notes 'the committee is convinced that the reserves of goodwill among its diaspora are deeply entrenched and waiting to be tapped if the right policy framework and initiatives are taken by India' (Ministry of External Affairs: xi). Two of its recommendations were for annual, international *Pravasi Bharatiya Divas* (Family of India Days) on 9 January each year (symbolically the day M.K. Gandhi returned from South Africa in 1915) and annual *Pravasi Bharatiya Samman* (Family of India Awards) for eminent PIOs and NRIs. Nevertheless, it also recommended that the Indian government develop a more pro-active stance towards attracting the diaspora to India for the purposes of tourism. As we shall see in the next section, although India's tourism policy has at one level mirrored its foreign policy, until very recently it has been very parochial and ambivalent.

## The structure and organization of tourism to India: a brief snapshot

Tourism to India faces an image problem due in part to poor accommodation, transport and sanitary conditions as well as tiresome bureaucracy and political uncertainty (Chaudhary 1996). Currently, just over half of India's foreign tourists come from Western Europe and North America, however the vast majority (over 90 per cent) are non-package 'backpacker' tourists who stay for at least one month. Tourist growth rates have now slowed and international arrivals have not kept pace with global rates of increase in spite of the economic liberalization that has taken place. In general, until very recently tourism has been accorded a relatively low priority and there has been a lack of urgency afforded to tourism development by the federal government of India (Raguraman 1998).

In terms of a central government apparatus, it was not until 1958 that the government of India created a separate Ministry of Tourism, albeit one that was attached to the Ministry of Aviation. It was and remains primarily a policy-making executive organization (Singh 2001), headed as it is by the Minister of Tourism, a politician and the Director-General (Tourism), a senior member of the Indian Administrative Service (IAS), a career civil servant. The Ministry of Tourism (renamed in 2000 as the Ministry of Tourism and Culture) has become the nodal agency for the development of tourism in India and seeks mainly to supplement the work of the constituent state governments in promoting tourism. The Ministry of Tourism concentrates its activities on international positioning and marketing, while the regional (state) governments and the private sector bear most of the responsibility for tourism development and management (Misra 1998). The publicly funded India Tourism Development Corporation (ITDC) spends most of its money on a relatively small number of flagship projects.

The first comprehensive national tourism policy in India was formulated as recently as 1982 and utilized the notion of promoting selective 'travel circuits' to maximize the economic benefits of tourism. This approach is encapsulated in the view that,

> The plan proposed to achieve intensive development of selected circuits, dispel the tendency of concentration in a few urban centers, encourage the diversification of tourist attractions and opening up economically backward areas which hold many tourist attractions.
>
> (Ministry of Tourism 2001: 2)

However, these circuits were based upon either heritage and/or religious grounds and reflected domestic tourism demands rather than foreign preferences (Richter 1989; Singh 2001). Set in its politico-cultural context, this strategy reflected India's focus on the nation-building and the internal integration of the 'nation-state' as its central priority.

In contrast, as a result of the economic liberalization reforms of the 1990s tourism was identified as a priority sector for economic investment and a new tourism policy was developed. In 1992, a National Action Plan for Tourism was established with specific aims to improve the basic tourism infrastructure; to develop more selective marketing strategies; and, more importantly, to increase foreign tourist arrivals and foreign exchange earnings (Sarkar and Dhar 1998). More recently, the current plan (1997–2002) has been interpreted as involving primarily the development of large-scale tourist resorts at selected destinations. The future plan (2002–2007) focuses more on international positioning, however (Ministry of Tourism and Culture 2003).

Despite these plans, in practice the work of the Ministry of Tourism in India has not been considered a high priority. Much of its budget has been and continues to be spent on international marketing. Developing foreign tourism arrivals is seen as a key focus, partly because of the need for foreign currency, but also because of India's changing priorities on the world stage (Hannam 2002). For example, in the wake of an estimated 15 per cent slump in foreign tourist arrivals after 9/11, it has recently developed a new slogan 'Incredible India' to back its first international television campaign aimed at attracting international tourists to visit its designated World Heritage Sites (Gantzer and Gantzer 2002; Shankar 2002). In reality, considerable responsibility is, in fact, devolved to the individual states throughout India's federal political system of government. Virtually all facilities and infrastructure, including the development and maintenance of the airports, is devolved with the national policies and plans in place practically as guidance notes to steer development and management. As Jayanth (2000: 2) argues,

> Even in the scheme for development of 'Special Tourism Areas' by the Centre, the initiative rests with the States. At least five such areas have been selected – Bekal in Kerala, Sindhudurg in Maharashtra, Mamallapuram (or Mahabalipuram) in Tamil Nadu, Puri in Orissa and Diu in Daman and Diu. Though the Centre may release some funds for infrastructure development, it has left it to the States and Union Territory to formulate the scheme and involve both foreign investors and Non-Resident Indians to take up specific projects for development of these areas. So, the responsibility will be with the State Governments to initiate a dialogue with the private sector, draw up a blue-print for the development of each of these areas and offer specific projects for investment.

## Strategic approaches to diaspora tourism in India?

As reflected in the previous comments, the High Level Committee on the Indian diaspora has made it known that the diaspora could make a significant contribution to the growth of tourism in India. This would be mainly in the form of providing the necessary investment for capital projects such as visitor attractions and resort complexes. In this light, the committee recommended that suitable schemes be devised both to attract members of the diaspora to India and to motivate PIO travel agents to promote tourism to India. The Committee noted that first-generation visitors from the UK to India are almost annual, subject to their financial status, while the succeeding generations tend to limit their visits to important family occasions or tourism. This is reinforced by Leonard's research (1999: 47) in which it was noted that,

> Second generation Hyderabadis abroad cheerfully told me that they had not liked Hyderabad at all when they visited it and saw no resemblance to the place described by their parents. Some who had not visited Hyderabad said they had the Taj at Agra and the historic buildings in Delhi on their list of places to visit ahead of Hyderabad. One organized visit of second generation youngsters from the United States was in some ways a disaster, filled with disappointed expectations and cultural misunderstandings: many of these youngsters said in the end that they preferred Bangalore. The homeland has in effect disappeared, so transformed that it is unrecognizable.

In spite of this evidence, the Committee on the Indian diaspora has recommended that there should be greater focus on promoting tourism among second generation PIOs. It argued that special tour packages, pilgrim packages and packages tailor made for this group of diaspora should be developed and publicized (Ministry of External Affairs 2001). In addition, it was argued that putting on special events for the diaspora would help to overcome some of the current negative perceptions that many members of the diaspora have of India. India would then, perhaps, become a more obvious and attractive tourist destination for the Indian diaspora. One of these events is the inaugural *Pravasi Bharatiya Divas* discussed below.

### Pravasi Bharatiya Divas

A significant example of the Indian government's recent attempts at engaging with its diaspora and the promotion of diaspora tourism is provided by the *Pravasi Bharatiya Divas* festival that has recently been inaugurated. The first of these events took place on the 9–11 January 2003 at an estimated cost of 15 Crore Indian Rupees (*c*.£2million) (Sengupta 2003). The dates chosen hold symbolic significance as those when M.K. Gandhi returned to India from South Africa in 1915 to lead India's struggle for freedom. Jointly organized by the Ministry of External Affairs and the Federation of Indian Chambers of Commerce and Industry (FICCI), it was advertised in the Indian Embassies, Consulates and High Commissions around the world as the largest and most prestigious gathering of the Indian diaspora (Consulate General of India 2002).

The objectives of the *Pravasi Bharatiya Divas* were stated as to:

1 Engage with all NRIs/PIOs to understand their sentiment about India, their expectations from India and to propose a policy framework for creating a more conducive environment for their sustained and productive interaction with India and her people;
2 Acquaint the Indian people with the depth, variety and achievements of the Indian diaspora and sensitize them on its problems and expectations from the mother country;

3   Provide an opportunity for the members of the Indian diaspora to network and build relationships that criss-cross 110 countries of the world;

4   Utilize this opportunity to develop closer synergies with the host countries through the Indian diaspora in view of the pre-eminent role the Indian diaspora has played in our relations with several countries.

(Indians meet Indians 2003)

In the end, the *Pravasi Bharatiya Divas* was more of an extremely high level conference held in New Delhi. Ironically, it was sponsored by Pepsi and Virgin amongst other global and national brand names. The Prime Minister of India, Shri Atal Bihari Vajpayee, inaugurated the event. However, for all the rhetoric and fine, noble intentions, the *Pravsasi Bharatiya Divas* was attended by just over 2,000 members of the Indian diaspora, much less than the expected 5,000 (Dutta 1993). Nearly 500 delegates came from the USA, followed by approximately 100 from Canada, Australia and the UK. In the future it is planned that a smaller event would be held annually with a larger one held every third year.

Among the general entertainment, short speeches on the position of the Indian diaspora were given by important and well-known members of the Indian diaspora such as the Right Honourable Sir Aneerood Jugnauth, Prime Minister of Mauritius; Sonia Gandhi, Leader of the Opposition in the *Lok Sabha* (the Indian Parliament); Lord Bhikhu Parekh and Lord Navnit Dholakia, members of the House of Lords in the UK; H.E. Dato' Seri S. Samy Vellu, a Minister of the Malaysian government; Mewa Rambgobin, a South African MP; Mahendra P. Chaudary, the former Prime Minister of Fiji; the Nobel Laureates, Professor Amartya Sen and Sir V.S. Naipaul; the musician Ravi Shankar; and other prominent authors, artists, industrialists, politicians, lawyers and academics.

A number of key themes were developed for discussion at the 'festival': science and technology in India; the global business matrix; culture, language, literature; the voluntary sector and development; knowledge-based industries; healthcare and pharmaceuticals; hospitality and tourism – branding strategies; entertainment and ethnic media; education; financial services; Indian states; and opportunities in defence and internal security research and development.

In practice, many who attended the *Pravsasi Bharitiya Divas* were openly critical of and disillusioned with both the organization of the event and with the Indian government's policies towards its diaspora (Sen 2003a). Lord Bihkhu Parekh reflected,

If India fails to develop a coherent diasporic policy, there is a grave danger that the long awaited reunion of the globally extended Indian family may break up in a much mutual recrimination. As Indians should know, the joint family can turn into a veritable hell if the patterns of interaction between its members are not clearly defined and charitably interpreted.

(NRI Speak 2003)

Many felt that there was still a large gap between the speeches made at this event and the actual policies that the government attempts to implement towards the diaspora (Sen 2003a). The Nobel Laureate Sir V.S. Naipaul was apparently disappointed that the event was more akin to a trade fair rather than a meaningful dialogue, a thought backed up elsewhere by industrialists (Srivastava 2003a). Many of the delegates felt that their voices were not being heard. NRIs in the Gulf were exceptionally critical of their treatment at the event and argued that once again the Indian government had ignored their interests. Indeed, only a few of the NRIs living in the Gulf States were invited to the *Pravasi Bharatiya Divas* despite them constituting the second largest Indian diaspora community in the world and representatives of the Indian government conceded that the diaspora from the Gulf States was not adequately represented (Thomas 2003). Goswami (2003) argued that the Indian policies tend to favour diaspora from the developed world and quoted Lord Parekh in this context: 'the big mistake that is being made is the treatment of the Indian diaspora as a homogeneous group'. Mody (2003) cited prominent delegates who also noted a religious bias in that there was a perception that the government was largely only interested in the Hindu diaspora. Although this accusation was refuted by the Deputy Prime Minister, it was evident that there was a sharp sense of divide amongst the delegates at the *Pravasi Bharatiya Divas*:

> The divide is between descendants of Indians who emigrated on compulsion to islands of the West Indies as landless labourers on sugar plantations and those or the children of those who moved abroad much later to much more comfortable means of livelihood . . . Gupta claims that he spent $15,000 in travelling to India with his spouse (the figures are inclusive of hotel bills). 'I decided to visit India only because a very senior official from among the organizers had rung me up personally. But, after coming here, I realize what a big mistake it has been . . . Here the organizers are only inviting politicians – of India or abroad – to talk. Instead of them, people who are self-made and people who have toiled hard in alien countries should be invited to speak. India can really benefit from the richness of the NRI experience. The Indian Prime Minister did mention this richness, but nothing is seen in actual terms'. Nodding at Gupta, S Sen from Atlanta joined the conversation. 'I completely agree with him. Already, some senior members are speaking of boycotting the seminar from next year. What's the point in coming here?'
>
> (Chatterjee 2003)

There were also criticisms that some of the Indian states themselves were not adequately represented at the event, particularly Uttar Pradesh, Bihar and West Bengal (Duttagupta 2003). Nevertheless, at a special session on the Indian states, tourism was higher up the agenda. For example, Narendra Modi, the Chief Minister of Gujarat, emphasized the importance of tourism, calling on every NRI to sell India as a tourist destination and convince at least five foreigners to visit India. However, his plan to identify a thousand families in Gujarat that could host tourists so that they could do away with hotels and give a real taste of Gujarati culture may seem rather optimistic in the least (Srivastava 2003b).

The need to boost tourism was further recognized at a special session on 'Hospitality and Tourism: Branding strategies for India' hosted by the Minister of Tourism. He argued that the Indian diaspora should contribute to marketing the right image of India overseas. He further argued that: '[t]he government is making its best efforts to remove bottlenecks faced by the tourism industry and has decided to launch a special campaign *Discover India, Discover Yourself* for the Indians living abroad' (rediff.com 2003). In reality, however, many of the delegates were disappointed by their return to India for this event. In particular, the state of India's airports and national airline were heavily criticized (Sen 2003b). Many also felt that the regulations imposed on many returning diaspora members were draconian were they to plan to stay for any length of time. These included an AIDS certificate, an income tax certificate, as well as a certificate to prove that they had not committed any crimes. Moreover, there were complaints of harassment at airports by customs officials. For example, Vetcha and Bhaskar (2003) cite the example of a man from England who was detained for twelve hours for bringing a television into India. Despite paying customs duty, it was argued that he was detained because he refused to bribe the officials and, as a direct consequence of this unfortunate event, he refused to return to India.

On the whole, events such as the *Pravasi Bharatiya Divas* can be viewed as spaces in which definitions of Indianness and questions of authenticity and continuity may be challenged (Van der Veer 1995b). Even as such events forge a bond with the past through a remembrance of origins, it seeks to extend transnational flows of capital via cultural patronage. The *Pravasi Bharatiya Divas* is clearly a much more global, commodified and business orientated event. However, of particular importance for those who maintain contacts with the homeland is that participation in this event announces to those in India that some kind of Indianness is being preserved by those who have settled outside India. It also helps to legitimize the act of migration and validates the status of the Non-Resident Indian (NRI) among resident Indians (Hansen 1999). Indeed, one critic noted the distinct clash between the economic and emotional interests present at the event (Sengupta 2003). Ultimately, however, the critical reception of the *Pravasi Bharatiya Divas* shows that there is still a great deal of work to be done if India is to develop a credible diaspora tourism. In the penultimate section of the chapter, I briefly explore one attempt at the sub-national level to develop diaspora-related tourism.

## Discover your roots?

As we have already seen, tourism has been largely devolved to the regions in India. As a result, it is perhaps no surprise that the first overt diaspora tourism policy has been set up at the regional (state) level in the form of the Uttar Pradesh 'Discover your Roots' scheme (Uttar Pradesh Department of Tourism 2002). Uttar Pradesh is one of the largest states within India and contains many of India's key tourist attractions such as the Taj Mahal. However, in spite of the wealth of its cultural heritages, it is simultaneously looking to develop new markets. This recent scheme has been set up by the publicly owned Uttar Pradesh State Tourism Development Corporation. It is largely web-based and has the following modus operandum,

> . . . on receiving inquiries from NRIs, Britishers and others whose ancestors had once lived/worked in Uttar Pradesh, [the Department of Tourism] will make efforts to locate their places of birth/origin through the old letters, school certificates, passports, land records and through personal inquiries. The department will also provide all necessary facilities in India for their visit to such places where they would be accorded a warm traditional welcome. It would be a great event for them to meet old acquaintances or their descendants. To get to see the remnants

of the house where their forefathers or ancestors lived, schools they visited, wells where they bathed, fields where they tilled and temple/mosque/church where they prayed would be a nostalgic experience for them. The NRIs whose roots are discovered which could be named after them [sic] may adopt the birth place/villages of their forefathers for development of activities thus immortalizing the memories of their forefathers for all times to come.

(up-tourism.com/roots 2002)

The state tourism department thus carries out research on behalf of the prospective tourist and attempts to locate their places of birth or origin through various documentary and oral evidence. The aim of the scheme is not just to increase tourist arrivals, but ultimately also to widen development investment from the Indian diaspora. A detailed application form is provided on the website, however, responses can be erratic. Although relatively simple in execution, the scheme is significant for a number of reasons: first, it provides a genealogical search of past places in the country of origin for interested potential tourists; second, it explicitly links tourism development to the wider investment potentials of the Indian diaspora; and third, it blurs the differences between NRIs, British colonialists and others who have had ancestors who lived and/or worked in India, thereby challenging conventional academic notions of transnationality, power and belonging.

However, for all its innovativeness, the scheme has not been a success. It has not been copied by other Indian states and response has generally been poor so far. Most first-generation NRIs who want to visit India as tourists already retain family connections and have little need of the service. Current evidence also suggests that second- and third-generation PIOs are generally disinterested in returning to India, but, in any case, would still turn to their families for advice on where to visit, rather than engage directly with this policy.

## Conclusion

This chapter has analysed and evaluated the success of India's recent tourism policies associated with the diaspora. More particularly, these have been posited in the context of Indian foreign policy. The idea of the diaspora being a 'hidden asset' in terms of its economic potential was not taken seriously by the Indian government until the late 1990s. At present, however, the approach of the Indian government to diaspora is contradictory; it is attempting to engage with its diaspora more explicitly and actively, but it remains by and large ambivalent. This is in spite of the potential (and actual) contributions diaspora members are able to make to the economy through their spending and (foreign) direct investment.

Conversely, diaspora members are as equally ambivalent about returning to India as tourists. For many, a visit to India has become not so much a vacation as an homage to the extended family; a break from the ordinary, everyday world by observing familial piety elsewhere. Notwithstanding, their visits tend to revolve around 'traditional' ancestral sites, although for second- and third-generation visitors, the allure of other visitor attractions, destinations and resorts rather than familial locations is difficult to resist. However, the ambivalence of the government towards both its diaspora and towards tourism as an economic sector are major concerns.

Clearly, Indian tourism's engagement with diaspora is at a crossroads. There are signs that India is becoming more confident internationally. Ultimately, time will be the judge of whether its tourism promotion, policy and actions will allow both parties to progress. At another level, should policy solutions fail, it will be interesting to see if future generations follow the trend of diffidence among contemporary younger generations towards familial destinations. Visiting friends and relatives, which has formed the mainstay of the Indian tourism economy, could be put under severe threat.

India faces critical issues in the development of its tourism sector, however it raises some pertinent themes for reflection and future research. In terms of a wider message, this chapter has revealed that it is crucial to understand the nature and condition of the diaspora and hence not to take it for granted when attempting to attract members to the homeland. In this case, though, the multiplicity of the divides within the diaspora suggest not only that it is practically impossible to know and to understand all constituencies, but also that, although an umbrella term may be a neat neutral solution for dealing with all, it masks tensions and contexts that satisfy few cohorts. There are limits to liminality and one is to be able to know large diasporic groups sufficiently to appeal to them. Another limit may be that the Indian federal government may only be able to provide a basic framework to engage with diaspora, but its role would become all the more significant if it could mobilize the individual states to appeal more vigorously to motivations among their diasporic populations. This is compatible with the Indian model of government. For their part, the individual states need to recognize that this process of engagement may involve traditional familial triggers, but increasingly a more creative approach will be required involving other, non-diasporic visitor attractions.

As research elsewhere on South and East Asia demonstrates (see Nguyen and King 1998, 2002, Ch. 11; Lew and Wong 2002, Ch. 13), diaspora and tourism can present economic and social opportunities. Both India and China have large overseas diasporic populations, the linkages between which are fostered through family and obligation. Starkly, the Indian state is attempting to mobilize the diaspora as a collective, as a function of the secular state, a 'secular diaspora' to add to other types of diaspora groupings (see Cohen 1997, Ch. 1). There has been very little comparative analysis of the propensity, performance or patterns of different types of diaspora groups. Thus, one interesting theme for future work remains, therefore, to untangle the travel behaviours of different types of diasporas and the constituencies within them. In states such as India, where investment in marketing is scarce, the value for money associated with marketing to particular groups assumes ever greater importance.

# References

Bahadur Singh, I. (1979) *The Other India: The Overseas Indians and their Relationship with India*, New Delhi: Arnold-Heinemann.

Bhachu, P. (1999) 'Multiple-migrants and multiple diasporas: cultural reproduction and transformations among British Punjabi women', in C. Petievich (ed.) *The Expanding Landscape: South Asians and the Diaspora*, New Delhi: Manohar.

Bhardwaj, S. and Madhusudana Rao, N. (1990) 'Asian Indians in the United States: a geographic appraisal', in C. Clarke, C. Peach and S. Vertovec (eds) *South Asians Overseas: Migration and Ethnicity*, Cambridge: Cambridge University Press.

Brah, A. (1996) *Cartographies of Diaspora: Contesting identities*, London: Routledge.

Chatterjee, A. (2003) 'Sharp sense of divide among delegates', *The Times of India*, 10 January.

Chaudhary, M. (1996) 'India's tourism: a paradoxical product', *Tourism Management* 17(8): 616–19.

Clarke, C., Peach, C. and Vertovec, S. (eds) (1990) *South Asians Overseas: Migration and Ethnicity*, Cambridge: Cambridge University Press.

Cohen, R. (1967) *Global Diasporas*, London: Routledge.

Consulate General of India (2002) *Pravasi Bharatiya Divas*. Chicago. Online. Available: http://chicago.indianconsulate.com/PRAVASI2002.htm (accessed 19 January).

Dutt, V. (2003) 'A window dressing, at PBD, they say', *The Times of India*, 15 January.

Duttagupta, I. (2003) 'For Keralites, it smacks of chhole bhature', *The Times of India*, 11 January.

Gantzer, H. and Gantzer, C. (2002) 'A prism called tourism', *The Hindu*, Sunday magazine supplement, 12 May.

Goswami, U. (2003) 'Govt unsure why it wants closer ties', *The Times of India*, 9 January.

Hannam, K. (2002) 'Tourism and forestry in India', paper presented to the Annual Conference of the Association of American Geographers, Los Angeles, March.

Hansen, K. (1999) 'Singing for the Sadguru: Tyagaraja festivals in North America', in C. Petievich (ed.) *The Expanding Landscape: South Asians and the Diaspora*, New Delhi: Manohar.

Indians meet Indians (2003) *Objectives of Pravasi Bharatiya Divas*. New Delhi, India. Online. Available: http://www.indiaday.org/nri/index1.htm (accessed 13 February).

Jayanth, V. (2000) 'Promoting tourism', *The Hindu*, 1 December.

Lall, M.C. (2001) *India's Missed Opportunity: India's relationship with the Non Resident Indians*, Aldershot: Ashgate.

Leonard, K. (1999) 'Construction of identity in diaspora: emigrants from Hyderabad, India', in C. Petievich (ed.) *The Expanding Landscape: South Asians and the Diaspora*, New Delhi: Manohar.

Lessinger, J. (1999) 'Class, race and success: Indian-Americans confront the American Dream', in C. Petievich (ed.) *The Expanding Landscape: South Asians and the Diaspora*, New Delhi: Manohar.

Ministry of External Affairs (2001) *The Indian Diaspora: Report of the High Level Committee on the Indian Diaspora*, New Delhi: Government of India.

Ministry of Tourism (2001) *Annual Report*, New Delhi: Government of India.

Ministry of Tourism and Culture (2003) *Annual Report*, New Delhi: Government of India.

Misra, S. (1998) 'Public–private partnerships: new ways of managing tourism in India', *Journal of Tourism* 3: 5–12.

Mody, A. (2003) 'Advani put on the defensive', *The Hindu*, 11 January.

NRI Speak (2003) 'Lord Bhikhu Parekh', *The Times of India*, 10 January.

Petievich, C. (ed.) (1999) *The Expanding Landscape: South Asians and the Diaspora*, New Delhi: Manohar.

Raguraman, K. (1998) 'Troubled passage to India', *Tourism Management* 19(6): 533–43.

Rediff.com (2003) *Govt launches special tourism campaign for NRIs*. Online. Available: http://www.rediff.com//money/2003/jan/10pbd11.htm (accessed 4 May 2003).

Richter, L. (1989) *The Politics of Tourism in Asia*, Honolulu: University of Hawaii Press.

Sarkar, A. and Dhar, P. (1998) *Indian Tourism: Economic Planning and Strategies*, New Delhi: Kanishka.

Sen, S. (2003a) '3-day meet ends, NRIs expect better times ahead', *The Times of India*, 11 January.

Sen, S. (2003b) 'Patriotism? At the cost of merit? Never!', *The Times of India*, 7 January.

Sengupta, R. (2003) *Lessons from the Pravasi Mela*. Online. Available: http://www.rediff.com//news/2003/jan/14ram.htm (accessed 4 May 2003).

Shankar, T. (2002) 'Tourism ministry coins new slogan', *The Hindu*, 27 November.

Singh, S. (2001) 'Indian tourism: policy, performance and pitfalls', in D. Harrison (ed.) *Tourism and the Less Developed World*, London: CAB International.

Srivastava, S. (2003a) 'Excel, that's your service to India', *The Times of India*, 10 January.

Srivastava, S. (2003b) 'States seek diaspora's help for development', *The Times of India*, 11 January.

Thomas, V. (2003) 'NRIs in Gulf lament partisan treatment', *The Times of India*, 11 January.

Uttar Pradesh Department of Tourism (2002) *Discover Your Roots*. India: Lucknow. Online. Available: http://www.up-tourism.com/roots/rootmain.htm (accessed 21 September).

Van der Veer, P. (ed.) (1995a) *Nation and Migration: The Politics of Space in the South Asian Diaspora*, Philadelphia: University of Pennsylvania Press.

Van der Veer, P. (1995b) 'Introduction: the Diasporic Imagination', in P. van der Veer (ed.) *Nation and Migration: The Politics of Space in the South Asian Diaspora*, Philadelphia: University of Pennsylvania Press.

Ved, M. (2003) 'Diaspora voices: a part, yet apart', *The Times of India*, 10 January.

Vetcha, R. and Bhaskar, T. (2003) 'Passage . . . back to India', *The Hindu*, Sunday magazine supplement, 12 January.

Vertovec, S. (ed.) (1991) *Aspects of the South Asia Diaspora*, Oxford: Oxford University Press.

# 17  Reinventing Tulip Time
## Evolving diasporic Dutch heritage celebration in Holland (Michigan)

*Deborah Che*

## Introduction

Michigan's Tulip Time Festival, which was established in 1927 to celebrate the heritage of one of the largest Dutch settlements in North America, has been one of the most successful events promoting the cultural traditions of a diasporic group in its new home country, community pride and tourism destination development. Tulip Time has been recognized as one of the top 20 events in the world by the International Festival and Events Association, as North America's third largest flower festival and one of its top 10 ethnic festivals (Tulip Time Festival 2001a). Tulip Time, however, is not a static Dutch festival on American soil. Like the evolving Dutch-American diasporic culture it celebrates, it too has evolved. As an American festival with a Dutch theme, it has incorporated wooden shoe-clad so-called '*klompen* dancers', tulips, Dutch foods and crafts and parades as well as American-style musical entertainment in order to meet tourist demands. This chapter will examine the Tulip Time phenomenon as the evolving manifestation of Dutch diasporic culture by first exploring how diasporic cultures change; Holland, Michigan as the diasporic setting; and finally the critical issues surrounding Tulip Time's development and evolution in order to meet changing tourist and community demands.

## Diaspora, cultural change and authenticity in the face of hybridization

Both continuity and change underpin the diaspora concept. On one hand, diaspora implies cultural maintenance and stability. Dispersed, diasporic groups hold on to a collective, idealized myth about their original homeland, to which they may envision returning even if they left it under traumatic conditions. Based on the homeland myth as well as on a sense of distinctiveness, a common history and belief in a common fate, diasporic groups hold a strong, sustained ethnic group consciousness and a sense of solidarity (Cohen 1997; Safran 1991). On the other hand, diasporas in new lands require a degree of cultural adaptation to new host societies, with the inevitable consequence of cultural hybridization. Diasporic cultures and identities are, therefore, the syncretic outcomes of a melange of influences.

Notwithstanding, diasporic cultural traditions result from an on-going interaction between inherited traditions and the demands of the host society (Schnapper 1999). Encounters and shared knowledges with the host societies and other external sources produce hybridized diaspora cultures. Gilroy (1993) views the cultures of the African diaspora as being produced by the flow of diasporic peoples, ideas and knowledges around the Atlantic Ocean (i.e. the black communities in the USA, UK, Caribbean, etc.) as well as by infusions from the Europeanized host societies. Diasporic cultures such as those of the Black Atlantic cannot be 'preserved' from 'mixing' but, according to Boyarin and Boyarin (1993), require mixing to be remade and to continue to exist (Isenberg 1997). The cultural forms of diasporic groups illustrate this hybridization. The popular music of 'black Britain', while Afro-Caribbean in its roots, has been heavily influenced by African-American culture and has also drawn strongly on English colloquial phraseology such as Cockney. Likewise, the African-American cultural form of jazz has changed as its practitioners spent time in Europe (Gilroy 1993). Given this hybridization, it is hard to define a pure or authentic culture. Cultural forms of both diasporic and non-diasporic groups constantly change as a consequence of external cultural influences (Bhattacharya *et al.* 2002). Such cultural forms, which also include tourism events produced by diasporic cultures, are constantly re-negotiated and re-made as both the diasporic and host cultures change.

Given the complexity and fluidity of the cultural influences that mediate diasporic cultural tourism events, special care should be taken in the assessment of their authenticity. Alternative conceptions of authenticity are necessary. Stemming from Boortstin, there has been a long-standing criticism of the cultural events produced for tourists as being inauthentic. Boorstin (1961) calls both embellished 'native' ancient rites and international expositions 'pseudo-events' that have ambiguous, sometime tenuous connections to reality (since such pseudo-events for tourist consumption are seemingly inauthentic they represent changes from practices produced for locals). Like Boorstin's pseudo-events, MacCannell's (1976) 'staged authenticity' mantra applied to seemingly authentic commoditized tourism events also apparently associates authenticity with a static, pre-tourism cultural form. These conceptions are problematic when referring to diasporic cultural tourism events, which are products of iterative, fluid hybrid cultures.

Recent heritage tourism work has sought to re-negotiate the concept of authenticity, pointing out the limitations of what Ning (1999) terms 'objective authenticity'; that is, a museum-based and professionally oriented approach, which associates authenticity with the static 'original'. Instead of tying authenticity to the tourism object observed, some new approaches link authenticity to the exponential aspects of tourists partici-pating in an event. Ning (1999) proposes the concept of 'existential authenticity', in which the authentic self is realized and an existential state of being as an outcome of tourists' performance in events outside the routine; that is, dominant institutions and institutionalized socio-economic and socio-political positions. As an alternative conceptualization it allows for authenticity even if the tourism objects themselves are inauthentic. McIntosh and Prentice (1999) likewise conceive of authenticity as an interactive, rather than passive concept, through their concept of 'insightfulness'. Tourists achieve insight as they assimilate the information presented in the attraction, which is filtered through their values and past experiences. Authenticity is realized as tourists attain personal insights and associations through their experiences of places. These new approaches, which temporally move authenticity away from the past into the present, facilitate an as yet overlooked framework through which to approach hybridized diasporic cultural tourism events.

Other approaches to authenticity also get around the fixation on the 'original' by focusing on the authenticity of experience from the view of the visited diasporic communities. Erik Cohen's (1988) concept of emergent authenticity, which allows for negotiability and change in defining cultural authenticity, may be more appropriate to diasporic cultural tourism. In Cohen's view, crafts and festivals that were initially produced for tourists and considered contrived or inauthentic can acquire new meanings for locals as a means of self-representation before tourists. Thus, these touristic events may eventually be recognized as 'authentic' local customs and products of an ethnic group or region. More recently, Marwick (2001) has reworked the emergent authenticity concept in light of tourism-driven commercialization. She notes that 'substitutive' commercialization involving the spontaneous reorientation of a declining craft can help preserve local traditions, while involving new or reworked cultural significance for locals. Such extant approaches have clear value for enhancing our understanding of the production of tourism spectacles by diasporic communities. Changing, hybrid cultural products, such as Holland's Tulip Time Festival, which celebrates the unfolding culture and history of a Dutch-American community, even in an increasingly commercialized way to meet tourist demands and revenue needs, may be considered to be delivering (emergent) authentic products, imbued with new meanings to the local culture.

## The Dutch diasporic community of Holland, Michigan

The Dutch settlement of Holland resulted from a common element of diasporas, a traumatic dispersal from the homeland. The community's founders led the dissident Seceders, or *Afgescheiden* (Dutch, literally, 'those who are cut off' – from the state church). This group opposed the liberalization of the State Reformed Church, which included changes from traditional Calvinism such as adding hymns which they found less holy than Psalm singing. State fines on, and imprisonment of, the Seceder leaders prompted emigration (Ten Harmsel 2002). Prominent among the leaders, Anthonie Brummelkamp and Albertus C. van Raalte formed a society and developed a constitution contained in the 'Landverhuizing' (Emigration) Memorial of 1846 that would govern all settlers in the proposed *Kolonie* (settlement). The 'Rules for the Society of Christians for the Holland Emigration to the United States of North America' advocated keeping the emigrants together for spiritual and economic reasons by buying land collectively to prevent the potentially detrimental intrusion of strangers (Lucas 1955). While economic issues such as the 1845 potato blight, agricultural restructuring and the high rate of taxation also promoted migration of the largely rural, agricultural peasants and working-class Seceders, the desire to practise traditional Calvinism was central to their emigration (Ten Harmsel 2002). From 1835 to 1880, Seceders, who made up 1.3 per cent of the Netherlands' population in 1849, made up 13 per cent of its emigrants (Swierenga 1985).

The Seceders chose the eventual site of Holland as the site for their *Kolonie* which would enable them to practice traditional Calvinism and maintain the strong group consciousness based on a sense of distinctiveness, common history and common fate characteristic of diasporic cultures (Cohen 1997). Van Raalte selected west Michigan for its railways, suitability for farming, distance to markets via the Great Lakes and for its lack of Dutch Catholics, Germans and other European immigrant groups that were

present in other states (Lucas 1955). In 1847, he specifically selected Holland as the
site for the *Kolonie* where the group could pursue its religious aims since the relatively
unpopulated areas beyond between the mouths of the Grand and Kalamazoo Rivers
enabled its development and growth (Figure 17.1). Van Raalte wrote,

> I could find no other place for our group where along inhabited rivers tens of
> thousands of our people could find work without danger of being scattered . . . I
> chose this region on account of its great variety, being assured that if immigration
> from the Netherlands should develop into a powerful movement we ought to
> remain together for mutual support . . . The object of my settling between these
> two rivers was economic and at the same time to secure a centre of unifying
> religious life and labour for the advancement of God's kingdom.
>
> (Cited in Lindeman 1972: 24)

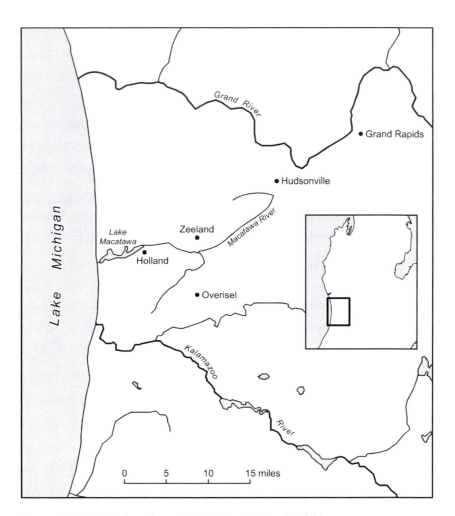

*Figure 17.1* The Dutch settlement (*Kolonie*) at Holland, Michigan.
Source: adapted from Bjorklund (1964).

Van Raalte's aim to attract like-minded believers and to solidify Holland and the surrounding area as a strict Calvinist stronghold was successful. The area established a concentrated Dutch settlement pattern within a 50-mile radius of southern Lake Michigan ranging from Muskegon, Grand Rapids and Holland to Chicago, Milwaukee and Green Bay, Wisconsin. In 1850, 16, or less than 1 per cent of all US counties accounted for 72 per cent of Dutch immigrants. Even by 1870, 56 per cent of Dutch immigrants resided in only 18 counties out of a total 2,295; 40 per cent lived in 55 townships out of a total 30,000 (Swierenga 1985). Compared to the Dutch Catholics who assimilated into multi-ethnic parishes, Dutch Calvinists have maintained a separate ethnic identity in America. As of 1990, the geographic pattern of Dutch Reformed churches still showed concentration around Lake Michigan, with newer establishments in Washington and southern California (Gaustad and Barlow 2001). The Dutch Calvinists' success in maintaining the diasporic culture is based on their preserving key institutions. Schnapper (1999) identified high rates of in-group marriage, schools and celebrations and specific religious and/or national demonstrations that refer to their history, culture and religion, such as Tulip Time.

While Holland was born out of a desire to maintain traditional Dutch Calvinism, the community has become hybridized. Bjorklund (1964) found the Dutch Reformed culture region centred around Holland retained core elements of the religious ideology and cultural traits that were continuations or modifications of practices in the Netherlands, while other cultural elements not needed or applicable in the USA were dropped. Many other cultural elements were Michigan creations which enabled the diasporic Dutch to succeed in America without compromising their religious principles. While the community quickly abandoned the Dutch dress, traditional housing forms and language and adopted North American agricultural practices, they retained Calvinist cultural elements that would foster the community's economic success and cultural continuity. Ideological beliefs of the centrality of the Calvinist family unit, productive activity to ward off sin borne of idleness, a selective Christian education and separateness from the irreligious, non-Dutch Reformed world were retained. In Michigan, they led to cultural expressions of single family, owner-operated farms with commercial and specialized agricultural production, seasonal work in nearby factories and Reformed Church-operated primary, secondary schools and colleges (Bjorklund 1964). This chapter will now turn to this evolving diasporic culture's largest public celebration, the Tulip Time Festival, which has retained a core emphasis on Dutch ethnicity and heritage in America while adapting to changing tourist and community demands.

## It's Tulip Time every year in May: the birth of a nationwide phenomenon

The Tulip Time Festival, which was established to celebrate Holland's heritage, has, like the diasporic Dutch Reformed, evolved over time while retaining its core. In 1927, local biology teacher Lida Rogers suggested a city beautification project and one-day festival oriented around tulip plantings given the city's ties to the Netherlands. At that time and into the 1950s, 90 per cent of Holland residents were of Dutch ancestry (Massie 1996), although Rogers, a member of the heritage-oriented Daughters of the American Revolution, was not one of them (Zingle 1995). The city, which viewed the tulip festival as a way to both promote community pride and businesses, adopted Rogers' proposal. The following year it purchased 100,000 tulip bulbs from the Netherlands, which bloomed in parks, along city streets and in private yards for the first Tulip Time in 1929 (Tulip Time History 2001). In 1930, Tulip Time, which had been a local West Michigan event, became an annual event with an organizing committee inviting visitors to come to Holland during the week of May 15 (Lindeman 1972). An estimated 45,000 to 50,000 visitors came for the 1930 festival (Massie 1996), which has been a tourist event ever since.

Key elements of Tulip Time, which followed Lida Rogers' emphases of flowers and Dutch heritage, were established during the 1930s. City tulip plantings, Dutch costumes, *klompen* dancing, street scrubbing, parades (Volks, Children, Band Review) and a flower show, which still comprise the core of today's festival, were initiated. Tulip plantings, the principal attraction of the festival, have grown into the city's famous tulip lanes. Residents wearing Dutch costumes during Tulip Time were popular and willing subjects for tourist photographs, according to press and promotional materials (Tulip Time Festival Inc. 1950). Over time, Tulip Time costumes have become more regulated. While some of the festival's first costumes, white bodices with Delft-blue skirts, had never been worn in the Netherlands (Sheeres 2000), subsequent Tulip Time Committees have urged that all new costumes be copies of one of the authentic costumes of the provinces of the Netherlands (Veen Huis 1953). *Klompen* (wooden shoe) dancers, who are predominantly local high school girls, perform routines that combine steps used in the folk dances in the Netherlands to special arrangements based on Dutch folk music. Parades form a central element of Tulip Time. The festival opens with the People's Parade, or *Volksparade*, which involves street scrubbing based upon the stereotypical cleanliness of the Dutch housewife and the Calvinist work ethnic. Hundreds of Holland's costumed citizens, armed with brooms, brushes and wooden yokes to carry the pails of water, 'scrub' the city streets clean for visitors (Plate 17.1). The Children's Parade, or *Kinderparade*, in which the city's school children parade in Dutch Costumes, came midweek. The festival's climax, the Parade of Bands, or *Musiekparade*, featured high school bands from Holland and from around the state of Michigan, costumed units and floats (Tulip Time Festival Inc. 1950). The popularity of these core festival elements, which sold the appeal of the old Netherlands and small-town America, led to Tulip Time becoming a national event by the end of the 1930s.

Incorporating non-Dutch entertainers into Tulip Time, which continues today, also helped to make it a national event. The Holland Furnace Company, which at the time claimed it was the world's largest installer of home heating equipment, brought Hollywood stars to Tulip Time. It also took Tulip Time to the nation via the national radio network broadcasts it sponsored from 1938 to 1941. The company flew in stars of the day such as Dorothy Lamour, Fay Wray, Richard Arlen and Pat O'Brien on a 'Hollywood to Holland Tulip Festival' American Airlines airplane decorated with painted tulips. Once in Holland, the stars gushed about the town's beauty and the spectacle of Tulip Time. Before a nationwide radio audience, Richard Arlen explained street scrubbing and the *Volks Parade* and exclaimed,

> This center of 30,000 Dutch Americans is filled to overflowing with merrymakers and lovers of beauty, because the Tulip Festival, like the Rose Bowl Game, the

*Plate 17.1* 'Street scrubbing' as part of the People's Parade at Tulip Time.
Source: Deborah Che.

Kentucky Derby and the New Orleans Mardi Gras, has become a great American institution.

(Van Reken and Vande Water 1993: 117)

These broadcasts spurred visits to Holland's flower festival as well as business for their sponsor, the Holland Furnace Company.

### Adaptation and reinvention of Tulip Time: adding the commercial to the cultural core

In the post-Second World War era of rising costs, an increasing gap between festival costs and revenues and the decline of the festival's paternalistic benefactor, the Holland Furnace Company (Wichers 1951; Bosman 1964), drove the need for revenue-generating events that would sustain the festival's long-term financial health and in particular, the non-revenue generating Dutch and tulip core. Only annual contributions and a 'reserve' fund, which was established in the early 1950s and largely made up of Holland Furnace Company donations, enabled the festival to pay its bills (Giles 1965a). Financial necessity drove the festival's expansion beyond the original core events and purpose which was,

> To promote, develop and publicize the tulip as a flower, to aesthetic appreciation of the tulip, to encourage and preserve the Dutch cultural heritage for the people

of the City of Holland as represented by the tulip as a flower and as a festival symbol; to stage and conduct a festival centered around the tulip and the Dutch heritage for the promotion of the interest of the City of Holland and its citizens.

(Tulip Time Festival Inc. 1948)

The inclusion of new revenue-generating events, especially musical ones, raised concerns about the direction, meaning, objectives and commercialization of the festival. Lida Rogers, the festival's founder, believed its continued success depended upon its focusing on the tulip and the Dutch motif. Mass tulip plantings, street scrubbing, Dutch dancing, the Dutch costume show, tour of the town and the Volks, Children's and Band Review parades should be included. Unrelated musical concerts, which could be staged by any town, should remain excluded. Lida argued that only Holland could really put on a Tulip Festival for which 'many visitors come hundreds of miles expecting to see large numbers of tulips'; that is, a spectacle preferably unspoiled by side shows and concessions and which allowed the visitor to enjoy a Dutch atmosphere (Rogers 1954: 2). Similarly William Wichers, Midwest Director of the Netherlands (Government) Information Office in Holland, Michigan, wanted the festival to stay true to its unique tulip and Dutch heritage roots. In 1951, Wichers wrote to the Board of Directors of the non-profit Tulip Time, Inc. that by basing 'our pageantry on the Dutch folklore which is the predominate background for our citizen we have created a pride in our heritage and given the visitor an insight into our character' (Wichers 1951: 1). Wichers particularly raised concerns about commercialism not linked to the festival's motifs, in the process noting that,

> Ours is a beautiful flower festival and through the years its distinctiveness and success was based on this factor. Commercialism was eliminated. Now by the urgency of being required to make the Festival pay its own way the committee is forced to schedule more and more revenue attractions. In the long run this policy will lead to excessive commercialism and be our undoing.
>
> (Wichers 1951: 2)

To balance financial concerns with those that Tulip Time was evolving from a flower and heritage festival to a musical one, he recommended retaining only those musical evenings involving local talent which generated revenue at little expense, provided an outlet for local musical talent and were popular with visitors. Instead of looking for revenue-generating events, Wichers felt the financial solution would be to solicit contributions from business firms that benefited from the festival (Wichers 1951). In the early 1960s, Tulip Time's organizers contacted local businesses for financial support since revenue events did not cover the costs of the non-revenue parades, awards, costumes and promotion (Giles 1965a). The festival, which had once been marketed nationally by the Holland Furnace Company, which contributed as much as US$30,000 in one year to promote the festival, had to turn down an invitation for a group of Dutch dancers to participate in a parade viewed by 500,000 people that would have only cost US$60 (Giles 1965b).

In response to financial problems that threatened the festival's long-term survival, Tulip Time developed more American headline entertainment shows and successfully targeted the growing group and package bus tour market in the 1960s and 1970s

(Tulip Time History 2001). The festival's clientele changed from predominantly independent travellers and families arriving by car from around the area to senior citizen, motorcoach groups who could visit during tulip-blooming May when schools were still in term (De Blecourt 1995; Koops 1995; Overbeek 1995; VandeWater 1995; Wichers 1995). From 1977 to 1995, the number of step-on (bus) guides grew from five guides servicing approximately 45 to 50 motorcoaches to as many as 50 to 100 guides available during the festival to service 500 motorcoaches (Koops 1995). Given the volume of the older motorcoach visitors, revenue events were geared towards their preferences. In 1979, the festival introduced its most popular show, the Stars of Lawrence Welk. Welk's ticket sales had at times accounted for about 60 per cent of Tulip Time's annual budget. Even with the reduction in the number of the shows booked, it still accounted for about 20 per cent of the festival's annual budget in 1995 (Duistermars 1995). While other headline American entertainers such as the Smothers Brothers, Jim Nabors, Debby Boone, Shirley Jones and the Oak Ridge Boys have been added to the festival's lineup, they also appeal to the older motorcoach clientele. This hybridization based on a mélange of Dutch influences and American popular music has made the festival viable.

These shows, which are crucial to the festival's financial health, however, supplement and support the tulip and Dutch core which bring tourists to Holland. According to a survey of 115 Group Tour (motorcoach) organizations, the tulip is still the queen, or the festival's number one draw (Tulip Time Festival Inc. 1999). The tulips complement the small-town atmosphere that allows tourists, as Appledorn (1995) puts it, to 'step back in time to a clean, safe, wholesome story book town with a wealth of community spirit'. The festival's appeal centred around its 'down-home face', illustrated by the 'church ladies cooking dinner, high school girls dancing in the streets, beautiful children and talented young people in bands'. The former Tulip Time director, Kristi Van Howe, encapsulated the festival's appeal. For her 'Tourists think they've arrived in a Dutch version of Prairie Home Companion, Lake Wobegone, or something . . .' (Van Howe 1995: 15). The Tulip Time Festival now draws over one million visitors annually and generates US$15 million in sales at hotels/motels, restaurants, area tourist attractions, gas stations, private property rentals to vendors and retailers in west Michigan. US$10 million of this spending occurs in the immediate Holland area (Tulip Time Festival 2001a).

## Further reinvention: addressing diversification of the tourism market and Holland, Michigan's population

As Tulip Time approaches its 75th birthday, it faces challenges as its once solidly dependable motorcoach market and Dutch community change. While Tulip Time is still ranked as one of the 'Top 10 Destinations for Motor Coaches' by several tourism associations (Tulip Time Festival 2001b), music meccas such as Branson, Missouri, and casinos are giving festivals competition for the senior motorcoach market (Martinez 2000). Many of these casinos have been constructed on Indian reservations nationwide, as part of what has proved to be geographically uneven economic development, with tribal casinos near big cities becoming major tourist attractions and generating millions of dollars in revenues, while other isolated facilities operate in the red (Lew and Van Otten 1998). Overall, however, Indian gaming is a huge industry;

in 2002, the nearly 300 Indian casinos generated US$13 billion in revenue (Bartlett and Steele 2002). For Tulip Time, Michigan casinos located mainly on Indian reservations offer inviting opportunities in intermediate locations for Chicago and Detroit seniors (Figure 17.2). Moreover, they also drive up entertainer fees since the casinos can pay twice as much the non-profit festival for loss leader concerts which lure gamblers (Duistermars 2001a). Tulip Time lost US$235,000 and US$103,000 in the 2001 and 2000 fiscal years respectively, a year after it recorded a surplus of US$64,200 in 1999. These losses were, according to Jesse (2002a), in part due to weather-related problems (rainstorms, early spring weather) that left Holland with stem, not tulip, festivals and thus contributed to declines in repeat motorcoach customers who complained about the lack of tulips (Duistermars 2001a).

Given the changing motorcoach market, Tulip Time is trying to attract younger people and families living within a 2 or 3 hour driving time from Holland (Duistermars 2001a; Van Kolken 2001). This drive-in family market had diminished in importance since the advent of the motorcoach era. To attract this weekend market as well as local youth and families on the weekdays, Tulip Time has added children's events, a carnival and concerts with acts popular with teens, with the most prominent being Christina Aguilera prior to her present adult 'incarnation' (Appleyard 1998). Inclusion of these new activities was aimed at changing perceptions that Tulip Time was only for old people and that one can't have fun (Duistermars 2001b). While the median age of Tulip Time customers has become younger, dropping from 78 years to the current one in the mid-60s (Duistermars 2001b), attracting independent, family, automobile-based travellers will be challenging for the May festival. Tourists who attended festivals or events on their most recent pleasure trip in Michigan were, compared with other tourists, more likely to begin their trips during the summer months (Yoon *et al.* 2000). Readers of *Michigan Living*, the magazine published by the American Automobile Association of Michigan, did not rank Tulip Time among its top five state festivals. All of these, with the exception of Detroit's Electronic Music Festival, took place during the summer when families could travel more easily (Tunison 2001). This relative weakness with auto travellers contrasts with the festival's name recognition among motorcoach passengers.

Tulip Time also faces challenges of producing a festival celebrating the tulip and the diasporic Dutch heritage in an increasingly multicultural community. From the time Tulip Time was born, into the 1950s, 90 per cent of Holland's citizens were of Dutch ancestry (Massie 1996). This percentage declined to 75 per cent by the 1970s, 38 per cent by 1990 and to approximately 30 per cent in the last 2000 Census (Yonkman 2002). Since the 1950s, Holland's Hispanic population has increased as agricultural migrant labourers from Texas and Mexico permanently settled in the area (Massie 1996; VandeWater 1995). In 1990, Holland had the highest percentage (14 per cent) of Hispanic citizens (predominantly of Mexican origin) of any Michigan city (Reens 2001). The changing ethnic shifts in Holland and the surrounding area can be seen in Table 17.1.

Like heritage tourism projects in other multi-ethnic cities (cf. Caffyn and Lutz 1999), Tulip Time faces the difficult challenge of integrating the needs of the present community and depicting the town's heritage so that it attracts tourists. Two of the top four ideas identified by the festival's long range planning board to support Tulip Time's mission which are to retain the Dutch flavour of the community and to offer a festival

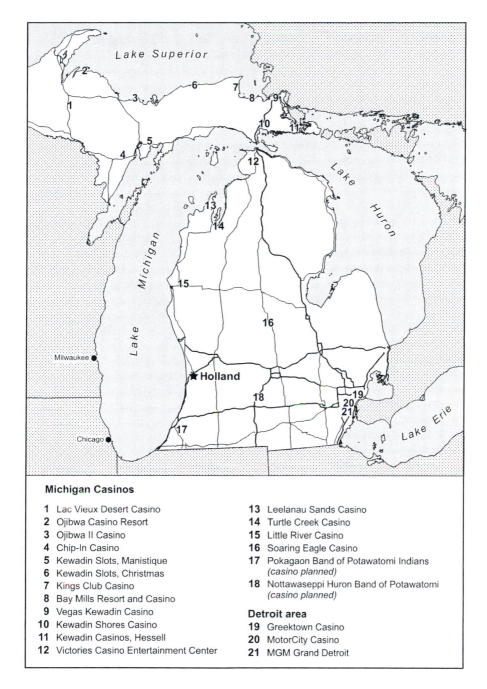

## Michigan Casinos

1 Lac Vieux Desert Casino
2 Ojibwa Casino Resort
3 Ojibwa II Casino
4 Chip-In Casino
5 Kewadin Slots, Manistique
6 Kewadin Slots, Christmas
7 Kings Club Casino
8 Bay Mills Resort and Casino
9 Vegas Kewadin Casino
10 Kewadin Shores Casino
11 Kewadin Casinos, Hessell
12 Victories Casino Entertainment Center

13 Leelanau Sands Casino
14 Turtle Creek Casino
15 Little River Casino
16 Soaring Eagle Casino
17 Pokagaon Band of Potawatomi Indians
   *(casino planned)*
18 Nottawaseppi Huron Band of Potawatomi
   *(casino planned)*

## Detroit area

19 Greektown Casino
20 MotorCity Casino
21 MGM Grand Detroit

*Figure 17.2* The location of casinos in Michigan.

*Table 17.1* Percentage of population claiming Dutch or Mexican ancestry in selected Michigan cities and their surrounding townships

|  | Dutch ancestry | | Mexican ancestry | |
| --- | --- | --- | --- | --- |
|  | 1990 | 2000 | 1990 | 2000 |
| Holland city | 38.7 | 27.4 | 13.4 | 17.7 |
| Holland township | 48 | 33 | 7.9 | 12.7 |
| Zeeland city | 72.3 | 58.6 | 2.2 | 2.6 |
| Zeeland township | 75.1 | 52 | 1 | 3.8 |
| Hudsonville city | 71.8 | 58.2 | 0.3 | 0.7 |
| Georgetown township | 53.9 | 42.7 | 0.5 | 1 |

Source: US Census Bureau (2002).

for all people – specifically relate to changing demographics (Allen Consulting Inc. 2000). To enhance the Dutch flavour, the festival has incorporated more Dutch cultural elements such as Dutch vocabulary/terms (Volks, Kinder and Muziek parades instead of Peoples', Children, Music/Band Review parades, respectively) (Robbins 2001); and revitalized its Marktplaats (market) featuring Dutch items (cheese, cocoa, Delft-ware, lace, prepared foods) for sale (Appleyard 1998). To offer a festival for all of Holland's people, Tulip Time is also working with Fiesta, Holland's annual celebration of Hispanic culture. In 2002, Fiesta and Tulip Time, which began a week earlier than normal to try to catch more tulip blooms, marked a shared Day of Festivals (Jesse 2002b). As Holland's school-aged population has diversified, Tulip Time has necessarily integrated the visibly non-Dutch in Dutch costumes into the core *klompen* dancing and parades (Bredeweg 1995; Van Howe 1995; Van Vyven 1995), which was not necessarily anomalous because, as Karsten (1995) has noted, black people from Curacao and Aruba speak perfect Dutch and cities in the Netherlands are increasingly ethnically and culturally diverse. But a diverse, multicultural community that may reflect present-day urban areas within the Netherlands and the USA may not be the 'step back in time' desired by tourists when they visit small-town Holland, Michigan (Appledorn 1995). Dealing with the community's cultural change will be one of Tulip Time's main challenges for the future.

## The evolving Tulip Time Festival: an (emergent) authentic tourist event

Tulip Time is an evolving cultural form with emergent authenticity. Like the diasporic Dutch culture in Michigan it celebrates, Tulip Time has maintained its core while adapting to ensure its long-term survival. The emergent authenticity concept, which allows for negotiability and change in defining cultural authenticity, is applicable to diasporic cultural tourism events. Tulip Time is an evolving product with meaning to the wider local community as it has included the growing, visibly non-Dutch population in core Tulip Time activities. The festival, which enables the event to acquire meaning for all in the community, regardless of heritage, thus mirrors the Dutch diasporic culture that has allowed for change necessary for its traditions to survive in America. While some tourists, as well as some locals, have found seeing

Hispanic and Asian schoolchildren in Dutch costumes disconcerting (Hoekstra 1995; Leenhouts and Leenhouts 1995; Overbeek 1995) and some visitors have even voiced that they would like to see strictly the blonde Dutch people participating to keep the festival 'pure' (De Blecourt 1995; Zwiep 1995), Sharon Koops, step-on motorcoach guide co-ordinator, has commented,

> I don't think Holland will ever let that happen. I hope not . . . I mean where else can you go and see a parade where they [people in Holland] put these crazy costumes on and scrub the streets? You don't get a parade like that very often and I think that is more important than worrying about who can be involved in it. If you want to – as my husband would say – if you want to be silly enough to put on a Dutch costume and go in the parade and scrub those streets, more power to you. Who cares what colour you are, or what your nationality is.
>
> (Koops 1995: 7)

Tulip Time can retain its celebration of Dutch heritage while remaining relevant to an increasingly multi-cultural population by reflecting Holland's diversity (Yonkman 2001).

The emergent authenticity concept also fits diasporic cultural tourism products in that it enables authenticity to be judged against the local culture, not that of the homeland. This distinction is important as the culture of diasporic groups has changed from that of the homeland in order to adapt to their new host societies. The Tulip Time Festival and its component activities are authentic west Michigan Dutch cultural forms. Ellen Van't Hof, a Calvin College physical education professor who has studied and participated in Tulip Time's *klompen* dancing, considered the dance to be authentic as it was 'genuine' and had 'an undisputed origin'. It was not an imported, intact Dutch dance as the Seceders who settled Holland did not approve of dancing and thus did not bring dances over. It was born in Holland in 1933 and regularized in 1953. It incorporated Dutch influences: costumes and footwear; folk music from the Netherlands (and Europe); and dance steps from Dutch, European and American folk traditions. In discussing the *klompen* dancing, Van't Hof evoked the essence of emergent authenticity, noting that the spirit of creation of a new unique dance which endures and thrives today should be the focus rather than 'bemoaning the fact that this dance did not somehow migrate directly from the shores of the North Sea to the shores of Lake Michigan' (Burdick 1998; Van't Hof 1998). Tulip Time's *klompen* dancing is an 'authentic' *local* product of the Michigan diasporic Dutch community. Similarly in response to implied criticism that Tulip Time is not authentic, or that there isn't much (Netherlands) Dutch in Tulip Time (De Boer 1998), Harry Hoekstra, the long-time director of the festival's Dutch Heritage Show, stressed that Tulip Time is an American festival with meaning or emergent authenticity to the local community, writing:

> Mr. de Boer, in the first place we are very happy that Tulip Time is an American festival. Holland, Michigan is an American city, rather small, but it has been able to organize a festival that cannot be found in the Netherlands. I lived many years in the Netherlands but the largest group of *klompen* dancers I've ever seen was maybe a group of 20–30 people and the dancing they did was rather tame compared to our *klompen* dancers here because our girls are all-American girls.

*Plate 17.2* When the Dutch met the Americans: *klompen* dancers pause for a moment in front of the bleachers and the assorted kiosks selling candy floss and fajitas.
Source: Deborah Che.

> I've watched them many times. Their smiling faces betray that they enjoy the American way to the old Dutch tunes . . . After all, that is the purpose of Tulip Time, to draw visitors, to make it a little Dutch. But we do keep in mind it is a Holland, Michigan festival'.
>
> (Hoekstra 1998)

Tulip Time is a celebration of diasporic Dutch heritage in Michigan (Plate 17.2), not the heritage of the Netherlands. Tulip Time is an (emergent) authentic event that has acquired meaning for the local community through self-representation before tourists. This aspect of emergent authenticity, its acquired meaning for the local toured communities, is critical to the long-term survival of community-based tourism events, especially where major demographic shifts have occurred. In addition to promoting local history, culture and the arts and providing locals with recreation and leisure opportunities, festivals like Tulip Time that have (emergent) authenticity can help enhance group (i.e. community) and place (i.e. regional, local) identity (De Bres and Davis 2001). They can engender 'pride of place' and a local positive self-identification as well as attract tourists. While Tulip Time has most appropriately been recognized by the US Library of Congress as a Local Legacy representing America's cultural heritage (Tulip Time recognition 2000), its longevity will depend on its appeal and continued (emergent) authenticity to local residents who make the festival possible through their participation in parades, musical and variety shows and tulip planting. While Tulip Time approaches its 75th birthday facing changing tourist and community demands, its adaptability, like that of the diasporic Dutch culture it celebrates, will help support its continuation. May it always be Tulip Time in May.

# References

Allen Consulting Inc. (2000) Summary Report of the Long Range (Tulip Time) Planning Board Retreat, held 10–11 January, Holmdel, NJ: Allen Consulting Inc.

Appledorn, B. (1995) Interview by J.V. Upchurch, in The Hope College Oral History Project for 1995, Holland, MI: *The Joint Archives of Holland*, 1–19.

Appleyard, K. (1998) 'Dutch Market Back at Festival', *Holland Sentinel*. Online. Available: http://www.hollandsentinel.com/stories/032898/new_dutch.html (accessed 27 July 2001).

Bartlett, D.L. and Steele, J.B. (2002) 'Wheel of misfortune', *Time*, 16 December: 44–8.

Bhattacharya, G., Gabriel, J. and Small, S. (2002) *Race and Power: Global Racism in the 21st Century*, London: Routledge.

Bjorklund, E.M. (1964) 'Ideology and culture exemplified in South-western Michigan', *Annals of the Association of American Geographers* 54: 227–41.

Boorstin, D.J. (1961) *The Image: A Guide to Pseudo-Events in America*, New York: Vintage Books.

Bosman, N. (1964) Letter sent to 360 recipients seeking contributions to the Tulip Time reserve fund, 16 December, Holland, MI.

Boyarin, D. and Boyarin, J. (1993) 'Diaspora: generation and the ground of Jewish identity', *Critical Inquiry* 19: 693–725.

Bredeweg, D. (1995) Interview by J.V. Upchurch, in The Hope College Oral History Project for 1995, Holland, MI: *The Joint Archives of Holland*, 1–14.

Burdick, J. (1998) 'Our Dutch Dancers' routines aren't authentic, Calvin scholar reports', *Holland Sentinel*. Online. Available: http://www.hollandsentinel.com/stories/040898/new_dancers.html (accessed 27 July 2001).

Caffyn, A. and Lutz, J. (1999) 'Developing the heritage tourism product in multi-ethnic cities', *Tourism Management* 20: 213–21.

Cohen, E. (1988) 'Authenticity and commoditization in tourism', *Annals of Tourism Research* 15: 371–86.

Cohen, R. (1997) *Global Diasporas: An Introduction*, Seattle: University of Washington Press.

De Blecourt, J. (1995) Interview by J.V. Upchurch, in The Hope College Oral History Project for 1995, Holland, MI: *The Joint Archives of Holland*, 1–28.

De Boer, K. (1998) 'Tulip Time has nothing for the young', *Holland Sentinel*. Online. Available: http://www.hollandsentinel.com/stories/121898/opn_letter2.html (accessed 27 July 2001).

De Bres, K. and Davis, J. (2001) 'Celebrating group and place identity: a case study of a new regional festival', *Tourism Geographies* 3: 326–37.

Duistermars, M. (1995) Interview by J.V. Upchurch, in The Hope College Oral History Project for 1995, Holland, MI: *The Joint Archives of Holland*, 1–11.

Duistermars, M. (2001a) Interview by author. Holland, MI, 3 August.

Duistermars, M. (2001b) Interview by author. Holland, MI, 27 September.

Gaustad, E.S. and Barlow, P.L. (2001) *New Historical Atlas of Religion in America*, New York: Oxford University Press.

Giles, R. (1965a) Letter to C.J. Marcus, 5 February.

Giles, R. (1965b) Letter to C.J. Marcus, 18 March.

Gilroy, P. (1993) *The Black Atlantic: Modernity and Double Consciousness*, Cambridge, MA: Harvard University Press.

Hoekstra, H. (1995) Interview by J.V. Upchurch, in The Hope College Oral History Project for 1995, Holland, MI: *The Joint Archives of Holland*, 1–13.

Hoekstra, H. (1998) 'Tulip Time is just fine as it is', *Holland Sentinel*. Online. Available: http://www.hollandsentinel.com/stories/122198/opn_tuliptime.html (accessed 27 July 2001).

Isenberg, N. (1997) '"Critical post-Judaism"; or, reinventing a Yiddish sensibility in a post-modern age', *Diaspora* 6: 85–96.

Jesse, D. (2002a) 'Tulip Time ready for change: new events and schedule part of effort to rebound from lack of blooms', *Holland Sentinel*. Online. Available: http://www.holland sentinel.com/stories/040702/loc_040702002.shtml (accessed 29 August 2002).

Jesse, D. (2002b) 'Festivals declare carnival a success: officials from Fiesta and Tulip Time plan to do it again next year', *Holland Sentinel*. Online. Available: http://www.holland sentinel.com/stories/052402/loc_05242007.shtml (accessed 29 August 2002).

Karsten, J. (1995) Interview by J.V. Upchurch, in The Hope College Oral History Project for 1995, Holland, MI: *The Joint Archives of Holland*, 1–14.

Koops, S. (1995) Interview by J.V. Upchurch, in The Hope College Oral History Project for 1995, Holland, MI: *The Joint Archives of Holland*, 1–10.

Leenhouts, J. and Leenhouts, T. (1995) Interview by J.V. Upchurch, in The Hope College Oral History Project for 1995, Holland, MI: *The Joint Archives of Holland*, 1–15.

*Holland Sentinel* (2000) 'Legacy: Tulip Time recognition', *Holland Sentinel*. Online. Available: http://www.hollandsentinel.com/stories/052700/new_15.html (accessed 28 November 2002).

Lew, A.A. and Van Otten, G.A. (eds) (1998) *Tourism and Gaming on American Indian Lands*, Elmsford, NY: Cognizant Communication Corporation.

Lindeman, M.C. (1972) 'Those hotheads who sought the freedom to be narrow', *Holland Herald* 7: 24–31.

Lucas, H.S. (1955) *Netherlanders in America: Dutch Immigration to the United States and Canada, 1789–1950*, Ann Arbor, MI: University of Michigan Press.

MacCannell, D. (1976) *The Tourist: A New Theory of the Leisure Class*, New York: Schocken Books.

Martinez, S. (2000) 'Festival, casinos fight for tourist dollar', *Grand Rapids Press*, 16 May 2000: A1, A4.

Marwick, M.C. (2001) 'Tourism and the development of handicraft production in the Maltese islands', *Tourism Geographies* 3: 29–51.

Massie, L.B. (1996) *Haven, Harbor and Heritage: The Holland, Michigan Story*, Allegan Forest, MI: Priscilla Press.

McIntosh, A.J. and Prentice, R.C. (1999) 'Affirming authenticity: consuming cultural heritage', *Annals of Tourism Research* 26: 589–612.

Ning, W. (1999) 'Rethinking authenticity in tourism experience', *Annals of Tourism Research* 26: 349–70.

Overbeek, L. (1995) Interview by J.V. Upchurch, in The Hope College Oral History Project for 1995, Holland, MI: *The Joint Archives of Holland*, 1–15.

Reens, N. (2001) 'Hispanic population focus at conference: city manager, mayor bring local experiences to national gathering', *Holland Sentinel*. Online. Available: http://www.hollandsentinel.com/stories/010601/new_6.html (accessed 27 July 2001).

Robbins, J.C. (2001) 'Ik hou van een parade; I love a parade: language keeps unique cultural flavour in festival', *Holland Sentinel*. Online. Available: http://www.hollandsentinel.com/stories/051901/loc_0519010014.shtml (accessed 19 July 2001).

Rogers, L. (1954) Letter to Mr Riemersma (president of the Tulip Time Committee), 17 December, Holland, MI.

Safran, W. (1991) 'Diasporas in modern societies: myths of homeland and return', *Diaspora* 1: 83–99.

Schnapper, D. (1999) 'From the nation-state to the transnational world: on the meaning and usefulness of diaspora as a concept', *Diaspora* 8: 225–54.

Sheeres, J.S. (2000) 'Klompendancing through America. Dutch enterprise: alive and well in North America', in L.J. Wagenaar and R.P. Swierenga (eds) *Proceedings of the 12th Biennial Conference of the Association for the Advancement of Dutch-American Studies*, Holland, MI: The Joint Archives of Holland, Hope College.

Swierenga, R.P. (1985) 'Dutch immigration patterns in the nineteenth and twentieth centuries', in R.P. Swierenga (ed.) *The Dutch in America: Immigration, Settlement and Cultural Change*, New Brunswick, NJ: Rutgers University Press.

Ten Harmsel, L. (2002) *Dutch in Michigan*, East Lansing, MI: Michigan State University Press.

Tulip Time Festival Inc. (1948) *Articles of Incorporation*, Holland, MI: Tulip Time Festival Inc.

Tulip Time Festival Inc. (1950) *Tulip Time in Holland (Michigan)*, Press Release, Holland, MI: Tulip Time Festival Inc., 17–20 May.

Tulip Time Festival Inc. (1999) *Group Tour Survey Results*, Holland, MI: Tulip Time Festival Inc.

Tulip Time Festival Inc. (2001a) *Tulip Time Festival 2001 Fact Sheet*, Holland, MI: Tulip Time Festival Inc.

Tulip Time Festival Inc. (2001b) *Tulip Time Festival Offers Visitors a Chance to Join in the Celebration of the Area's Dutch Heritage*, Press Release, Holland, MI: Tulip Time Festival Inc.

Tulip Time History (2001) *Holland: Tulip Time Festival Inc.* Available: http://www.tuliptime.org/history/index.html (accessed 9 May 2001).

Tunison, J. (2001) 'Flowers fading? AAA readers don't rate Tulip Time among state's top festivals', *Grand Rapids Press*, 10 April 2001: A1, A4.

US Census Bureau (2002) *American Fact Finder, 1990 and 2000*, Washington, DC: US Census Bureau. Online. Available: http:factfinder.census.gov (accessed 25 November 2002).

Van Howe, K. (1995) Interview by J.V. Upchurch, in The Hope College Oral History Project for 1995, Holland, MI: *The Joint Archives of Holland*, 1–19.

Van Kolken, P. (2001) Chairman: 'Festival dates can't be moved. Tulip Time leader hopes weather will improve floral display. Survey shows visitors want more Dutch items', *Holland Sentinel*. Online. Available: http://www.hollandsentinel.com/stories/052101/loc_0521010003.shtml (accessed 19 July 2001).

Van Reken, D. and Vande Water, R.P. (1993) *Holland Furnace Company*, Holland, MI: D.L. Van Reken.

Van't Hof, E. (1998) 'Klompen article omitted key facts', *Holland Sentinel*. Online. Available: http://www.hollandsentinel.com/stories/041098/opn_letter1.html (accessed 27 July 2001).

Van Vyven, M. (1995) Interview by J.V. Upchurch, in The Hope College Oral History Project for 1995, Holland, MI: *The Joint Archives of Holland*, 1–22.

VandeWater, R. (1995) Interview by J.V. Upchurch, in The Hope College Oral History Project for 1995, Holland, MI: *The Joint Archives of Holland*, 1–27.

Veen Huis, E.R. (1953) *Historic Dutch Provincial Costume Adapted to Klompen Dancers of the Tulip Festival in Holland, Michigan*, Fort Collins, CO: Colorado Agricultural and Mechanical College.

Wichers, N.E. (1995) Interview by J.V. Upchurch, in The Hope College Oral History Project for 1995, Holland, MI: *The Joint Archives of Holland*, 1–12.

Wichers, W.C. (1951) Letter to Tulip Time Festival Inc. Board of Directors, 5 January, Holland, MI.

Yonkman, D. (2001) 'Dutch with diversity: Festival holds to tradition amid multicultural changes; US Census numbers show an increasing minority population living in the Holland area', *Holland Sentinel*. Online. Available: http://www.hollandsentinel.com/stories/051501/loc_0515010002.html (accessed 14 March 2002).

Yonkman, D. (2002) 'Dutch impact dwindles locally; Percentage of Holland residents with roots from the Netherlands declining', *Holland Sentinel*. Online. Available: http://www.hollandsentinel.com/stories/060402/loc_060402003.shtml (accessed 29 August 2002).

Yoon, S., Spencer, D.M., Holecek, D.F. and Kim, D.-K. (2000) 'A profile of Michigan's festival and special event tourism market', *Event Management* 6: 33–44.

Zingle, M. (1995) Interview by J.V. Upchurch, in The Hope College Oral History Project for 1995, Holland, MI: *The Joint Archives of Holland*, 1–9.

Zwiep, S. (1995) Interview by J.V. Upchurch, in The Hope College Oral History Project for 1995, Holland, MI: *The Joint Archives of Holland*, 1–11.

# 18 Selling diaspora

## Producing and segmenting the Jewish diaspora tourism market

*Noga Collins-Kreiner and Dan Olsen*

## Introduction

The commodification of national identity and ethnicity has been one of the most distinctive features of tourism development in the last decade. One motivation for this commodification process has been the increased interest among communities and individuals in uncovering more about their collective pasts and identities by discovering family roots and by improving their awareness of past historical events and places. This rapid growth in heritage-related travel has also seen an increase in travel by people belonging to diasporic communities. As such, diaspora-related tourism has grown into a significant market in recent years and many destinations now design and market tourism products to such 'hyphenated' communities around the globe. For example, one of the results of these new levels of interest in people's familial past is the growth of genealogy-related travel wherein people travel in search of their roots and the communities of their ancestors, whether to genealogical libraries or to monuments or buildings that symbolize individual and collective pasts (Timothy 1997; see also Ch. 9).

Research on the tourism component of diasporic communities has invariably focused on the concepts of pilgrimage, authenticity and experience, focusing on the motivations and subjective experiences of travellers to diasporic sites of nationalist or heritage significance. However, there is little research on how those experiences are mediated, commodified, marketed and sold. While there has been recent work on diasporic destinations as attractions and the mediation of their symbolic meanings through souvenirs and tour guides (cf. Shenhav-Keller 1995; Cohen 1999, 2002), there have been few attempts to examine the actual delivery of diaspora tourism products from the producers to the consumer as delivered through intermediaries (e.g. tour operators and agencies) (Klemm 2002). In addition, existing studies have tended to overlook the extent to which retailers and tour agents manipulate and transform authenticity, experiences, education and spirituality into commodities for sale. This is particularly true of the diaspora market, where tourism suppliers and agencies stress particular themes or motives that are calculated to induce tourist consumption.

Therefore, this chapter attempts to go beyond the limited research to date and to survey the different market patterns within Jewish diaspora tourism, paying particular attention to how the motivations and experiences of Jewish diaspora travellers are mediated, marketed and sold by tour operators and agencies. In particular, it covers the ways in which Jewish diasporic experiences and identities are manipulated, commodified and subsequently sold to members of the Jewish diaspora, focusing on marketing slogans and regulation of travel experiences. The chapter begins by reviewing the history and current literature on Jewish diaspora, then turns to the identification and description of the different niche market segments within the Jewish diaspora as identified and commodified by tour operators and agencies. Finally, the niche market segments within the context of tourism policy and marketing are discussed, particularly focusing on the commodification strategies of tour operators and agents working within the formal tourism production system (see Britton 1991).

## Jewish diaspora, travel and tourism

What constitutes 'diaspora' has changed over time. Werbner (2002: 120–1) notes that postcolonial and postmodernist interpretations of diaspora are challenging the paradigm that diasporas are 'scattered communities yearning for a lost national homeland, whether real or imaginary,' and points out that diasporas 'need to be grasped as deterritorialized imagined communities which conceive of themselves, despite their dispersal, as sharing a collective past and common destiny'. These new interpretations have caused the term to encompass 'a motley array of groups such as political refugees, alien residents, guest workers, immigrants, expellees, ethnic and racial minorities, [and] overseas communities' (Shuval 2000: 42; see also Korom 2000) rather than referring specifically to ethnic groups (e.g. Jews and Africans) who have been exiled from a homeland.

Diaspora includes displaced persons that maintain and revive a strong sentimental and/or material connection with a homeland based on ethnicity and/or religion (Sheffer 1986; Safran 1991). As a social and conceptual construct, diasporas are 'founded on feeling, consciousness, memory, mythology, history, meaningful narratives, group identity, longings, dreams, [and] allegorical and virtual elements' (Shuval 2000: 43). As Gordon and Anderson (1999: 288) note, diasporas 'denote a certain kind of identity formation, the feeling of belonging to a community that transcends national boundaries'. Ben Ari and Bilu (1997: 10) claim that diasporas are movements of peoples and experiences that leave trails of collective memories about other places and other times and in this way create new maps of desire and attachment. These 'trails of collective memory' have recently inspired members of communities to travel to seek out more about their collective past, identity and their position within the wider diaspora group. In their typology of sacred space, Jackson (1995) and Jackson and Henrie (1983) argue that homelands are endowed with a sense of sanctity because of the collective emotional attachment diasporic groups feel towards them.

Shapiro (2001) notes that the Jewish diaspora experience is a classic example of diaspora and is usually the 'ideal type' to which other forms of diaspora are compared. The Jewish diaspora is characterized by displaced Jews sharing common ethnic, cultural and religious roots, which in turn creates a common identity, as well as a sentimental and material connection to their ancestral homeland: modern-day Israel. Part of this shared identity comes from Jews' history as a persecuted people, starting in AD 70 with their scattering throughout Europe and Asia by the Romans and continuing through the Holocaust during the Second World War. Even with the creation of the State of Israel in 1948, most Jews today still live scattered throughout the world. The estimated strength of world Jewry today is little more than 13 million, though an exact count is difficult because of the faith's traditional reluctance to count its people and the impossible task of reaching out to so many unaffiliated Jews in the diaspora (World Jewish Congress 2002). Some 5 million Jewish people live in Israel (Israel Bureau of Statistics 2001) and another 6 million live in the USA and Canada. The remaining 2 million Jews are located in various places around the world, with the majority located in Western Europe. The cities with the largest Jewish population in the diaspora include New York City (1,900,000), Los Angeles (585,000), Miami (535,000) and Paris (350,000).

The relationship between Israel and the Jews abroad is unique. On the one hand, the creation of the State of Israel signified for the Jewish world a major change in the Jewish condition, with Israel becoming an object of pride and identity and worthy of active support. The Jews of the diaspora, especially American Jewry, have provided financial, moral and political support for Israel (World Jewish Congress 2002). These factors commonly lead Jews from all over the world to view Israel as their ancestral home and subsequently travel there, particularly, as Orenstein (1994: 370) writes, 'Jews no longer necessarily discover Judaism through Jewish texts, rituals and traditions. Often, Jews discover Judaism through their personal quests and journeys'. However, focusing on the modern-day Jewish diaspora experience in North America, Shapiro (2001), citing Cohen (1991) and Safran (1991), argues that although support for Israel by North American Jews places a central role in the public life of North American Jewry, within the private religious sphere of individual Jews it is peripheral. As Shenhav-Keller (1995: 151) notes, they face a 'duality of worlds', where they live life in the periphery while still feeling a personal indebtedness to Israel, their elective centre. Because of this, Shapiro (2001: 24) questions whether 'North American Jewry really possess the double consciousness of diaspora life, the profound attachment to a prior home, the serious ambivalence about life outside Israel that one expects from the term "diaspora"'.

While there are no tourism statistics concerning diaspora tourism among different groups, it is estimated that over half a million Jews travel to Israel each year, comprising approximately 20 per cent of all visitors to Israel (Israel Ministry of Tourism 2001). In 2001, most visitors came from Europe (Italy 18.3 per cent; Sweden 17.3 per cent; France 19.0 per cent; UK 23.2 per cent and Germany 17.4 per cent). Only 3.24 per cent of the Jewish population of the USA visited Israel that year. It is likely that at least the same number, if not more visitors to other Jewish heritage sites in the world, particularly in Eastern Europe. Many Jewish travellers will visit sites concerned with Jewish heritage and/or religion even if the main reason for their travel is business or other form of leisure holidays. For example, in a survey of 3,000 Jewish households in greater London, Waterman (2002) found that 24 per cent of the respondents reported having visited a Jewish museum outside the UK in the previous 12 months. Destinations in Eastern Europe have increased in popularity since 2000 because of the political tensions in the Middle East, which have steered many Jews to seek alternative, safer destinations (Bauer 2002). Shackley (2001) notes this pattern at sacred sites and indeed all tourism sites, located in areas of socio-political instability are difficult to manage and at times can experience a low visitor rate.

## Segmenting the Jewish diaspora market

Most research on Jewish diaspora tourism relates to Holocaust sites (e.g. Gruber 1992, 1994; Kugelmass 1993; Kirshenblatt-Gimblett 1998; Ashworth 1999, 2003; Szkółka 2000; Feldman 2001; Lehrer 2001). Several researchers have examined other patterns of Jewish tourism, including Ioannides and Cohen Ioannides (2002; see also Ch. 6) and Kugelmass (1993), who investigated patterns of Jewish travel in the USA and Collins-Kreiner (2000, 2002) and Epstein (1995), who have written about different kinds of pilgrimage in modern times to 'new' holy sites in Israel. Other researchers have published on elements of Jewish youth summer trips to Israel and especially on the impact of 'Birthright Israel', which provides a gift of educational trips to Israel for young Jewish adults aged 18 to 26 (e.g. Chazan 1997; Post 1999; Kelner *et al.* 2000). Other works include Goldberg's (1995) ethnographic perspective on visits to Israel; Shapiro's (2001) study of *Livnot u'lehibanot*, a three-month work-study programme; Heilman's (1995) research on the visits made by participants of the movement 'Young Judea'; and Cohen's (1999) work on the 'Israel Experience' (see also Ch. 8).

As noted previously, the focus of this chapter is the patterns and market segments of Jewish diaspora tourism, particularly focusing on the supply side of the Jewish diaspora market. While there are many guidebooks, tour operators and travel agencies that cater to Jewish tourists living in the diaspora, the focus of this study is the Internet as a supplier of images and narratives that promote particular destinations and types of experiences. This was done for two reasons. First, the Internet is quickly becoming widely accessible to potential tourists and as such, more and more people are using it as a decision-making tool, using the constructed images and personal narratives to decide which destinations they wish to visit and activities they wish to experience. Second, while the Internet has been used for tourism marketing and promotion and this has been studied a great deal in recent years, little research has been done on the content of these images and narratives and what market niches various tour operators and agencies have targeted.

In this study, 50 tourism websites oriented towards the Jewish market were analysed. North American websites were the focus, as approximately 75 per cent of diaspora Jews reside in the USA and Canada. Search engines were used to select and access the sites. The main themes sought from the Internet included the types of tours offered, stated motivations or reasons people should take a particular type of tour, images associated with tours, market definitions and itineraries. In addition, three tour operators dealing with Jewish travel in Toronto, Ontario (Canada) were interviewed to provide additional insight into the findings from the Internet. The results of this exercise show that the main reasons for visiting Israel and other places connected to Jewish diaspora heritage, as cited by tour operators, were oriented towards what may be regarded as pull motives. The major reasons mentioned included: a feeling of belonging; religious/spiritual motivations; searching for roots; 'pure' vacation; vacation with 'feelings'; visiting friends and relatives (VFR); solidarity with Israel; celebrations of the Bar/Bat Mitzvah; and association with other Jewish people. Through a content analysis, several types of Jewish tours were identified, which reflect these various motives. The majority of websites (over forty) promoted heritage tours to sites in Eastern Europe and Israel. Some of the sites focused more on exotic or adventure tours, heritage tours outside Eastern Europe and Israel, Bar/Bat Mitzvah tours, youth educational tours to Israel, solidarity missions and Jewish singles tours. A minority of sites (less than ten) offered pilgrimage tours specifically, Jewish cruises and interaction tours. These different types are discussed in more detail below (see Table 18.1).

### Heritage tours

The majority of websites focused on Jewish heritage tours in Eastern Europe and Israel. These tours concentrated on presenting a 'different' point of view of the sites related to Jewish history, culture and identity. The most popular region for Jewish heritage tours was Eastern Europe, especially Prague, Budapest, Vienna and Krakow. Other heritage tours are organized to Spain, Netherlands, Belgium, France, Italy, India, China and Morocco. Heritage tours are also offered to other countries, but these are fewer in number and, according to interviews with tour operators, do not have as many participants as the tours to Eastern Europe and Spain. For example, one heritage tour to India was advertised as a 'Judaic tour of India' including visits to major centres of Judaism in India, meeting Jewish leaders, visiting synagogues and observing unique customs and traditions.

Jewish tourists visit cities in the 'old countries', such as Prague and Vienna, or the country of origin of their parents or grandparents. The former motherlands were called *Der Heim*, which in Yiddish means 'the home'. These visits are a way for Jews to discover how their ancestors lived. As part of these nostalgic experiences, there are visits to old Jewish neighbourhoods, homes of famous Jewish personalities, synagogues and graveyards and the death camps of the Second World War. This 'pining for the past' leads to a kind of tourism that provides a way for Jews of the diaspora to get in touch with their roots.

*Table 18.1* Different types of Jewish tours available on the Internet and indicative URLs

| Type of tour | Tour operator | URL |
| --- | --- | --- |
| Heritage Tours | Spiritual Journeys(r), Inc. | www.SpiritualJourneys.net |
| Interaction Tours | The Jewish Travel Network | www.jewishtravelnetwork.com/ |
| Solidarity Missions | UJA | toronto.ujcfedweb.org/travel_home.html |
| | Keshet Israel | www.keshetisrael.co.il/solidarity.htm |
| Bar/Bat Mitzvah Tours | Israel Tour.com | www.israeltour.com/11dayTour.asp |
| Jewish Cruises | Kosherica Cruises | www.kosherica.com |
| Jewish Singles Tours | United Jewish Appeal | ujcfedweb.org/travel_home.html |
| Exotic/Adventure Tours | Jewish Travel.com | www.totallyjewishtravel.com |
| Youth Educational Tours | Birthright-Israel Experience | www.israelforfree.com |
| | Israel Experts.com | www.israelexperts.com/trips.html |
| Pilgrimage Tours | United Synagogue Youth (USY) | www.usy.org/programs/pilgrimage/ |
| General Tours | Jewish Travel.com | www.jewishtravel.com |
| | Totally Jewish Travel.com | www.totallyjewishtravel.com/tours |

## Interaction tours

Jewish interaction tours include people who travel to a particular location and as part of the overall experience stay in the homes of other Jews as a way of making new friends and visiting new places. For example, The Jewish Travel Network is a member-supported hospitality and home exchange, which finds and offers homes worldwide and provides specialized travel information. Its listings and information are provided as a service to its members. Members register either as hosts or as contacts offering hospitality and information about their community. They may also indicate a special interest in contacting singles and/or students and youth.

The home exchange or 'home swap' service provides a connection for people to find others who may be interested in exchanging homes during a mutually convenient time. All the home exchange listings are published on the website and individuals/families/groups contact each other and make their own arrangements. Israel is the most popular destination, particularly for home exchanges because many families have relatives in Israel. It should be noted that there are usually special websites concerned with these types of tours and that most of the visits are offered during Jewish Holidays. A typical advertisement might include:

> Travel while enjoying the warmth and hospitality of a Jewish home. . . . When you travel, have you ever wished you could meet and interact more with Jewish people who are living there?; Imagine yourself on a Friday and knowing that you can look forward to a Sabbath observance with a Jewish family.
>
> (The Jewish Travel Network 2002)

## Solidarity missions

Solidarity missions involve Jews travelling to the Holy Land to show their solidarity with Israel. These trips are organized to Israel only and are especially popular during times of war or other difficult political situations. Since Israel was established in 1948, there have been many Jewish solidarity missions to that country to show identification and support. The missions focus on gaining an understanding of the events and in particular of the dilemmas faced by Israeli leaders and citizens in the aftermath of political crises. Those who participate in solidarity missions recognize the importance of being physically present in Israel at times of conflict and operators attempt to bring as many visitors as possible. Most solidarity missions are arranged by Jewish organizations with tour operators arranging the technical aspects of the tour, such as transportation and lodging. These missions aim to give participants a positive, patriotic experience. These trips include meetings with political leaders, journalists, academics and 'regular people' in the style of hands-on encounters. Programme highlights include panel discussions with Israeli and foreign journalists, leaders and political figures on the roots of the current crisis and visits to the Jewish Quarter of Jerusalem, the Western Wall of the Second Temple and other Jewish sites.

## Bar/Bat Mitzvah tours

A Bar Mitzvah tour might be classified as both a pilgrimage and a heritage tour because of its ritual characteristics and motives. Celebrating a Bar Mitzvah is one of the obligations of every Jewish boy when he reaches the age of 13 and a similar Bat Mitzvah celebration is often observed for 12-year-old girls. Today there is a tendency to make the Bar/Bat Mitzvah tour a deeply meaningful experience for all family members and especially for the young. There are companies that specialize in Bar/Bat Mitzvah tours including sites and activities of special interest for young people. Many of the tours visit Israel and celebrate at the Western Wall. In recent years, there has been a trend among Jewish families from around the world to participate in Bar/Bat Mitzvah tours to visit other countries owing to the current tense political situation in Israel.

### Jewish cruises

Jewish cruises were advertised as combining vacations with a 'Jewish flavour' experience. The vacation aspect of the cruises includes regular recreational activities (e.g. scuba diving, basketball, tennis, aerobics, gambling, shopping, shows). Quality Jewish entertainment (e.g. lectures on Judaism and Jewish topics and performances by Jewish musicians), religious services and strict kosher foods epitomize the Jewish flavour of the cruises. These cruises in a way combine elements of sun–sea–sand leisure with heritage tourism. Cruises are offered through many websites, but there are also specific specialized Jewish companies such as Kosherica Cruises. This type of tour is especially interesting because of its ability to combine heritage with traditional leisure at a non-specific site. The cruise ships are sites that originally had nothing to do with Judaism, but become Jewish because of their synagogues, lectures, food, entertainment and interaction with other Jews.

### Jewish singles tours

Some tours aim to get Jewish singles together. Most of these have been instituted to prevent marriage outside the faith. Such tours are combined with either heritage tours or cruises. Most of these are not organized by tour operators but by Jewish organizations such as the United Jewish Appeal (UJA), with travel agents and tour operators making the logistical arrangements. The most popular tours connected to singles travel are the cruises, probably because of their ability to combine Jewish heritage and leisure tourism, as well as the exotic feeling of being away from the travellers' traditional surroundings.

### Exotic or adventure tours

Tours were also found and labelled as exotic or adventure tours aimed at Jewish people who would like to have a Jewish connection during their travels. These tours are, for example, designed for people who want to travel to exotic places but formerly could not, because there were no kosher facilities or other special arrangements. Now kosher meals may be provided anywhere in the world. Another reason could be a wish to celebrate a Jewish holiday during the tour. Some of the advertised tours included an African Kosher Safari Tour, Passover in Peru and the Silk and Spice Tour in China.

### Youth educational tours

The goal of youth educational tours is to provide participants with 'a stimulating encounter with Israel – and by extension, with their own identity' (Post 1999: 54). The best known of these tours is *Birthright Israel*. Initiated by philanthropists Charles R. Bronfman and Michael Steinhardt and supported by North American Jewry's communal institutions, the Israeli government, the Jewish Agency and various other private philanthropists, the US$210 million, multi-year programme entailed what is so far the largest single mobilization of resources to address the challenge of adult Jewish identity and involvement (Chazan 1997; Kelner *et al.* 2000). Being in Israel contributes directly to the programme's Jewish identity-building goals. The itineraries capitalize on this by explicitly linking educational themes with physical settings. This capacity to tap the reservoir of core elements of diaspora Jewish folk religion (a sense of history, religion, peoplehood, family, heroism, insecurity and power) is experienced most vividly at evocative settings like the Western Wall, the Golan Heights and the Jewish National Fund (JNF) forests.

All the educational tours, including *Birthright Israel,* tours of youth movements such as Betar, Young Judea, UJA Youth movements and tours coordinated by The Jewish Agency are mainly aimed at teenagers and organized by different Jewish or Israeli organizations. For example, an estimated 500,000 Jewish youngsters have visited Israel over the past 25 years through the Israel Experience programme organized and sponsored by the Joint Authority for Jewish-Zionist Education (Cohen 1999; see Ch. 8 for a fuller discussion). There is no free or individual choice with these tours (except actually signing up for the tour) and the role of tour operators is minimal.

### Pilgrimage tours

Pilgrimages, or religious visits to sacred sites, were not mentioned in the websites as a 'pull' factor for travel. Although all of the tours to Israel offered by the different tour operators included a visit to sites of religious significance, the tours themselves were not offered as religious tours *per se.* A reason offered by the operators interviewed was that there is no demand for such tours (cf. Ch. 6). It might be that religious Jews in the diaspora do not require organized tours in order to visit religious sites and are capable of making their own way. Another possible reason may be that since the Holy Temple (*Bet Hamikdash*) has been destroyed, no real religious pilgrimage according to ancient custom can take place. The act of pilgrimage that takes place nowadays is only a social act in memory of the ancient pilgrimage. However, it may be interesting to note that the word 'pilgrimage' and the phrase 'pilgrimage tour' were rarely offered to the Jewish public. It appeared only once in the 50 websites analysed.

## Conclusion

In the last decade there has been evidence of a large and expanding travel industry catering to Jewish travellers, particularly those living in North America (Ioannides and Cohen Ioannides 2002). There is a growing number of guidebooks, tour operators and travel agencies for secular or moderate Jewish travellers wishing to visit Jewish sites. Based on this examination of websites it seems the expansion of Jewish diaspora-oriented tourism leans heavily towards travel related to Jewish heritage, culture, religion and leisure. In fact, the appearance of several motives on a single webpage shows a mixing of religious, heritage and cultural reasons for travelling within the context of diaspora tourism. While Jews in the diaspora may or may not feel a private affinity with Israel, their cultural and spiritual centre, travel patterns suggest that their culture and religion have a strong influence on their travel patterns and activities. Therefore, it is argued that the niche markets within diaspora tourism that were identified fall under one or more of the categories of traditional leisure, culture/heritage, or religion (cf. Figure 18.1). The findings also demonstrate that diasporas are not internally homogenous, as individuals within diasporic groups have individual tastes and preferences concerning both touristic activities undertaken during travel and their attachment to their diasporic heritage, religion and culture. Because of this, tourism agencies put together various products and diversify their market bases to attract additional consumers.

As noted previously, the most popular form of Jewish tour is the heritage tour. Many of the images and slogans associated with the marketing of Jewish diaspora show that the emphasis of tour operators and marketers is on a narrative of Jewish identity through the ideal of belonging. For example, the government of Israel advertises to Jewish and non-Jewish tourists alike with slogans such as, 'Israel is calling. It is time to answer', or 'Israel: No one belongs here more than you!' It makes Jewish travellers and other tourists feel a part of something larger than themselves, creating in a sense a symbolic attachment to Israel. Whereas the former refers to the substantive meanings

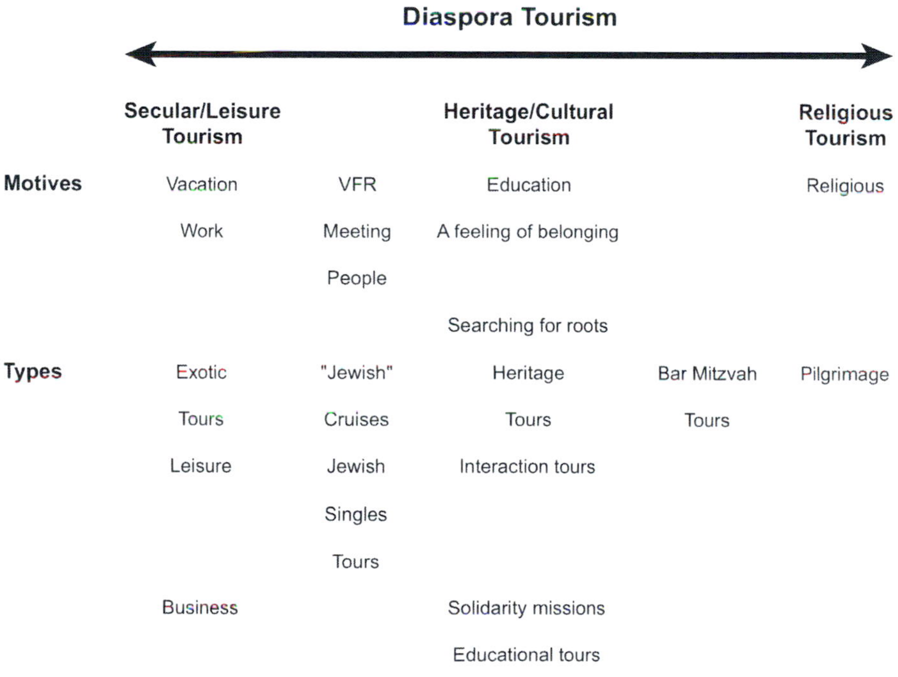

## Diaspora Tourism

| | Secular/Leisure Tourism | | Heritage/Cultural Tourism | | Religious Tourism |
|---|---|---|---|---|---|
| **Motives** | Vacation | VFR | Education | | Religious |
| | Work | Meeting People | A feeling of belonging | | |
| | | | Searching for roots | | |
| **Types** | Exotic Tours | "Jewish" Cruises | Heritage Tours | Bar Mitzvah Tours | Pilgrimage |
| | Leisure | Jewish Singles Tours | Interaction tours | | |
| | Business | | Solidarity missions | | |
| | | | Educational tours | | |

*Figure 18.1* Niche markets within Jewish diaspora tourism.

and to the positive or negative feelings attached to the symbol, the identification dimension refers to the relationship positioned between the symbol's reference and the self (Kelner *et al.* 2000). This can be seen most clearly through the Israel Experience, which is fast becoming a normative feature of American Judaism. Calling such programmes 'agents of re-ethnification', Mittelberg (1999) argues that they are providing North American Jews with a new set of shared experiences that foster group cohesion. Conceived as a programme to influence Jewish youth, *Birthright Israel* may have much more far-reaching effects, potentially reshaping the North American Jewish community as a whole by enshrining a pilgrimage to Israel as a prevailing rite of American Jewish passage.

However, with the rise in heritage tourism over the past few decades, there is beginning to be an over-saturation or congestion of heritage tours, especially within a diaspora tourism context where there is a limited customer base. To become more competitive and to attract more Jewish tourists, tour operators and agencies are attempting to diversify their heritage-type tours by combining Jewish patrimony and sense of belonging with other tourist activities, as seen in the exotic or adventure tours, the Jewish cruises and singles tours. Many of the tour companies also offer kosher menus in an attempt to persuade more orthodox Jews to participate, particularly when the tours coincide with Jewish holidays. In addition, tour operators emphasize the opportunity to travel with other Jews, since many Jews desire this because they can relate better in terms of language, culture and religion, they can develop lasting friendships and in the case of the Jewish singles tours, life-long relationships. This is akin to the interaction tours, with the exception that tour agencies and operators have had little to do with the creation of this particular type of experience. In this case, individual travellers are responsible for establishing linkages between themselves and the host community they wish to visit.

While peddlers of the Jewish travel experience have created various types of tours and market segments, as described above, there has been no research on the efficacy or potency of the themes and motives identified here. An extension of this work might include estimating tourist participation in the different diaspora tourism niche markets to measure the potency or effectiveness of the themes and motives, as the effectiveness of the messages of tourism suppliers is ultimately measured by how many tourists consume their products. Even though different themes and motivations are created and marketed by tourism producers, there is no guarantee of these new efforts attracting different types of tourists. This chapter has specifically focused on the Jewish diaspora tourism market, taking the promotional slogans and themes produced by tourism suppliers and agencies and identifying different markets and travel experiences they have targeted within Judaism. Indeed, one aspect of diaspora tourism marketing in general that has not been studied in detail is whether the themes or motives stressed by suppliers to induce tourism consumption are successful or not.

## References

Ashworth, G.J. (1999) 'Heritage dissonance and Holocaust tourism: some cases from European planning', paper presented at the Annual Meeting of the Association of American Geographers, Honolulu, Hawaii, 5 March.

Ashworth, G.J. (2003) 'Heritage, identity and places: for tourists and host communities', in S.

Singh, D.J. Timothy and R.K. Dowling (eds) *Tourism in Destination Communities*, Wallingford, UK: CAB International.

Bauer, A. (2002) Manager, Dan Tours, personal communication, 1 April, Toronto.

Ben Ari, E. and Bilu, Y. (1997) *Grasping Land: Space and Place in Contemporary Israeli Discourse and Experience*, Albany: SUNY Press.

Britton, S. (1991) 'Tourism, capital and place: towards a critical geography of tourism', *Environment and Planning D: Society and Space* 9: 451–78.

Chazan, B. (1997) *Does the Teen Israel Experience Make a Difference?* New York: Israel Experience, Inc.

Cohen, E.H. (1999) 'Informal marketing of Israel Experience educational tours', *Journal of Travel Research* 37(3): 238–43.

Cohen, E.H. (2002) 'A new paradigm in guiding: the *Madrich* as a role model', *Annals of Tourism Research* 29(4): 919–32.

Cohen, S. (1991) 'Israel in the Jewish identity of American Jews: a study in dualities and contrasts', in D. Gordis and Y. Ben-Horin (eds) *Jewish Identity in America*, Los Angeles: Wilstein Institute.

Collins-Kreiner, N. (2000) 'Pilgrimage holy sites: a classification of Jewish holy sites in Israel', *Journal of Cultural Geography* 18(2): 57–78.

Collins-Kreiner, N. (2002) 'Is there a connection between pilgrimage and tourism? The Jewish religion', *International Journal of Tourism Studies* 2(2): 1–18.

Epstein, S. (1995) 'Inventing a pilgrimage: ritual, love and politics on the road to Amuka', *Jewish Folklore and Ethnology Review* 17(1/2): 25–32.

Feldman, J. (2001) 'B'ikvot nitzol hashoah hyisraeli: Mishlahot noar l'Polin v'zehut le'umit' *Teoria Ubikoret* 19: 167–90.

Goldberg, H. (1995) *A Summer on a NFTY Safari, 1994: An Ethnographic Perspective*, Montreal and Jerusalem: CRB Foundation.

Gordon, E.T. and Anderson, M. (1999) 'The African diaspora: toward ethnography of diasporic identification', *Journal of American Folklore* 112(455): 282–96.

Gruber, R.E. (1992) *Jewish Heritage Travel: A Guide to East-Central Europe*, New York: Wiley.

Gruber, R.E. (1994) *Upon the Doorposts of Thy House: Jewish Life in East-Central Europe, Yesterday and Today*, New York: Wiley.

Heilman, S.C. (1995) *A Young Judea Israel Discovery Tour: The View from the Inside*, Montreal and Jerusalem: CRB Foundation.

Ioannides, D. and Cohen Ioannides, M. (2002) 'Pilgrimages of nostalgia: patterns of Jewish travel in the United States', *Tourism Recreation Research* 27(2): 17–26.

Israel Ministry of Tourism (2001) *Tourism to Israel 2000*, Jerusalem: Ministry of Tourism.

Israel Bureau of Statistics (2001) *Statistical Abstract of Israel*, Jerusalem: Bureau of Statistics.

Jackson, R.H. (1995) Pilgrimage in an American religion: Mormonism and secular pilgrimage, in D.P. Dubey (ed.) *Pilgrimage Studies: Sacred Places, Sacred Traditions*, vol. 3, Allahbad, India: The Society of Pilgrimage Studies.

Jackson, R.H. and Henrie, R. (1983) 'Perception of sacred space', *Journal of Cultural Geography* 3(2): 94–107.

The Jewish Travel Network (2002) *Interaction Tours*. Online. Available: www.jewishtravelnetwork.com (accessed 4 March 2003).

Kelner, S., Saxe, L., Kadushin, C., Canar, R., Lindholm, M., Ossman, H., Perloff, J., Phillips, B., Teres, R., Wolf, M. and Woocher, M. (2000) 'Making meaning: participants' experience of birthright Israel', Maurice and Marilyn Cohen Center for Modern Jewish Studies, Brandeis University, *Birthright Israel Research Report* 2.

Kirshenblatt-Gimblett, B. (1998) *Destination Culture: Tourism, Museums and Heritage*, Berkeley, CA: University of California Press.

Klemm, M.S. (2002) 'Tourism and ethnic minorities in Bradford: the invisible segment', *Journal of Travel Research* 41(1): 85–91.

Korom, F. J. (2000) 'South Asian religions and diaspora studies', *Religious Studies Review* 26(1): 21–8.

Kugelmass, J. (1993) 'The rites of the Tribe: the meaning of Poland for American Jewish Tourists', *Going Home: YIVO* 21: 395–454.

Kugelmass, J. (1994) 'Why we go to Poland', in J. Young (ed.) *Holocaust Monuments: The Ruins of Memory*, New York: Prestel.

Lehrer, E. (2001) 'The only Jewish bookshop in Poland', *Pakn Treger* 36: 34–7.

Mittelberg, D. (1999) *The Israel Connection and American Jews*, Westport, CT: Praeger.

Orenstein, D. (ed.) (1994) *Lifecycles: Jewish Women on Life Passages and Personal Milestones*, vol. 1, Woodstock: Jewish Lights Publishing.

Post, M. (1999) 'Don't bash birthright', *The Jerusalem Report,* 20 December: 54.

Safran, W. (1991) 'Diasporas in modern societies: myths of Homeland and Return', *Diaspora* 1(1): 83–99.

Shackley, M. (2001) *Managing Sacred Sites,* London and New York: Continuum.

Shapiro, F.L. (2001) 'Learning to be a diaspora Jew through the Israel Experience', *Studies in Religion* 30(1): 23–4.

Shenhav-Keller, S. (1995) 'The Jewish pilgrim and the purchase of a souvenir in Israel', in M.F. Lanfant, J.B. Allcock and E.M. Bauer (eds) *International Tourism: Identity and Change*, London: Sage.

Sheffer, G. (ed.) (1986) *Modern Diasporas in International Politics*, Sydney: Croom Helm.

Shuval, J.T. (2000) 'Diaspora migration: definitional ambiguities and a theoretical paradigm', *International Migration* 38(5): 41–55.

Szkółka, A. 2000. 'Pilgrimages to Auschwitz', in N.-J. Chaline and A. Jackowski (eds) *Pergrinus Cracoviensis: Selected Research Problems in the Geography of Pilgrimage*, vol. 10, Cracow, Poland: Institute of Geography, Jagiellonian University.

Timothy, D.J. (1997) 'Tourism and the personal heritage experience', *Annals of Tourism Research* 24: 751–4.

Waterman, J. (2002) Director of Research, Institute for Jewish Policy Research, personal communication, 29 August, London, UK.

Werbner, P. (2002) 'The place which is diaspora: citizenship, religion and gender in the making of chaordic transnationalism', *Journal of Ethnic and Migration Studies* 28(1): 119–13.

World Jewish Congress (2002) *Jewish Communities in the World.* Online. Available: http://www.wjc.org.il (accessed 4 April 2002).

# 19 Tourism and diasporas

## Current issues and future opportunities

### *Dallen J. Timothy and Tim Coles*

Political geographers define nation-states as nations, or groups of people with common cultural characteristics, that are housed within a state or country where the one national group dominates (Glassner 1996: 48). Among the most common examples are Japan, Egypt and Sweden. Today, however, the world is more complex than this traditional definition suggests in political, economic and socio-cultural terms. The threshold between pure nation-states and multi-ethnic states is no longer as clear as it once was.

People who have been forced from their homes as political refugees, migrants who move for better working opportunities and other diaspora groups have created a huge global melting pot of transnational societies that defy the traditions of nation-statehood. Even the countries in Western Europe that have traditionally held very conservative immigration policies and therefore remained relatively homogenous in population terms, such as Finland and Norway, have become home to many migrants during the past ten years from Africa, Asia and Eastern Europe. Likewise, political trans-formations of the past decade, such as the collapse of the Iron Curtain, have allowed many people of various diasporas of Eastern Europe to travel to their homelands. These conditions, coupled with the fact that more people are travelling abroad now than ever before, have created a form of travel that is motivated by a desire to visit relatives, see the homes and communities they left behind and fulfil feelings of nostalgia for places of their familial heritage.

This book has aimed to advance the nascent discussion in tourism studies about the relationships between tourism, migration and diasporas. The individual contributions have highlighted many of these relationships, although this book is only a foundation of a potentially much more fruitful area of research for scholars of tourism and cultural studies. The collision between diaspora and tourism is clearly a complex one that encompasses countless perspectives on race, migration, colonialism, persecution, power, tradition, conflict, choice (or lack thereof) and culture. While this book focuses on many of the dynamics of individual diasporas, several major concepts may be drawn from its contents.

One of the most notable themes generated in this volume is that of contestation of diasporic spaces. Maddern's chapter notes that particular spaces and their meanings and narratives of migration are contested by different individual groups. Ellis Island and the Statue of Liberty are not the only examples of this. The role of diaspora and migration is undeniable in all sorts of places that have distinctive ethnic heritages that are commonly commodified for tourism purposes. Even in completely contrived places, ethnic/diaspora heritage commonly provides an important, if somewhat thinly disguised, element of the tourism product. One prime example is Leavenworth (Washington). This is a mining and railroad town, that turned itself into a 'Bavarian' community tourist destination. In the 1960s, in the face of rapid economic decline, Leavenworth was successfully reinvented as an authentic-looking southern German village where tourists can experience 'authentic' German foods, music, architecture and shopping. The efforts were successful despite the fact that neither Leavenworth nor its residents were of Bavarian descent (Price 1996; Timothy and Boyd 2003).

Inasmuch as diaspora travel can be viewed as a form of heritage tourism, there is an inherent notion of contestation and power. The complexities of community life, particularly in racial and ethnic terms, nearly always create some form of dissonance and contestation regarding the treatment of heritage and community identity. Often questions arise regarding which ethnicity of many is being favoured, preserved, or interpreted, who represents the community and whether or not the dominant group's interests are in conflict with those of other groups (Ashworth 2003). Contestation among cultural groups generally results when multiple groups share the same heritage and historic places but view them differently, when there are divergent factions within one group causing a heterogeneous view of a cultural past and the existence of parallel, but disconnected, pasts (Olsen and Timothy 2002; Timothy and Boyd 2003).

The individual experiences of white and black American tourists at slave heritage sites in West Africa, for instance, are notably different, despite their occurrence at the same location (Teye and Timothy 2004). Likewise, the strength of Singapore's varied ethnic groups has brought about various types of cultural contestation. This is particularly so among the Chinese and Indians in the Little India locale of the city. Singapore law allows any person to live and set up shop in Little India, which has resulted in many Chinese merchants moving into the area that the South Asian community considers its own. Indian merchants are generally opposed to this population shift, because they feel Little India is their ethnic space. According to Chang (1999), this attitude stems from the notion that Singapore's majority population is Chinese and Little India is seen as a haven for the minority Indians. Tensions have emerged between the two groups in recent years resulting from contradictory ideas about who represents the 'true insider'.

A related issue is authenticity. Ning (1999) challenges the traditional views of authenticity and the visitor experience by suggesting that people can have double authentic experiences. First are those that are relevant to the individual and his/her construction of what current diasporic identity is and hence what should or should not be relevant. The second experience has relevance to the unfolding diasporic group identity as it emerges; the ways in which first- and second-generation migrants and their progeny depict 'their' cultures to tourists necessarily mingle elements of truth and falsehoods. When diaspora groups and their cultural heritage become the attraction such as in Holland and Frankenmuth (Michigan), Little Italies and Jewish Ghettos, decisions must be made regarding which elements will be modified by experiences in the new country, preserved from the homeland and performed for tourists. In some cases, areas have become tourist destinations based on their previous association with another specific ethnicity. Many continue to be an attraction even after the ethnic majority has scattered. One example of this is the Little Italies of New York and Boston. Non-Italian Americans have become the overwhelming majority in America's Little Italies and are responsible for keeping the Italian image alive (Conforti 1996). Timothy and Boyd (2003) call this phenomenon 'ethnic intruders', which is a form of inauthenticity wherein cultural events, places, or people are demonstrated for tourists by people who have little or no socio-ethnic connection to the place or event itself (cf. Adams 2002).

When diaspora groups travel to the homeland, authenticity may become a purely relative notion, conditioned by factors in the new country and limited contact with the original homeland through intervening years. Conditions do not remain stagnant in the homeland, so it is common for hyphenated peoples to have travel experiences in the motherland that are very different from their expectations. In the words of one commentator, who visited the childhood home of his father (Shanghai) after many years of hearing about it,

> For all the times my father told me about Shanghai in my 30-plus years, my feel for it was much like . . . an image that didn't quite seem real. I'd heard about the city so often and the huge role it had played in . . . my father's life and shaping his character, that it became almost mythical to me, until Sept. 25, when I stepped back in time and saw it for myself.
>
> (Compart 1999: 1)

All diasporas are different and reflect the specificities of their conditions, histories, new homeland immigration policies and population sizes. It appears that some diasporas are more readily predisposed to attract tourists and to undertake travel themselves. Conversely, it was noted earlier in this volume that some diasporas are more difficult to penetrate from a tourism perspective. The latter category appears to be found among groups that migrated as a result of political oppression rather than for economic reasons.

This book has demonstrated that in diaspora discussions there is a worrying function of expediency and ease in describing the scattered populations of the world under broad categories, such as the notion of diaspora itself. Cohen (1997) notes that not all members of every group that migrates internationally should necessarily be called a diaspora. Despite the tendency to generalize, there are some subtle and not so subtle, differences between various diasporas. The generalization of 'Pacific Islander', for example, does not take into account distinguishable linguistic, religious and social structural differences between various island nations and individual islands within national archipelagos. Likewise, the term African-American does not distinguish between the descendants of slaves or more recent first- and second-generation migrants from Africa. As noted in the introductory chapter, there are several different types of diaspora, which broad generalizations typically do not address or recognize.

Perhaps this is a result of the Western world view that has dominated discussions and scholarship in both tourism and diaspora studies. Chow (2003) argues for the need to read diaspora from a wider range of perspectives. Even further, Hollinshead (1996) questions the ability of Western researchers to articulate truly different perspectives and embrace differences in various diasporas and he calls for informed sensitivity in the cultural tourism marketplace. In general, society refers to groups, such as 'African-American', 'Asian-American', or even more specifically South Asian or British-Indian. However, as this volume clearly demonstrates, there are noteworthy differences associated with various groups within individual countries or regions. For example, the religious and cultural differences associated with Sikhs, Punjabis and Tamils, to name but a few groups, make this perfectly clear in the Indian context.

Another important point is the production and operationalization of diaspora tourism experiences. As alluded to in a few places in this book, it is difficult to measure the size of the diaspora and hence its potential value to the market. This raises multitudinous methodological and conceptual challenges, not least of which is defining who diaspora tourists really are and where they are scattered. Just as there are different types of cultural and heritage tourists (McKercher and du Cros 2002; Timothy and Boyd 2003), there are no doubt as many different forms of diaspora tourists, ranging from people visiting their birthplaces to people attending mass clan reunions. The Scottish Tourist Board echoes this belief in noting three types of visitors driven by diasporic motives to Scotland (Fowler 2003; STB 2001). The Welsh Tourist Board, the Irish Tourist Board, the Hungarian National Tourism Office and many other national tourism organizations have begun promoting diaspora-based tourism, although some have realized only limited success. Some observers have questioned whether ecotourism is really an important global tourism niche or simply a form of consumption that is more hype than reality (Chalker 1994; Hill 2000; McKercher 1993; Timothy and Ioannides 2002; Wall 1994; Wheeller 1994). Perhaps diaspora tourism will succumb to the same questions and criticism if researchers do not begin to understand its size and strength in the travel experience.

Several other issues were hinted at in various chapters of this book that may be fruitfully brought to bear on future research in the study of tourism and diasporas. One issue that needs to be examined in much more depth is the relationships between various players, or stakeholders, in the production of the diaspora travel experience. The travel intermediaries themselves most certainly play a role in disseminating certain types of information and creating experiences that may or may not be authentic and relative to the needs of diaspora travellers. For many people, a visit to the homeland is akin to a spiritual quest and a search for meaning in their own lives (Lowenthal 1985). It then becomes important to understand how the producers of the experience mitigate the feelings, occurrences and desires of diaspora travellers in an effort to satiate whatever motives drive people to experience the land of their ancestors. For those engaged in the search for tangible artefacts of their forebears, how important are the sympathies, sensitivities and performances of archivists, librarians, consultant genealogists and tour guides?

While several observers have noted the role of community members in producing the experience of ethnic heritages in the new homeland (e.g. Holland, Michigan), the views and perceptions of the people who live in the homeland have been practically ignored. In many cases, conflict ensues because visitors, regardless of their progenitors' origins, are seen as outsiders. In Ghana, for example, there is a considerable tension between Ghanaians and black Americans. Many African-Americans view Africa as their homeland and many are disappointed to learn that to modern-day Ghanaians, they are simply 'foreigners'. For many communities in the old country, it may be difficult to accept the claims to the homeland by these alien strangers.

This volume has focused overwhelmingly on international diasporas, but it must also be noted that domestic mass migrations among ethnic groups may be an important, albeit ignored, element of diaspora studies. 'Domestic diasporas' are especially notable when they involve the movement of certain ethnic groups from internal homelands to other regions of a state. A good example of this is the Lapps (Sami) of northern Finland, many of whom have been dispersed throughout the cities of Finland but have won domestic 'homeland' status in Lapland, educational equality and linguistic rights (Modeen 1999). Just as some international diasporas do, domestic diasporas involve many issues related to equality, representation and sometimes social defiance.

This book has also made clear that there are many diaspora groups involved in tourism that have not been well documented in the literature. Future research should address the issues and experiences involved in these other diasporas as well, since each group has its own set of experiences, values and traditions that determine its own travel patterns, as well as the type of attraction it becomes. The Irish and Italian diasporas are among the largest in North America and Australia, for instance and have formed a significant part of the population foundations of many urban areas. In the Caribbean region and Mexico, Lebanese and Syrian migrants play a very important, albeit ardently contested, role as merchants and the same is true of the Chinese in Southeast Asia. The Turks living in Germany have become one of that country's most important labour sources, with Berlin often described under the soubriquet of 'the third largest Turkish city in the world'. Another concept that deserves additional attention by tourism researchers is the notion of stateless nations and the issues they face surrounding travel, identity, citizenship and prejudice as visitors and as the visited. Among many other groups, these problems and experiences are particularly notable among the Kurds, Palestinians and European Gypsies (Al-Qudsi 2000; Brearly 2001; Crowe and Kolsti 1991; Kenrick 1993; Ross 1999; Sayigh 2001; Yavuz and Gunter 2001).

Examining these less-studied diasporas and quasi-diasporic peoples in the tourism context is important and can be enhanced by undertaking comparative studies between groups. For example, researchers know little about how life has emerged and how conditions differ between Lebanese-Jamaicans and Lebanese-Mexicans, Greek-Americans and Greek-Australians and Japanese-Canadians and Japanese-Peruvians. Another idea that should be explored through further research is how diaspora members from the homeland and elsewhere in the diaspora respond to diasporic spectacles. For instance, how do the Irish react to St Patrick's Day celebrations among Irish-Americans in Boston? How do Germans view the attractions and cultural events of the German-Brazilians? Researchers have also failed at this point to demonstrate how diaspora groups travel and incorporate visits to friends and relatives in the diaspora beyond the homeland. Why do British Sikhs visit Canadian Sikhs or Singaporean Sikhs, but not those in Punjab? Finally, we need to understand far more about the involvement of diasporas in tourism production for both diaspora members and non-members. Scattered diasporas in south-east Asia are responsible for the delivery of tourism infrastructure, distinct aspects of the tourist experience such as retailing and gastronomy and particular spaces as attractions (Hitchcock 1999). How do groups such as overseas Indians, Sikhs and Chinese valorize their roles in the tourism systems of major global cities such as Singapore, Bangkok, Kuala Lumpur and Jakarta? In comparison to one another, how do they perceive and understand their performance of the tourist experience? As is widely acknowledged, tourism constitutes a performance not just by the tourist, but also by the producer (Crouch 1999; Franklin and Crang 2001).

As noted much earlier, this book was motivated by the belief, which we retain now more than ever, that diasporas are a major global constituency active in the production and consumption of tourism. There has been much progress in understanding diasporic relationships with tourism. These new major strands of research are vital but so is a scholarly community that is willing to tackle these complex issues and which is sensitively attuned to the diasporic condition. Diasporas may provoke all kinds of dilemmas, they may be fluid and time-and-space specific, but as Hollinshead correctly observes, perhaps the most delicate problem for the future of research in this area is whether the research community is ready and willing to take up the challenge.

# References

Adams, J. (2002) 'The 'Bavarianization' of German Texas: ethnic tourism development in Neu Braunfels', paper presented at the Annual Conference of the Association of American Geographers, Los Angeles, March.

Al-Qudsi, S.S. (2000) 'Profiles of refugee and non-refugee Palestinians from the West Bank and Gaza', *International Migration* 38(4): 79–107.

Ashworth, G.J. (2003) 'Heritage, identity and places: for tourists and host communities', in S. Singh, D.J. Timothy and R.K. Dowling (eds) *Tourism in Destination Communities*, Wallingford, UK: CAB International.

Brearly, M. (2001) 'The persecution of Gypsies in Europe', *American Behavioural Scientist* 45(4): 588–99.

Chalker, B. (1994) 'Ecotourism: on the trail of destruction or sustainability?', in E. Cater and G. Lowman (eds) *Ecotourism: A Sustainable Option?*, Chichester: Wiley.

Chang, T.C. (1999) 'Local uniqueness in the global village: heritage tourism in Singapore', *Professional Geographer* 51: 91–103.

Chow, R. (2003) 'Against the lures of diaspora: minority discourse, Chinese women, and intellectual hegemony', in J.E. Braziel and A. Mannur (eds) *Theorizing Diaspora. A Reader*, Malden, MA: Blackwell Publishing.

Cohen, R. (1997) *Global Diasporas*, London: Routledge.

Compart, A. (1999) 'Genealogy travel: TW writer's road to Shanghai', *Travel Weekly* 58(99): 1, 18.

Conforti, J.M. (1996) 'Ghettos as tourism attractions', *Annals of Tourism Research* 23: 830–42.

Crouch, D. (1999) 'Introduction: encounters in leisure/tourism', in D. Crouch (ed.) *Leisure/ tourism Geographies: Practices and Geographical Knowledge*, London: Routledge.

Crowe, D. and Kolsti, J. (1991) *The Gypsies of Eastern Europe*, Armonk, NY: M.E. Sharpe.

Fowler, S. (2003) 'Ancestral tourism', *Insights*, March: D31–D36

Franklin, A. and Crang, M. (2001) 'The trouble with tourism and travel theory', *Tourist Studies* 1(1): 5–22.

Glassner, M.I. (1996) *Political Geography*, 2nd edn, New York: Wiley.

Hill, L. (2000) 'Questioning ecotourism: assessing ecotourism operations – case studies from Bali, Indonesia', in J. Cukier and E. Dixon (eds) *Tourism Resources, Impacts and Planning*, Hamilton: University of Waikato, Department of Geography.

Hitchcock, M. (1999) 'Tourism and ethnicity: situational perspectives', *International Journal of Tourism Research* 1(1): 17–32.

Hollinshead, K. (1996) 'Marketing and metaphysical realism: the disidentification of Aboriginal life and traditions through tourism', in R. Butler and T. Hinch (eds) *Tourism and Indigenous Peoples*, London: Thompson I.B.P.

Kenrick, D. (1993) 'The Romany gypsies of Europe', *Jewish Quarterly* 40(4): 8–10.

Lowenthal, D. (1985) *The Past is a Foreign Country*, Cambridge: Cambridge University Press.

McKercher, B. (1993) 'The unrecognised threat to tourism: can tourism survive "sustainability"?', *Tourism Management* 14: 132–6.

McKercher, B. and du Cros, H. (2002*) Cultural Tourism: The Partnership between Tourism and Cultural Heritage Management*, New York: Haworth.

Modeen, T. (1999) 'The Lapps in Finland', *International Journal of Cultural Property* 8(1): 133–50.

Ning, W. (1999) 'Rethinking authenticity in tourism experience', *Annals of Tourism Research* 26: 349–70.

Olsen, D.H. and Timothy, D.J. (2002) 'Contested religious heritage: differing views of Mormon heritage', *Tourism Recreation Research* 27(2): 7–15.

Price, T. (1996) *Miracle Town: Creating America's Bavarian Village in Leavenworth, Washington*, Vancouver, WA: Price and Rogers.

Ross, J. (1999) 'Uncommon humanity: travellers, ethnicity and discrimination', *Contemporary Issues in Law* 4(1): 1–30.

Sayigh, R. (2001) 'Palestinian refugees in Lebanon: implantation, transfer or return?', *Middle East Policy* 8(1): 94–105.

Scottish Tourist Board (STB 2001) *Genealogy Tourism Strategy and Marketing Plan*, Edinburgh: STB.

Teye, V.B. and Timothy, D.J. (2004) 'The varied colours of slave heritage in West Africa: White American stakeholders' *Space and Culture* 7(2).

Timothy, D.J. and Boyd, S.W. (2003) *Heritage Tourism*, Harlow: Prentice Hall.

Timothy, D.J. and Ioannides, D. (2002) 'Tour operator hegemony: dependency, oligopoly and sustainability in insular destinations', in Y. Apostolopoulos and D.J. Gayle (eds) *Island Tourism and Sustainable Development: Caribbean, Pacific and Mediterranean Experiences*, Westport, CT: Praeger.

Wall, G. (1994) 'Ecotourism: old wine in new bottles?', *Trends* 31(2): 4–9.

Wheeller, B. (1994) 'Egotourism, sustainable tourism and the environment: a symbiotic, symbolic or shambolic relationship?', in A.V. Seaton (ed.) *Tourism: The State of the Art*, Chichester: Wiley.

Yavuz, M.H. and Gunter, M.M. (2001) 'The Kurdish nation', *Current History* 100(642): 33–9.

# Index